ENVIRONMENTAL INFORMATION: RESEARCH, ACCESS & ENVIRONMENTAL DECISIONMAKING

by Sarah Lamdan

ENVIRONMENTAL LAW INSTITUTE
Washington, D.C.

1730 M Street NW, Washington, DC 20036

Cover design by Davonne Flanagan.

Published June 2017.

Printed in the United States of America
ISBN 978-1-58576-185-2

Contents

About the Author

Sarah Lamdan is an Associate Professor and librarian at the City University of New York School of Law. She received her J.D. from the University of Kansas School of Law with a Certificate in Environmental and Natural Resources Law, a master's degree in Legal Information Management from Emporia State University's Graduate School of Library and Information Management, and a B.A. in English from the University of Kansas. Sarah specializes in government information access and transparency with a focus on environmental agency records, and she teaches legal research, advanced legal research, and lawyering skills courses. She previously served as research specialist for several international law firms and at Bloomberg Law. Her writing on the Freedom of Information Act (FOIA), information access, and environmental justice issues have been published in the *Kansas Journal of Law & Public Policy*, *Vermont Journal of Environmental Law*, *Government Information Quarterly*, and other legal and information science publications. Sarah also facilitates FOIA workshops and seminars on participating in the federal and New York administrative rulemaking process.

About the Editor

Taryn L. Rucinski is a Branch Librarian for the U.S. Court of Appeals for the Second Circuit for the Southern District of New York Libraries and the former Environmental Law Librarian for the Pace University School of Law. She graduated *cum laude* with her J.D. and certificate in environmental law from Pace in 2010, received her M.L.I.S. with a concentration in law librarianship from St. John's University in 2012, and her B.A. *cum laude* in history from The College of New Jersey in 1999. A former Editor-in-Chief of the *Pace Environmen-*

tal Law Review, she serves as an adjunct professor of law specializing in environmental legal research for Pace and is an adjunct professor of library science for St. John's. Professor Rucinski is also co-author of the Environmental Law Lexicon and has authored a number of articles and research guides related to environmental law. Her research specialties include environmental and cultural heritage law, legal history, and New York law.

Acknowledgements

Thank you to everyone who supported this first ever text dedicated to helping people find and use environmental information and data—this book is the product of years of hard work and insight provided by legal scholars and practitioners, professional researchers, and wonderful friends, family, and students. To my scholarly soulmate, dedicated editor, and all around greatest fellow law librarian partner in environmental research, Taryn Rucinski, thank you for the countless hours you spent wading through facts and citations to make sure that this is the most complete and accurate book possible. I couldn't have written this without you.

Thank you to Susan Nevelow Mart and the *Boulder Conference on Legal Information: Scholarship and Teaching* in San Antonio (2014), Philadelphia (2015), and Chicago (2016). This environmental information text was born at Boulder and incubated there over several years under Susan's leadership and the collaboration, input, and feedback provided by every Boulder attendee. The Boulder Conference facilitates new and innovative legal information writing, and I am proud that this book is a product of such a wonderful collaborative event.

Thank you to the many people who edited and provided feedback to the chapters within this volume. Thank you to the Environmental Law Institute and their publications staff for their help improving and enriching this text. Thank you also to Martha Merson of TERC for her advice on making sense of environmental information. To everyone else who edited this book—Susan Azyndar, Paul Callister, Susan deMaine, Laura Jensen, Nicole Sasaki, and Gail Whittemore—your insight and input made this book better than it was before, and I'm so grateful for the time you spent on this effort.

Thank you to my excellent research assistants Maria Brinkmann, Jonathan Cantarero, Robert Feliu, and April Whitehead. Your help over the years of research and drafting was invaluable, and your enthusiasm and willingness to engage in environmental information concepts encouraged me to create a more practical, educational book.

And a special thank you to my wonderful CUNY School of Law community. I am honored to work at a law school dedicated to law in the service of human needs, and proud to participate in that mission every day. You fostered my vision and gave me space to share my work, lent your ears to hearing out my ideas, and provided invaluable feedback and support throughout the process. To my fellow library faculty and staff, and to my co-librarians Doug Cox, Raquel Gabriel, Yasmin Sokkar Harker, Jonathan Saxon, and our wonderful director Julie Lim: you are the best. I am lucky to call you my colleagues.

And last but certainly not least, I dedicate this book to my awesome family. Mom and Dad, you support all my ideas, no matter how flighty. Eyal, you are the greatest partner I could ask for. Thanks for your patience as I spent days poring over environmental research and writing, and for offering your help and encouragement every step of the way. Also love and thanks to Ben and Evelyn, who are not old enough to read this yet, but will someday know that their Mom likes to write stuff about environmental law things.

Foreword

It is an honor to provide the foreword to this book. A chance conversation between law librarians passionate about environmental legal research at the sixth annual *Boulder Conference on Legal Information: Scholarship and Teaching* in San Antonio hatched the idea for this book, and I have watched it evolve ever since. The Boulder Conferences offer a yearly opportunity for law librarians to receive feedback and advice on their scholarship from senior reviewers and peers. The author and editor took that chance conversation and turned it into a book proposal, presenting four draft chapters and an outline of a table of contents as a work-in-progress at the seventh annual Boulder Conference in Philadelphia. Two more draft chapters and a revised table of contents were workshopped at the eighth annual Boulder Conference in Chicago. The author framed her work for the conferees, law librarians all, as providing a step beyond traditional legal research. The book you have before you is the result of the hard work Sarah has put into its drafting and Taryn editing, the time for research and writing wrested from the competing demands of day jobs as law librarians and from private life.

The very title of this book acknowledges that, in the regulatory state, those who wish to set environmental agendas, influence policy decisions, take direct political action, seek access to detailed environmental information, or follow traditional litigation strategies against agency actions or inactions need access to more than the traditional legal materials that most research guides provide. This book is for anyone who seeks to participate in local, national, or international environmental conversations. The aim is to collect, in one place and at an accessible level, all of the tools necessary to participate in those conversations. While the book does address traditional legal research in one chapter, the book's overall focus is on obtaining environmental information produced and collected by federal, state, local, and international governments in order to promote environmental decisionmaking.

The book begins with an acknowledgment of the complexity of the task it is seeking to accomplish. Chapter 1 concedes that the breadth of the very concept of environmental information, as that term has been defined, makes the task daunting. Compressing the European Union's definition, set forth in full in Chapter 1, gives some hint of the scope of the information the book discusses:

> Environmental information shall mean any information in any form about the state of the elements of the environment, any substances that are released into the environment, any legal measures of any kind that might affect the environment, any reports on the environment, any cost-benefit or economic analyses used in any legal measures affecting the environment, and any information on the state of human health and safety that might be affected by any of these environmental elements.

Each chapter of the book uses a real environmental information-seeking story to set the stage for the ensuing discussion. Chapter 2 gives the reader an overview of the United States' reactionary statutory framework and provides a useful taxonomy of environmental laws: information and planning laws; pollution control laws; resource management laws; chemical tracking laws; and liability laws, discussing the laws with each category.

Chapter 3 presents an overview of the legal research process. First, the stage is set with an environmental information seeker's question about permitting procedures for cement plants and air pollution. After a review of the legal research process, with special attention to the benefits and detriments of both free and paid resources, the chapter applies that process to the cement plant problem. Legal research is complex. This chapter does an excellent job of clarifying the legal research process for non-law researchers. Chapter 4 addresses accessing environmental information related to the U.S. environmental statutory framework, starting with a primer on searching the U.S. Environmental Protection Agency's (EPA's) website, EPA.gov, a notoriously difficult task. The chapter discusses separately and in depth how to access environmental information related to prospective emergency plans and overviews, manifests for hazardous material, pollution reports, and public notice and mandatory labeling for potentially dangerous materials and events. For example, under "Access to Hazardous Materials Tracking Information," the information sources relating to the Resource Conservation and Recovery Act (RCRA) list and explain the *RCRA Biennial Report* and RCRAInfo. Creating the taxonomy and then describing the sources of law for each category and the relevant databases makes it easier to locate the information needed.

Chapters 5 and 6 look at federal transparency laws. Chapter 5 discusses what to do when the material an information-seeker wants to find is not available through an agency's affirmative disclosure of environmental information. A discussion of the Freedom of Information Act (FOIA), its history, its limitations, and the exemptions to disclosure follows. Readers are given a guided tour of the request process, the appeals process, and FOIA resources.

Chapter 6 discusses the other federal transparency laws, including federal meeting laws, laws governing the management of agency information, laws requiring finding aids for researchers, laws restricting access to information, records classification, FERC requests, and the Privacy Act.

Chapter 7, using a 549-page proposed rule by the U.S. Department of Transportation's (DOT's) Pipeline and Hazardous Material Safety Administration as the starting point, takes readers through the rulemaking process. The chapter outlines the entire process, starting with publication of a proposed rule in the *Federal Register*, the procedures for public notice and comment, the final rule, and ending with codification in the *Code of Federal Regulations*, thoroughly discussing the general government and specific agency websites involved at each stage. Most importantly, the chapter discusses the records that are generated before a proposed rule is published and identifies what sources about the pre-rulemaking process are available. And because the regulatory process does not end after a final rule is issued, the chapter also introduces guidance documents as an information resource.

Chapter 8 takes up the important category of grey literature, broadly defined as "not official government records," and material not published by mainstream publishers. Grey literature relevant to environmental issues includes some surprises. For example, corporations must disclose certain environmental information in their filings with the Securities & Exchange Commission (SEC). This chapter starts by illustrating how to locate such filings and guidance documents. In addition to the SEC, DOT, the U.S. Department of Justice, the Federal Aviation Commission, and the Food and Drug Administration are other agencies identified as having environmental materials. This chapter addresses how to find congressional documents and Congressional Research Service reports. Researching like a reporter and using power search techniques on Google to find this information round out the tools provided.

Chapter 9 takes up the very difficult research task of finding state and local environmental information, which may not be easily available. Where would an environmental information-seeker go, for example, to find out what fluids are in the fracking chemicals used in oil extraction projects in a particular state? Information that must be collected by states can be different than the information collected by the federal government, and can vary based on zoning, land use, building codes, or sanitation laws. States differ in their freedom of information laws, so there is little uniformity in what types of environmental information is available from state to state. In addition, federal law frequently delegates implementation to the states, within certain guidelines. This is known as cooperative federalism. The chapter also has citations to each state's environmental information access laws.

Chapter 10 introduces the environmental information researcher to foreign and international environmental resources. Environmental issues know no political borders. Foreign law is the domestic law of other countries, and the chapter lists the online resources for locating those laws. There is also a discussion of how to find and use other nation's information access laws. International law, meanwhile, governs the relationships between countries, especially treaties. The chapter has a comprehensive list of sites where treaties and other international environmental documents can be located.

Chapter 11 ties many of these disparate pieces together by discussing how to make sense of scientific environmental information. As does each chapter, Chapter 11 uses a real scenario to illustrate the problem. After discussing resources to help environmental information researchers understand the jargon, the chapter addresses understanding the formats, structures, and functions of environmental testing reports, environmental impact statements, and permit records. The authors recommend using the ADEPT method of analyzing environmental information: Analogy, Diagram, Example, Plain English, and Technical Definition. The final pieces of the chapter discuss how to get help in assessing environmental risk and how to move forward with advocacy.

Finally, two appendices list the major environmental statutes and their amendments and provide contact information for EPA regional offices.

Taken together, the chapters in this book offer traditional legal researchers—as well as anyone else interested in having a say in the conversation about local, national, and international environmental issues—all the tools they need to find and use the information they want within this vast and messy landscape. I predict that this book will become a staple for environmental advocates.

Susan Nevelow Mart
Director of the Law Library and Associate Professor
University of Colorado Boulder School of Law

Chapter 1: Introduction to Environmental Information

Environmental information is all around us. It permeates almost every aspect of our daily lives. Every time a politician, scientist, news reporter, or other decisionmaker discusses an issue like contaminated drinking water, climate change, or fracking and says—"look at the evidence"—they are referencing environmental information. In the simplest terms, environmental information is any data, resource, or material that informs us about the biotic world as a whole, or the various plants, animals, organisms, and ecosystems that comprise it. This definition encompasses an unknowable amount of information that is constantly changing and evolving.

In our modern-day world, environmental information is the backbone of environmental law and regulation, playing a critical and necessary role at every level of environmental decisionmaking. Navigating this arena and dealing with the divergent interests of the different groups that participate in environmental activities and policymaking can often make accessing environmental information a very thorny and complicated venture. A 2014 account of a U.S. Environmental Protection Agency (EPA) battle over environmental information in the agricultural context underscores the complexities of environmental information and its access. As one reporter, frustrated with EPA's attempts to balance environmental information access and personal privacy, cynically described the events:

> The Environmental Protection Agency (EPA) walked into court in Minnesota last week and asked a judge to issue an order to stop the federal agency from sharing information about some of the biggest polluters of our nation's waterways with the American public. That's right, the agency that is charged with protecting your environmental health is working hard to make sure . . . that you don't have access to information. It would be shocking if it weren't so sadly commonplace.[1]

The Minnesota courtroom in the reporter's story is the U.S. District Court for the District of Minnesota, where a skirmish between public interest envi-

1. Scott Edwards, *Snatching Defeat From the Jaws of Victory*, HUFFINGTON POST, Feb. 9, 2015, http://www.huffingtonpost.com/scott-edwards/snatching-defeat-from-the_3_b_6645502.html?utm_hp_ref-green (last modified Apr. 11, 2015).

ronmental groups and private farming organization put EPA in the center of an argument over the release of factory farm data. The data in question reflected the effects of concentrated animal feeding operations (CAFOs) on discharges of nutrients, pathogens, heavy metals, and pharmaceuticals into waterways.[2] In 2012, EPA granted a Freedom of Information Act (FOIA) request about CAFO-related pollution, and provided records to several environmental advocacy groups including the Natural Resources Defense Council, Pew Charitable Trusts, and Earthjustice. After releasing information to the advocacy groups, EPA tried to put the proverbial cat back in the bag, attempting to force the groups to return the information to the federal agency because of personal information contained in the released records.

The release of the information put corporate farmers on the secrecy side of the information access tug-of-war, with the American Farm Bureau (Farm Bureau) and National Pork Producers Council trying to protect their members' personal information from public exposure, and the environmental organizations on the other side, trying to obtain information that would show which pollutants were in their members' waterways. EPA hovered between the two groups, delicately trying to balance the farmers' privacy interests with the public's right to know about pollution threats. The Farm Bureau called EPA's FOIA disclosure a "betrayal of trust"[3] and a security threat, while the environmental organizations claimed that the failure to disclose the information threatened environmental safety and allowed under-regulated agricultural companies to run amok, causing environmental damage without consequence.

EPA is often put in a difficult position, information-wise, recognizing both the importance of protecting the privacy of the information that polluters report to the agency and also the key role of public participation in the Agency's programs.[4] EPA recognizes that every environmental issue and effort benefits from public participation, and that environmental information access is vital to that participation.[5] Citizen awareness is crucial to all facets

2. *See* Motion to Intervene, American Farm Bureau Fed. & Nat'l Pork Producers Council v. EPA, No. 13-cv-01751 (D. Minn. Sept. 5, 2013), http://www.foodandwaterwatch.org/sites/default/files/Memorandum_of_Law_in_Support_of_Motion_to_Intervene_%289-5-13%29.pdf.

3. Perry Chiaramonte, *Ranchers, Farmers Fear Eco-Terrorists After EPA Releases Private Info*, Fox News, Jan. 17, 2014, http://www.foxnews.com/us/2014/01/17/ranchers-farmers-fear-eco-terrorists-after-epa-releases-private-info/ (last visited Nov. 10, 2016).

4. *See* U.S. EPA, *Public Participation in the NPDES Permit Issuance Process*, Sept. 2013, https://www3.epa.gov/npdes/pubs/publicparticipation.pdf (last visited Nov. 10, 2016).

5. EPA's very mission statement declares one of its purposes as ensuring that "all parts of society—communities, individuals, businesses, and state, local and tribal governments—have access to accurate information sufficient to effectively participate in managing human health and environmental risks." U.S. EPA, *Our Mission and What We Do*, https://www.epa.gov/aboutepa/our-mission-and-what-we-do (last updated Aug. 18, 2016).

of environmental preservation and the progress of environmental safety and welfare. Simultaneously, laws like FOIA and the Privacy Act of 1974 emphasize the importance of privacy, carving out privacy exemptions, and crafting privacy guarantees for U.S. citizens.

The balance between information privacy and access, as well as the complexities of environmental data access creates a challenging landscape for those seeking environmental information. Although environmental information is a critical part of environmental decisionmaking, there is not much discussion about access and usability issues that burden environmental information in the United States and the complexities of obtaining, understanding, and utilizing environmental information.[6]

I. Why This Book?

This book is designed to help environmental information seekers locate, obtain, and make sense of environmental records, documents, and pieces of data. It serves as a manual for those wishing to obtain and understand environmental information. Environmental information is abundant, but in order to find what you need, you must often first know how to traverse the organizational landscape of local, state, and federal agencies and organizations. You must know where the information is located, understand what environmental information is gathered pursuant to the different state and federal environmental laws, and be able to ascertain the types of information available to you in accordance with the laws governing records access and open meetings.

This is a practical book for legal practitioners, advocates, students, and enthusiasts interested in seeking environmental information. The term "environmental information seeker" is used throughout the book because the tips and concepts the book introduces expand beyond legal research or general research and into the broader realm of information-gathering.[7] Environmental information seeking goes beyond merely conducting online searches in familiar databases for scholastic and legal sources, to include methods for obtaining information from nontraditional sources like government offices and open meetings. This book incorporates traditional research skills with an understanding of environmental and transparency laws, demonstrating how

6. Sarah Lamdan, *Beyond the FOIA: Improving Environmental Information Access in the U.S.*, 28 Geo. Envtl. L. Rev. (forthcoming 2017).

7. Notice the term "researcher" is used in the chapter (see Chapter 3) explaining traditional legal research methods.

to find environmental information through a combination of research skills and government transparency provisions.

This book is not a traditional legal research guide or theoretical look at environmental and information access laws. Rather, it is the ultimate how-to book: an in-depth guide through the complex and multilayered environmental information access schemes. There is a chapter dedicated to traditional legal research methods, but there are also chapters guiding information seekers through the use of environmental laws, transparency and open government laws, and general administrative law tools to obtain environmental records and data. This book also contains a 50-state survey on information and environmental information access laws; an overview of foreign and international environmental law resources; and a chapter focused on finding and using nongovernmental information sources. The appendices provide legislative history information for most major environmental laws, among other environmental information research references. The final chapter explains how to make sense of the information you gather, breaking down the complex mathematical and scientific data, and parsing the jargon that is ubiquitous in environmental information. In short, this book is the perfect companion for anyone who wants to find and understand complex environmental issues. Although in an ideal world, environmental information would be easy to find and use, the current state of environmental information access requires additional knowledge and expertise—the kind that this book provides.

II. What Is Environmental Information?

Before environmental information seekers embark on their journey using the tools and tips in this book, it is useful to consider the meaning of environmental information and what types of information fall into that category. Beyond scientific testing results (the results of tests analyzing air, soil, and water, for example), there are broader concepts like climate change that cannot be captured by singular data points. There are also environmental policy issues, including food safety, noise abatement, and the ownership and use of public lands. All of these data points, concepts, and policy issues have associated bodies of information, studies, and data associated with them. The overall scope of environmental information in the United States, compared to other nations, is amorphous, lacking both an official definition and set parameters.

So far, the United States has not provided guidelines on the scope of environmental information under its laws—there is no official definition of what "environmental information" means in the United States. The international community, however, has well-defined parameters and definitions for the term "environmental information." International treaties like the regional Aarhus Convention and sovereign nations including those in the United Kingdom have developed official definitions of environmental information.[8] The United Kingdom borrowed its structural definition from the European Union, whose definition reads:

> 1. "Environmental information" shall mean any information in written, visual, aural, electronic or any other material form on:
>
> (a) the state of the elements of the environment, such as air and atmosphere, water, soil, land, landscape and natural sites including wetlands, coastal and marine areas, biological diversity and its components, including genetically modified organisms, and the interaction among these elements;
>
> (b) factors, such as substances, energy, noise, radiation or waste, including radioactive waste, emissions, discharges and other releases into the environment, affecting or likely to affect the elements of the environment referred to in (a);
>
> (c) measures (including administrative measures), such as policies, legislation, plans, programmes, environmental agreements, and activities affecting or likely to affect the elements and factors referred to in (a) and (b) as well as measures or activities designed to protect those elements;
>
> (d) reports on the implementation of environmental legislation;
>
> (e) cost-benefit and other economic analyses and assumptions used within the framework of the measures and activities referred to in (c); and
>
> (f) the state of human health and safety, including the contamination of the food chain, where relevant, conditions of human life, cultural sites and built structures inasmuch as they are or may be affected by the state of the elements of the environment referred to in (a) or, through those elements, by any of the matters referred to in (b) and (c).[9]

The EU's definition is broad, encompassing many different pieces of the environmental information spectrum and capturing a variety of media and

---

8. Info. Comm'r's Office, United Kingdom, *What Is Environmental Information? (Regulation 2(1))* (updated Nov. 11, 2016), https://ico.org.uk/media/fororganisations/documents/1146/eir_what_is_environmental_information.pdf.
9. Council Directive 2003/4/EC, art. 2, of 28 January 2003 on Public Access to Environmental Information and Repealing Council Directive 90/313/EEC, 2003 O.J. (L 41), 28, http://eur-lex.europa.eu/LexUriServ/LexUriServ.do?uri=OJ:L:2003:041:0026:0032:EN:PDF (last visited Nov. 10, 2016).

document types under the broad umbrella of information access. From the elements of the environment to factors that cause pollution, environmentally focused plans and projects, cost-benefit analyses that touch on environmental factors, and human health and safety information, the European Union believes all of that information is vital to public participation in the environmental process.[10]

While the United States has not defined environmental information, a collection of U.S. statutes, regulations, and practices provides access to many types of environmental information defined by the European Union, as the chapters of this book demonstrate. For instance, in the United States, FOIA is pervasive part of environmental information access—note that the statute is frequently mentioned throughout the book. In the realm of government transparency in America, most roads lead to FOIA.

Additionally, much of the environmentally focused government information in the United States comes through EPA. This book may seem rather EPA-centric at times, but that is largely because EPA controls and oversees the majority of environmental programs in the nation, delegating power to the states to oversee local efforts in concert with the federal agency.[11] Despite the EPA-heavy nature of this book, be aware that EPA is not the sole agency dealing in environmental matters: in truth, most federal and state agencies, to some extent, engage in activities that touch upon environmental concerns. Just a few of the many examples of non-EPA agencies participating in environmental work include the U.S. Food and Drug Administration's regulation of pesticides in foods, the Occupational Safety and Health Administration's design of work safety procedures for hazardous waste treatment, and the U.S. Department of Justice's Environment and Natural Resources Division bringing enforcement actions against environmental law breakers.

Just as environmental governance and oversight expands beyond EPA, the research tips within this book are useful in other contexts beyond environmental work. Many of the processes and information seeking tips in this book can be used in nonenvironmental contexts as well. Open meetings laws and freedom of information laws apply to a plethora of agencies, governing all meetings and records—not just ones with environmental focus. Many areas of law have their own disclosure requirements. For instance, securities laws proscribe many financial disclosure requirements on corporations that make their stock available for purchase and trade. Though the examples

10. *Id.*
11. This state delegation is discussed further in Chapter 9.

and illustrations in this book pertain to environmental issues, the prescribed methods for pursuing the information are applicable across many disciplines.

III. Types of Environmental Information

Environmental information takes a multitude of forms: scientific data reflecting environmental facts and theories; legally binding rules made by courts, legislatures, and agencies that impact the natural environment; and writings of scholars, advocates, artists, educators, and other interested parties about environmental issues and ideas.

A. Legal Information

Federal, state, and local governments produce volumes of environmental information. From legally binding statutes, regulations, court decisions, and agency directives to scientific data gathering and organizing, much of the environmental information people use as the basis for environmental action and activity originates in a government agency, legislative body, court, or other executive office.

Legal information is generally divided into two categories: primary sources and secondary sources. Primary sources are legally binding rules contained in statutes, regulations, constitutions, or case law. These sources tell people what they can and cannot do, and what enforcement measures await those who fail to abide by the rules. They officially delegate power and responsibilities to both the regulators and the entities they regulate. Secondary sources include all other types of information: guidance documents, explorative studies, informational pamphlets—basically, any information created by government entities that is not characterized as a primary source. These two sources of legal information, and their various forms, are discussed in greater detail in Chapter 3.

There are also U.S. (or federal) laws that grant public access to government information—specifically environmental information. These laws are the focus of much of this book.

B. Grey Literature

Grey literature is a library science term that refers to information that is not controlled by commercial publishers, like blog posts and other social media entries, pre-prints, reports (for example: progress reports, statistical reports, technical reports, market research reports, etc.), memoranda, conference proceedings, technical standards, noncommercial translations, bibliographies, official documents not published commercially, etc.[12] Due to its diverse origins and lack of commercial dissemination, grey literature can be difficult to find, and in some circumstances, it can be difficult to determine whether a government document is grey literature or something with more regulatory authority. In fact, some legal information science scholars have argued that government agencies disguise official directives as grey literature in order to avoid time-consuming public notice and participation requirements of administrative rulemaking processes.[13]

The Internet has greatly enhanced the accessibility of grey literature. Often times, grey literature can be found on the websites of the agencies or institutions that produced the documents, and sometimes it may also be located in library collections or other online information repositories. Although grey literature does not have binding legal authority or the backing of a commercial publisher, it can be an excellent source of scientific, technical, administrative, or policy information important to environmental decisionmaking.

Another aspect of environmental information that falls under the grey literature category is the vast amount of data compiled about natural resources. Scientific data is at the heart of environmental information, giving environmental decisionmakers the factual foundations for many of their positions. To understand whether a waterway is polluted, it is important to determine whether maximum permissible contaminant levels have been exceeded, which requires experts to determine and analyze pollutant measurements in parts per million. The quantification of environmental danger requires scientific and technical calculations and information sets.

Additionally, because much of the U.S. environmental regulation scheme requires regular tracking of pollutants in the nation's air, water, and soil, as well as data on flora and fauna, much of the scientific and technical information needed for environmental decisionmaking in the country is either generated or collected by government agencies. Nevertheless, private academic, for-profit, and nonprofit institutions run scientific environmental studies of their own. The information generated by nongovernment organizations can also contribute to the environmental decisionmaking process, keeping in mind potential perspectives, interests, and even bias of the organizations

12. Fourth International Conference on Grey Literature in Washington, D.C. (Oct. 4-5, 1999), *What Is Grey Literature?*, Grey Lit. Rep., Feb. 2000 (New York Academy of Medicine), http://www.greylit.org/about (last visited Sept. 2, 2016).

13. Mary Whisner, *Some Guidance About Federal Agencies and Guidance*, 105 L. Libr. J. 385 (2013), http://www.aallnet.org/mm/Publications/llj/LLJ-Archives/Vol-105/no-3/2013-19.pdf (last visited Nov. 10, 2016).

funding the research. Access points for nongovernmental scientific data and information are discussed in Chapter 8.

IV. Environmental Information's Role in Environmental Decisionmaking

Environmental information is key to environmental decisionmaking and crucial for managing environmental issues. Governments, industry, and the public all need comprehensive, trusted, and timely environmental information to make decisions about environmental issues.[14]

Lack of access to this information makes it impossible for individuals to know about environmental threats or opportunities. In short, without information, people cannot properly take stock in the state of their environmental surroundings. Access to environmental information not only provides individuals with a view of the environmental status quo, it provides the means to formulate predictions about the state of the environment in the future. Thus, environmental information forms the foundation of effective decisionmaking.

Knowing about environmental projects, which are often government-facilitated ones, and access to environmental data, which is often collected and maintained by the government, also honor the notion that U.S. citizens have a fundamental right to know about what their government is doing.[15] Federal, state, and local governments steward many of the environmental projects and studies that occur in the United States, and it is the obligation of those governments to provide information transparency and access to the public. Providing the public with environmental information empowers citizens to participate in the nation's environmental decisionmaking processes, which are usually led or overseen by government entities. Equal information creates a more equal playing field, and promotes democratic decisionmaking. Providing more information to the public puts citizens in a better position to bargain with private entities as well as the government.[16]

Environmental information access is also important because transparency can substantially affect government behavior. U.S. Supreme Court Justice Warren Brandeis declared that shining sunlight on information is the "best

14. Bureau of Meteorology, Aus. Gov't, *Environmental Information*, http://www.bom.gov.au/environment/ (last visited Nov. 10, 2016).
15. Nat'l Archives & Records Admin. v. Favish, 541 U.S. 157, 171 (2004) (quoting Department of Justice v. Reporters Comm. for Freedom of the Press, 489 U.S. 749, 773 (1989)).
16. *See* Clifford Rechtschaffen & Eileen Gauna, Environmental Justice: Law, Policy, and Regulation (Carolina Academic Press 2003).

of disinfectants," exposing secrets and bringing issues of national importance to light. He explained that "[p]ublicity is justly commended as a remedy for social and industrial diseases."[17] When more transparency is offered in environmental undertakings, it is far more difficult to hide information about those undertakings, and poor decisionmaking is easily visible to all. Thus, environmental information access measures incentivize careful environmental decisionmaking.[18] Using transparency as a preventative measure for environmentally detrimental practices lays bare transgressions before they permanently damage the environment.

Environmental information empowers citizens to assume an adversarial role when necessary, since it is a critical component when filing lawsuits against parties violating environmental statutes and regulations. For example, an informed citizen can sue a company under the Clean Water Act for polluting a river near their home or sue for an injunction to stop or prevent the pollution from happening in the first place. Citizens can also sue government entities that fail to perform their duties under environmental laws. An individual or group of citizens could sue EPA for failing to enforce rules promulgated under an environmental law. When environmental information is available to the public, individuals can file citizen suits to prevent the damage before it occurs through injunctive relief, rather than engaging in litigation only after the environmental damage has been done. Because access to environmental information is strained in places with scarce resources where populations are often disproportionately affected by environmental hazards, information access needs are all the more critical in those places.[19]

Information access is also an important gateway to more complex issues about the balance of power between citizens and the government. Disputes about information access in the United States are often subplots to, and surrogates for, a broader national discussion on public participation government operations and policymaking that lies at the heart of U.S. democracy and environmental progress.[20] For example, a 2004 lawsuit in which environmental nonprofit organizations trying to access records from then-Vice Presi-

17. Louis D. Brandeis, Other People's Money, and How the Bankers Use It 92 (1914), https://archive.org/details/otherpeoplesmone00bran.

18. Gidiere III, *infra* note 20, at 23.

19. This phenomenon of information scarcity is discussed in environmental justice and the not in my backyard (NIMBY) movement literature. *See e.g.*, N.Y. Times Editorial Bd., *Who Bears the Most Risk From Fracking?*, N.Y. Times, May 31, 2016, http://www.nytimes.com/2016/05/31/opinion/who-bears-the-most-risk-from-fracking.html?emc=edit_th_20160531&nl=todaysheadlines&nlid=69056167&_r=0 (discussing environmental justice and fracking).

20. Philip Stephen Gidiere III, The Federal Information Manual: How the Government Collects, Manages, and Discloses Information Under FOIA and Other Statutes 13 (2d ed. 2013).

dent Dick Cheney's Energy Task Force opened up a substantive debate about executive privilege and public access to special task force documents.[21] The organizations used the argument as the legal hook to then expose these central policy issues in the lawsuit.

Environmental information access acts as a catalyst to environmental problem solving and decisionmaking on a daily basis. New information tends to elicit public reaction. For instance, in 2011, an internal EPA list containing hundreds of major Clean Air Act (CAA) violators, who were caught spewing dangerous levels of toxins into the air, including mercury and formaldehyde, was released to the public in a FOIA response. Journalists were able to report the situation to concerned citizens, and those citizens were able to start thinking critically about solutions for the air pollution. News investigations not only provided statistics and facts about the air polluter watch lists, but also identified factors for the continuing air pollution issues, including outdated rules and technology, battles between EPA and state environmental agencies and authorities, and self-reporting issues among the polluters themselves.

Armed with this information, local news and environmental organizations posted the story far and wide as local reporters listed specific CAA violators in the vicinity,[22] and law firms alerted parties in violation of the CAA about potential scrutiny and even lawsuits that may result from their inclusion on the watch list.[23] Awareness of the secret watch list and the CAA violations reported therein led to a 2013 investigation of the watch list by EPA's Inspector General, and an indictment of the federal agency's failure to consistently enforce the environmental law.[24]

---

21. Cheney v. U.S. Dist. Court for D.C., 542 U.S. 367, 373, (2004).
22. This story appeared in many newspapers and news blogs across the nation in the week following the initial National Public Radio (NPR) story, including: in Tuscon, Arizona: Corbin Hiar, *EPA Posts Secret "Watch List" That Includes Chronic Polluters*, THE SENTINEL, Nov. 22, 2011, http://www.tucsonsentinel.com/nationworld/report/112211_watch_lists/epa-posts-secret-watch-list-that-includes-chronic-polluters/; in Los Angeles, California: *EPA's Secret List Shows Pollution Unchecked*, L.A. TIMES, Nov. 7, 2011, http://latimesblogs.latimes.com/greenspace/2011/11/epas-secret-list-shows-pollution-unchecked.html; and in New Orleans, Louisiana: *29 Louisiana Plants Show Up on Secret EPA Watch List of Air Polluters*, TIMES-PICAYUNE, Nov. 7, 2011, http://www.nola.com/business/index.ssf/2011/11/louisiana_companies_show_up_on.html.
23. Arnold & Porter LLP, *Advisory: EPA's Clean Air Act Watch List*, Nov. 2011, http://files.arnoldporter.com/advisory%20epa_caa_watch_list.pdf.
24. Jim Morris, *Inspector General to Review EPA's "Watch List,"* CTR. FOR PUB. INTEGRITY, Feb. 6, 2013 (updated May 19, 2014), https://www.publicintegrity.org/2013/02/06/12144/inspector-general-review-epas-watch-lists. Here is the report that resulted from the investigation: Office of Inspector General, U.S. EPA, *Report: The EPA Should Assess the Utility of the Watch List as a Management Tool*, https://www.epa.gov/office-inspector-general/report-epa-should-assess-utility-watch-list-management-tool (last modified Sept. 26, 2015) (referencing report number 13-P-0435).

As our understanding of environmental risks grows more nuanced with ever-increasing scientific discovery and innovation, and as environmental issues become more widely discussed in policy and public health forums, the barriers to environmental information access will become ever more apparent and obtrusive. Environmental information shapes the future of our natural landscape because information is the starting point for environmental protection and public health measures. The following chapters of this book will focus on the specific sources of environmental information available in the United States, and offer practical solutions for locating and obtaining the information needed for effective environmental decisionmaking.

Chapter 2: Overview of Environmental Law

Law in Action

Jerome Barker, a chemistry teacher at Mission Valley High School in Wichita, Kansas, is creating an Earth Day research project for his junior chemistry students. Mr. Barker is asking his students to assess the properties of active chemicals in various household cleaning products. After the assessment, he wants his students to design fact sheets for the public, warning people of any hazards related to the chemicals and instructing them on the safe use and application of the products. He is also directing his students to find any environmental laws that may affect the sale, use, or disposal of the products. Mr. Barker wants to know what kind of background information his students would need to understand and research environmental laws about chemical hazards.

Environmental information seekers can benefit tremendously from a basic understanding of U.S. environmental law and policy. Environmental law is a unique and relatively new body of doctrine and legal authority. Like most bodies of law, environmental jurisprudence does not consist of a singular, streamlined, monolithic plan for dealing with ecological issues. Rather, environmental law in the United States can best be characterized as a set of disparate reactions to various public health issues. Environmental laws have evolved ad hoc over the course of the past century. New environmental laws tend to arise on a reactionary basis, materializing only when environmental problems can no longer be ignored.

The reactionary nature of environmental law in the United States is apparent in the very formation of the nation's environmental law scheme. In the 1950s, in the wake of the industrial revolution and an explosion of post-World War II consumerism,[1] industrial development along U.S. waterways caused disasters like the fire in the Cuyahoga River near Cleveland, Ohio.[2] As people realized that water pollution threatened human

1. Robert F. Blomquist, *Government's Role Regarding Industrial Pollution Prevention in the United States*, 29 Ga. L. Rev. 349, 357 (1994-1995).
2. Ohio History Central, *Cuyahoga River Fire*, http://www.ohiohistorycentral.org/w/Cuyahoga_River_Fire?rec=1642 (last visited Nov. 13, 2016).

health and safety, industries were forced to stop using waterways as a pollution "sink," or tool for washing away chemicals, because laws were passed limiting waterway pollution. To comply with the water pollution regulations, polluters began evaporating contaminated wastewater into the air instead of feeding it into U.S. waterways. By the 1960s, as more particulate matter was released into the air, pollution increased, leaving a layer of soot on surfaces and even people.[3] When air pollution regulations were strengthened to decrease soot in the air, the remaining convenient and yet unregulated pollution "sink" was land. Industries started placing their wastes in barrels, underground storage tanks, and lagoons. This land storage eventually caused landfill chemical leaching disasters like Love Canal, where an industrial chemical dumping site caused injuries, cancer, birth defects, and led to waste disposal and toxics regulations in the 1970s and 1980s.[4] By looking more closely at the development of U.S. environmental laws, you can see why and how the nation's major environmental programs are the way they are.

I. History of Environmental Law in the United States

Before pollution became a topic of federal legislation, the United States effectively managed the environment as a profitable portfolio of natural resources that could be reduced to fungible products; including lumber, metals, and minerals.[5] While the federal government arranged for public land and water management and funded major infrastructure like highways and dams, the federal government did not create systems for governing environmental conservation or public health.

Theodore Roosevelt Jr. was one of America's first champions of the natural environment during his tenure as president from 1901 to 1909. He sought to conserve the American wilderness, and was influenced by writers like Henry David Thoreau and John Muir. He used his presidential power to pass the laws that established the nation's first national parks.[6] These are considered by many

3. In Chicago, civic leaders referred to the smog in the air as the "smoke horror." Stephan Benzkofer, *For Much of Its History, Chicago Covered by Smoke, Soot*, Chi. Trib., June 5, 2015.
4. S. Rep. No. 96-848, at 7 (1980) (stating that "propelled the problems of inadequate hazardous chemical waste disposal into the national spotlight").
5. Richard N.L. Andrews, Managing the Environment, Managing Ourselves: A History of American Environmental Policy 227 (1999).
6. Norman Tyler et al., Historic Preservation 27-28 (2d ed. 2009). Henry David Thoreau was an American writer who left regular society to live in nature in a cabin by Walden Pond. He wrote the book, *Walden*, a book that is widely embraced by environmentalists as a foundational text. John Muir was an environmental philosopher who founded the Sierra Club and petitioned Congress for the 1890 national parks bill in the United States.

to be the first environmentally focused laws.[7] During Roosevelt's presidency, the federal legislature also created public health-focused laws that safeguarded natural resources like drinking water.[8] Several decades later, in the 1930s, the Civilian Conservation Corps was established by the New Deal.[9] Under the New Deal, projects were launched to improve the nation's infrastructure, including its drinking water and sewage systems.[10] Much of the focus of these early laws was on the environmental needs of growing populations and urban centers faced with waste disposal and potable water access issues.

In the 1950s, the fast growth of chemical, plastics, petroleum, automotive, aviation, and munitions industries created more highly visible forms of pollution with the proliferation of industrial plants and large-scale chemical operations.[11] The federal government developed several federal programs to establish air and water pollution standards to curb the soot, smog, and polluted waterways.[12] These pollution control laws, which included statutes like the Federal Water Pollution Control Act[13] and the Air Pollution Control Act,[14] focused on public health rather than environmental improvements. These anti-pollution programs were administered by the Public Health Service.[15]

In 1962, Rachel Carson published her book, *Silent Spring*, warning the world about the environmental impacts of pesticide use.[16] Several years later, in 1966, the Federal Water Quality Administration separated from the Public Health Service and incorporated with the U.S. Department of the Interior (DOI). This administrative change shifted environmental governance from a

---

7. Jack Lewis, *Looking Backward: A Historical Perspective on Environmental Regulations*, 14 EPA J. 26 (1988).
8. Robert A. Shanley, *Franklin D. Roosevelt and Water Pollution Control Policy*, 18 PRESIDENTIAL STUD. Q. 319 (1988).
9. Robert Fechner, *The Civilian Conservation Corps Program*, 194 ANNALS AM. ACAD. POL. & SOC. SCI. 129 (1937).
10. Robert D. Leighninger Jr., *Cultural Infrastructure: The Legacy of New Deal Public Space*, 49 J. ARCHITECTURAL EDUC. 226 (1996).
11. Lewis, *supra* note 7.
12. *Id.* (The Federal Water Quality Administration of 1965 and the National Air Pollution Control Administration originated as a research body in 1955, but it was not named that until 1968.)
13. Federal Water Pollution Control Act of 1948, Pub. L. No. 80-845, 62 Stat. 1155.
14. Air Pollution Control Act of 1955, Pub. L. No. 84-159, 69 Stat. 322.
15. Frank J. Barry, *The Evolution of the Enforcement Provisions of the Federal Water Pollution Control Act: A Study of the Difficulty in Developing Effective Legislation*, 68 MICH. L. REV. 1103, 1105 (1970) (explaining that the Federal Water Pollution Control Act of 1948 was led by the U.S. surgeon general who, at the time of the statute, was the head of the Public Health Service); *Advisory Committee on Community Air Pollution*, 72 PUB. HEALTH REP. 645 (1957). The Air Pollution Control Act was a law that enabled the U.S. surgeon general, as head of the Public Health Service, to research air pollution control, and the Public Health Service created the National Advisory Committee on Community Air Pollution to carry out those research goals.
16. RACHEL CARSON, SILENT SPRING (1962).

public health administration initiative and into the realm of natural resource regulation. In the natural resource governance scheme of the 1960s, each type of natural resource was considered separately, or by medium, rather than holistically.[17] This single-media separation was the result of administrative convenience in an era that did not yet officially recognize the interrelatedness of ecological issues.

Environmental law in its modern form first emerged in the late 1960s. Congress broke from its traditional tendency to create piecemeal, media-by-media legislation to create the first system of holistic environmental regulation in the nation. The National Environmental Policy Act (NEPA), the first major modern environmental law in the United States, was signed by President Richard M. Nixon on New Year's Day in 1970, and the first Earth Day was celebrated several months later, on April 22, 1970.[18]

During this era, other forms of social and legal change were also underway as the Vietnam War continued and the civil rights movement changed the nation's political landscape.[19] In 1972, the Watergate scandal unfolded. The anti-government skepticism that culminated in the civil rights and anti-war movements of the 1960s spurred the creation of government transparency laws and oversight laws that assured a more open government. This book deals with both environmental laws and government transparency or "sunshine" laws as access points for environmental information and sources of environmental records transparency. It is useful to note the historical confluence of increased government transparency and increased environmental legislation in the civil rights-focused era. The rise of both transparency and environmental measures occurred during a time rife with government distrust and a real push from the public to increase government accountability.

The proliferation of environmental laws in the 1970s gave the federal government a much larger role in environmental protection.[20] To fill that role, President Nixon issued an order establishing the U.S. Environmental Protection Agency (EPA), setting parameters also for the creation of the National

---

17. Richard J. Lazarus, *The Greening of America and the Graying of United States Environmental Law: Reflections on Environmental Law's First Three Decades in the United States*, 20 Va. Envtl. L.J. 75, 79 (2001). In 1948, toxic smog from a Donoro, Pennsylvania, zinc smelting operation asphyxiated 20 people and sent 7,000 others to the hospital. Incredibly, in 1952, the Cuyahoga River actually caught fire.
18. Adam Rome, *The Genius of Earth Day*, 15 Envtl. Hist. 194 (2010).
19. Andrews, *supra* note 5, at 237.
20. *Id.* at 229 (noting "[b]oth Democrats and Republicans responded with an outpouring of federal legislation to protect the environment. . . . By the end of 1980, bipartisan majorities had passed an unprecedented body of federal environmental protection legislation.").

Oceanic and Atmospheric Administration (NOAA).[21] The creation of a federal agency dedicated to regulating and enforcing environmental laws brought together the various governmental efforts to reduce pollution. Despite the clear intentions of the federal government to create a new environmental agency, the process of consolidating the various environmental programs into a singular agency in the federal government was quite complicated.[22]

II. Creation of EPA

Following decades of public health legislation that only considered ecological concerns in limited ways, EPA was created to focus on the environment, integrating all of the federal government's federal research and regulatory programs dealing with pollution under the umbrella of a single agency.[23] President Nixon's Reorganization Plan No. 3 of 1970, creating EPA, was a major step in centralizing and developing environmental programs in the United States.[24] While the plan created an agency to focus on environmental issues, it did not establish an overall mission for EPA. Instead, the plan merely reallocated various federal government offices and programs, grouping them together to form the Agency. For this reason, EPA's power was much like the disparate laws and programs it oversaw, "fragmented across a complex patchwork of legally separate programs."[25]

Although EPA was designed to streamline the programs, integrating the scattered, often conflicting public health governance with ecological goals was so burdensome that creating a cohesive environmental agency proved challenging.[26] This is apparent today when trying to maneuver through EPA's various environmental law projects and programs on the Agency's website. The many environmental programs under EPA's authority vary in their organization, structure, and design, separated into a web of offices and sub-offices, forming a jumbled collection of regulation and guidance.[27]

21. EPA Order 1110.2 (Dec. 4, 1970); *see also* U.S. EPA, *The Origins of EPA*, https://www.epa.gov/aboutepa/origins-epa (last updated Sept. 7, 2016). Note that this order came out in the same year that NEPA and the Clean Air Act (CAA) were also signed into law.
22. Andrews, *supra* note 5, at 227.
23. *Id.*
24. After its proposal, Reorganization Plan No. 3 was sent to the U.S. Senate and the U.S. House of Representatives pursuant to 5 U.S.C. ch. 9. In the summer of 1970, the plan went through extensive congressional hearings and was ultimately approved.
25. Andrews, *supra* note 5, at 230.
26. Lewis, *supra* note 7.
27. Eventually, EPA did create a mission statement, which is available on their website, U.S. EPA, *Our Mission and What We Do*, https://www.epa.gov/aboutepa/our-mission-and-what-we-do (last visited Nov. 16, 2016).

With the new agency and its mandates, the environmental law scheme was extended beyond the state and local prerogatives that existed prior to 1970 to enter the federal government's domain.[28] The new agency's policy tools—federal standards and permit requirements for private entities—had not previously been used to regulate the environment. This put the federal government in an adversarial position against industry, as polluters were required to obtain permits, limit their pollution outputs, and satisfy reporting requirements in order to operate. While the newly regulated industries may have been displeased with these developments, these laws have been a boon for environmental information seekers.

The creation of EPA was a large step toward a unified, holistic national environmental policy, but in form and practice, the agency was imperfect. Like the programs the agency oversaw and enforced, the agency itself had a jumbled structure, with different administrators from various agencies lumped together to work on a hybrid mix of "media" programs (air and water); "categorical" programs (pesticides, radiation, and solid wastes); and functional tasks (enforcement, research, policy analysis, and administration).[29] EPA was also decentralized. Nixon implemented a "new federalism" model by having one-half of EPA's members spread across 10 regional offices, making it one of the most decentralized federal agencies. Each regional office had substantial autonomy in permitting, enforcement, and program development, leading to inconsistencies in environmental law application across the nation.[30]

Today, EPA is a bit more unified, and many more environmental laws exist. Nevertheless, the different regional offices still operate in a somewhat autonomous fashion, so there continues to be variations in environmental information access across the nation.

With the proliferation of environmental laws and regulations since EPA's inception in 1970, EPA creates and gathers more data than ever.[31] Federal law requires that the Agency collect volumes of environmental information. As a result, EPA[32] creates, collects, and catalogs a large amount of environmental data through its regulatory programs and environmental studies.

---

28. E. Blaine Rawson, *Overfiling and Audit Privileges Strain EPA-State Relations*, Nat. Resources & Env't, Winter 1999, at 483 (discussing the new federal environmental programs created in the 1970s as a response to the inconsistent, and often ineffective, state pollution control measures).
29. Andrews, *supra* note 5, at 232.
30. *Id.*
31. According to EPA's website, it currently administers the implementation of 33 environmental law programs. *See* U.S. EPA, *Laws and Executive Orders*, https://www.epa.gov/laws-regulations/laws-and-executive-orders#majorlaws (last updated Sept. 8, 2016).
32. Although EPA is not the only agency that promulgates regulations affecting the natural environment—most agencies deal with environmental issues in one way or another—for instance, the U.S.

EPA's data collection could benefit people who seek all kinds of environmental information, because so much information from scientists, engineers, and regulated entities is now stored within a single agency. EPA serves as a pseudo-repository for environmental information, with EPA's libraries and databases providing access to many environmental records. Further, government transparency laws provide statutory rights to access government documents increasing environmental information access. The advent of both EPA and the Freedom of Information Act (FOIA) has given the public unprecedented access to environmental administrative records and the right to challenge EPA in court in order to obtain those records.[33] EPA's creation and its power to enforce and administer environmental law programs fundamentally changed the relationship of citizens to environmental information in the United States.

The improved information access is not without its flaws. Like the structure of EPA itself, the information created by the agency is scattered throughout regional offices across the country rather than held in a centralized location. The breadth, depth, and availability of particular types of data also vary depending on the environmental program and the office that administers it. The programmatic and geographic separations within EPA add an additional layer of challenges to the task of locating agency records.

III. Types of Environmental Laws

With the diversity of types and topics of environmental law that exist today,[34] it is useful to consider environmental laws categorically. Historically, different phases of environmental law creation resulted in different types of environmental laws: the first wave of environmental laws tended to be media-based laws (dealing with specific types of environmental media—land, water, etc.) and the second wave of laws focused on particular pollutants (toxics laws, etc.).[35]

These categories are imperfect. As environmental law jurisprudence continues to grow, attempts to discuss or categorize it become more complicated.[36] For one, it is difficult to isolate complex environmental laws and their

---

Department of Transportation, the U.S. Food and Drug Administration, and the U.S. Army Corps of Engineers regularly engage in regulatory schemes and undertake government projects that impact the environment.

33. ANDREWS, *supra* note 5, at 228.
34. *See infra* app.1 (note this appendix does not include laws related to national parks or federal land acquisitions, which also govern issues relevant to the environment to some extent).
35. RICHARD J. LAZARUS, THE MAKING OF ENVIRONMENTAL LAW 169-70 (2004).
36. ELI, LAW OF ENVIRONMENTAL PROTECTION §1:4 (2016).

regulatory programs into single categories. Furthermore, many environmental laws incorporate several different types and styles of environmental regulation into a single statutory and regulatory scheme.[37] It is also impossible to treat the various environmental media laws uniformly. For example, improving air quality is a completely different exercise than protecting forestlands. Also, major preexisting structures within the government, like federal wastewater management systems, and institutions beyond the government's control, like the automobile trade, cannot be obliterated in order to serve environmental interests, and instead, must be respected as immutable parts of the environmental policy landscape.

With those major caveats, and noting the inexact nature of these categories, environmental law schemes tend to conform to one or several of these models. The following are descriptions of the broad categories of environmental laws, including basic characteristics and information about the types of laws that fall into each category:

- Informational and Planning Laws. These are laws like NEPA and the Emergency Planning and Community Right-To-Know Act (EPCRA). The primary purpose of these laws is information gathering intended to better inform environmental decisionmaking and planning.
- Pollution Control Laws. These laws include some of our earliest environmental regulatory schemes and include statutes such as the Clean Air Act (CAA) and the Clean Water Act (CWA). These laws aspire to balance commercial productivity with environmental concerns, allowing for controlled discharges of minimal pollutants within a larger permitting and monitoring framework.
- Resource Management Laws. Public land management laws fall into this category; however, the most notable resource management statute is the Endangered Species Act (ESA). These types of laws manage the health, population, and habitat of living organisms. These laws employ permitting processes to allow for limited use or "taking" of organisms.
- Chemical Tracking Laws. This category contains a triumvirate of laws including the Toxic Substances Control Act (TSCA), the Resource Conservation and Recovery Act (RCRA), and the Federal Insecticide, Fungicide, and Rodenticide Act (FIFRA). Each law addresses a different category of chemicals and tracks the chemicals from creation,

37. For instance, the Emergency Planning and Community Right-To-Know Act (EPCRA) is an informational and planning law, but it also requires some chemical tracking and listing in the required inventory reports filed by the facilities within EPCRA's scope.

through their use and eventual disposal to limit their harm to human health and potential damage to the environment.

- Liability Laws. The most well-known liability law is the Comprehensive Environmental Response, Compensation, and Liability Act (CERCLA). This law is the source of many complex lawsuits that assign and parse liability among parties responsible for cleaning up pollution.

These categories of environmental law demonstrate the different ways that environmental regulation can be achieved: carrying out environmental goals and preserving industrial interests and their commercial value. Different permitting programs have different ways to weigh the various pollution limitations—some of the laws are designed to not just prevent pollution, but to improve the environmental efficacy of industrial technologies by forcing polluters to implement the best technology available. There are laws that assess environmental and public health risks and balance them against the commercial benefits of environmentally harmful activities. Some environmental statutes employ market incentives like tax rebates and the ability to buy and sell pollution rights to urge polluters to make environmentally beneficial decisions.[38] These statutory models attempt to strike a balance between conservation and commercial interests, allowing the U.S. economy to continue to grow, but also curbing the massive environmental degradation that has occurred in other nations.

Historically, different types of laws have been used in various ways by different administrations. Different political eras have had disparate views on the value of conservation and the best forms of environmental governance. Policymakers create and shape the nation's environmental laws, determining who manages U.S. natural resources and how much funding the subsequent environmental programs receive. For example, President Ronald Reagan wanted to select an EPA administrator that would bring the agency "to its knees" to minimize environmental law and administration. Major environmental decisions in the United States are determined by the government, making access to government information all the more critical.

38. Robert W. Hahn & Robert N. Stavins, *Incentive-Based Environmental Regulation: A New Era From an Old Idea*, 18 Ecology L.Q. 1, 42 (1991).

IV. Summary of Major Environmental Laws

This section provides a brief overview of several major U.S. environmental statutes. Some of these statutes fit neatly into a single category, but many actually contain elements from several different types. For example, some laws require best technologies and while also requiring agencies to complete an assessment balancing risks and benefits as they promulgate standards. Chapter 4 provides specific information about how to access the environmental information generated and collected by federal agencies through the environmental programs created to implement each of these laws. This chapter simply provides historical context and a brief overview of select major statutes to provide a deeper understanding of how the U.S. environmental law scheme operates.

Although this chapter discusses some major environmental laws, many more laws exist affecting aspects of the natural environment that do not appear in this chapter. Laws like the Oil Pollution Act, National Marine and Fisheries Act, and dozens more, are not discussed in this chapter but are important environmental law provisions, nonetheless.[39] Appendix 1 of this book provides citation and legislative history information for many environmental laws in addition to those discussed here.

A. *Informational and Planning Laws*

These are broad, overarching environmental laws that focus not on particular natural media, but on environmental planning and awareness. They facilitate environmental decisionmaking by forcing agencies to consider the environmental consequences of their actions; and forcing entities handling hazardous materials to participate in designing safeguards and working with municipalities to prevent the detrimental effects of hazardous materials-related emergencies.

39. For more information about major federal environmental laws, see David M. Bearden et al., Cong. Research Serv., Environmental Laws: Summaries of Major Statutes Administered by the Environmental Protection Agency (Dec. 20, 2013), https://www.fas.org/sgp/crs/misc/RL30798.pdf.

National Environmental Policy Act (NEPA)

Codification:
- 42 U.S.C. §§4341-4370h, ELR STAT. NEPA §§2-209

Regulations:
- 40 C.F.R. §§6, 1500-1508 (2016)
- 7 C.F.R. §520 (2016)
- 10 C.F.R. §1021 (2016)

Select History:
- National Environmental Policy Act of 1969,[40] Pub. L. No. 91-190, 83 Stat. 852 (1970)
- An Act to Amend the National Environmental Policy Act of 1969 in Order to Clarify the Procedures Therein With Respect to the Preparation of Environmental Impact Statements, Pub. L. No. 94-83, 89 Stat. 424 (1975) (amending 42 U.S.C. §4332)
- An Act to Authorize Further Appropriations for the Office of Environmental Quality, and for Other Purposes, Pub. L. No. 94-52, 89 Stat. 258 (1975)

Agency:
- Council on Environmental Quality (CEQ)
- All federal agencies

NEPA was introduced in 1969 and signed into law on January 1, 1970, several months before the first observance of Earth Day. The text of this statute is short, simple, and broad, creating a national policy that requires all federal agencies to consider environmental protection in their plans and projects.[41] NEPA also established the Council on Environmental Quality (CEQ) and charged the CEQ with overseeing NEPA's administration.[42]

NEPA's policy goals are to force agencies to "stop and think" about the environmental impact of all federal and federally funded activities, and to take a "hard look" at the potential environmental ramifications of government activities.[43] The law requires agencies to draft environmental impact statements (EIS) when an agency activity may cause significant environmental consequences. These EIS are made available to public officials and citizens, and the agency drafting the EIS must provide comment periods for the drafts and final versions of the documents.

Many agency projects, however, do not require federal agencies to research and compile an EIS. If an agency determines that an activity will not have a

40. A discussion of NEPA's legislative history can also be found in the case of Pacific Legal Found. v. Andrus, 657 F.2d 829, 833-37 n.5, 11 ELR 20871 (6th Cir. 1981).
41. Lazarus, *supra* note 17, at 77-78.
42. Alvin K. Alm, *NEPA: Past, Present, and Future*, 14 EPA J. 32 (1988).
43. Natural Resources Defense Council v. Morton, 458 F.2d 827, 838, 2 ELR 20029 (1972).

significant environmental impact, they can make a finding of no significant impact (FONSI). If an agency makes a FONSI, the agency can draft a brief environmental assessment (EA) in lieu of conducting a full environmental study needed to draft an EIS. Thus, if a project is not deemed environmentally significant, an agency will provide much less information about the potential environmental outcomes of the project.

NEPA's provisions caused a sea change in agencies' decisionmaking processes, forcing the agencies to pause before engaging in an activity in order to prioritize environmental considerations in their decisionmaking processes. The courts have upheld NEPA's "stop and think" philosophy since the law's inception. In a 1971 U.S. Court of Appeals for the District of Columbia Circuit case, the court required the Atomic Energy Commission to rigorously and with good faith, carry out NEPA review before granting a construction permit for a nuclear power plant.[44]

The main critique of NEPA is that it is a "toothless tiger," lacking any enforcement mechanism.[45] Beyond drafting an EIS or EA, agencies do not have to do anything to prevent or ameliorate the environmental harm caused by their activities. Although NEPA successfully forces agencies to consider the environmental impacts of their projects and activities, the statute does nothing to force agencies to make beneficial environmental decisions. Although the public can look at an EIS and provide feedback to the agency about the environmental impacts of a proposed project, the law does not prevent agencies from ignoring public advice or best environmental practices, and agencies that move forward with environmentally detrimental projects suffer no ill consequences.

44. Calvert Cliffs Coordinating Comm. v. U.S. Atomic Energy Comm'n, 449 F.2d 1109, 1 ELR 20346 (D.C. Cir. 1971).

45. *See generally* Roger C. Cramton & Richard K. Berg, *On Leading a Horse to Water: NEPA and the Federal Bureaucracy*, 71 Mich. L. Rev. 511, 517 (1973) (noting "NEPA is not a toothless tiger").

Emergency Planning and Community Right-To-Know Act (EPCRA)

Codification:
- 42 U.S.C. §§11001-11050, ELR Stat. EPCRA §§301-330

Regulations:
- 40 C.F.R. §§3.1 et seq.

Select History:
- Superfund Amendments and Reauthorization Act (SARA) of 1986,[46] Pub. L. No. 99-499, 100 Stat. 1728

Agency:
- EPA
- Toxics Release Inventory (TRI) Program
- U.S. State and Local Governments
- Local Emergency Planning Committees (LEPCs)
- State Emergency Response Commission (SERC)

EPCRA was enacted in 1986 as section III of the Superfund Amendments and Reauthorization Act (SARA). The law requires federal, state, and local agencies to provide prospective environmental information to the public when there is the potential for a hazardous waste disaster. Specifically, the law forces government entities to develop emergency plans for the public in case of environmental catastrophe.

The law, created in the wake of the Bhopal, India, disaster,[47] was designed to provide the public with information about potential environmental danger in their municipalities to improve disaster preparation and response. EPCRA has four major requirements: emergency planning[48]; emergency release notification[49]; hazardous chemical storage reporting[50]; and the toxic release inventory (TRI).[51]

Under EPCRA, facilities must create Occupational Safety and Health Administration (OSHA) material safety data sheets (MSDS) and provide those information sheets to: state emergency response commissions (SERC);

46. *See id.* tit. III.
47. Tim Edwards, *What Happened*, The Bhopal Medical Appeal, http://bhopal.org/what-happened/ (last visited Nov. 15, 2016). The Bhopal Disaster was the world's worst industrial disaster. In 1984, a gas leak at the Union Carbide India Limited pesticide plant in Bhopal, India, exposed more than 500,000 people to deadly chemicals, killing thousands of people and blinding and burning the respiratory tracts of thousands more.
48. 42 U.S.C. §§11001-11003.
49. *Id.* §11004.
50. *Id.* §§11021-11022.
51. *Id.* §11023.

local emergency planning committees (LEPC); local fire departments; and any additional entities required by state law if those facilities have a certain number of specific types of "extremely hazardous" substances on site. These data sheets must be made available to the public, as well. EPCRA does not focus on environmental threats that are not immediately catastrophic, limiting its focus to data regarding operations where catastrophic disasters could occur.

EPCRA's establishment of the TRI is the most well-known aspect of the law.[52] The TRI compiles data about toxic releases—times when the chemicals that threaten human health and the environment are discharged into the air or water or placed in some type of land disposal. Unlike NEPA, which lacks an enforcement mechanism, EPA imposes fines for regulated industries that fail to report TRI data in a timely manner.

B. *Media-Specific Pollution Control Laws*

Both the CWA and the CAA were introduced soon after NEPA to provide media-specific goals to protect specific resources—namely air and water. These natural element-based laws focus on a particular aspect of the environment and require polluters to apply for permits to build and run their stationary point-sources of pollution. Stationary point sources include things like electrical plants and manufacturing plants that output pollution into the air and water. These regulated entities must regularly report their polluting activities to the federal government. Unlike NEPA, EPCRA, and other laws that focus on specific hazardous chemicals and their cleanup and containment, the CAA and the CWA focus on limiting pollution in our air and water, and reporting pollutants that enter both of those natural media. Following NEPA's general EIS requirements, the CWA and the CAA reflect the increasing complexity of environmental laws, as these statutes are full of particularity and details.

52. *Id.*; 40 C.F.R. pt. 372.

Clean Air Act (CAA)

Codification:
- 42 U.S.C. §§7401-7671q, ELR STAT. CAA §§101-618

Regulations:
- 40 C.F.R. §§49-97, 122-24, 144-45, 270-71 (2016)

Select History:
- Clean Air Act of 1970,[53] Pub. L. No. 91-604, 84 Stat. 1676
- 1977 Amendments to the Clean Air Act of 1970, Pub. L. No. 95-95, 91 Stat. 685
- 1990 Amendments to the Clean Air Act of 1970, Pub. L. No. 101-549, 104 Stat. 2399

Related:
- Air Pollution Control Act of 1955, Pub. L. No. 84-159, 69 Stat. 322
- Clean Air Act of 1963, Pub. L. No. 88-206, 77 Stat. 392
- Air Quality Act of 1967, Pub. L. No. 90-148, 81 Stat. 48
- Motor Vehicle Control Act of 1960, Pub. L. No. 86-493, 74 Stat. 162
- Motor Vehicle Air Pollution Control Act of 1965, Pub. L. No. 89-272, 79 Stat. 992

Agency:
- EPA
- EPA-Approved State Delegated Programs

The CAA of 1970 was the first federal environmental law that proscribed best practices for preventing pollution, seeking to actually reduce particulate matter in the air, rather than merely proscribing rules about resource management. The CAA did four major things to control air quality. First, it authorized EPA to set national ambient air quality standards (NAAQS) based solely on human health risks (without regard to cost) for the six most common air pollutants.[54]

Second, it required states to develop state implementation plans (SIPs) to achieve NAAQS. It also required every new point source to obtain federal permits prior to construction and operation. Permits were based not on risk but on technology. Thus, to get a permit, a point source has to show that it is meeting the current new source performance standards and using the maximum achievable control technology (MACT) for controlling air pollution.

53. Sometimes referred to as the 1970 Amendments referencing the 1967 Air Quality Control Act.
54. The six common air pollutants, called the Criteria Air Pollutants, are ground-level ozone, particulate matter, carbon monoxide, lead, sulfur dioxide, and nitrogen dioxide. They are discussed in depth at *Criteria Air Pollutants*, EPA (last updated Oct. 19, 2016), https://www.epa.gov/criteria-air-pollutants.

Third, it required automobile manufacturers to reduce air pollution from new vehicles by 90% in 1975. This requirement was a technology-forcing requirement, obligating the automobile industry to develop new technology to achieve a prescribed result.

Fourth, it authorized EPA to develop risk-based standards for control of hazardous air pollutants, those that were less common than the primary six but might cause serious hazards in local areas where they were emitted.

The CAA represented a major policy shift in U.S. environmental governance, moving the onus of preventing industrial air pollution from states, instead creating uniform federal standards and a national technology-based permit system for industries polluting the air.[55] The CAA's regulatory program thresholds were based on medical science rather than a balancing of health considerations against compliance costs, which meant that some of the CAA's standards created costly obligations for industrial polluters. In 1977 amendments to the CAA, EPA provided some leeway to industries, allowing new industrial operations to purchase "offsets" (emissions rights among firms in regional air sheds) if they planned to operate in a nonattainment area so polluters could bank pollution rights to sell to later users. In 1979, the CAA was amended to allow businesses to collect excess emission reduction credits to trade with other sources that need credits. Now polluters can trade and balance pollution credits as though they are all enclosed under one regional "bubble," sharing the same net pollution reduction requirements.

55. 42 U.S.C. §7416 (preempting states from adopt or enforce any emission standard or limitation which is less stringent than standards or limitations under the CAA).

Clean Water Act (CWA)

Codification:
- 33 U.S.C. §§1251-1387, ELR STAT. FWPCA §§101-607

Regulations:
- 40 C.F.R. §§104-136, 140, 144-45, 230-33, 401-71, 501-03 (2016)

Select History:
- Federal Water Pollution Control Act of 1948, Pub. L. No. 80-845, 62 Stat. 1155
- Federal Water Pollution Control Act Amendments of 1972, Pub. L. No. 92-500, 86 Stat. 816
- Clean Water Act of 1977, Pub. L. No. 95-217, 91 Stat. 1566
- Water Quality Act of 1987, Pub. L. No. 10-4, 101 Stat. 7

Agency:
- EPA
- EPA-Approved State Delegated Programs
- U.S. Army Corps of Engineers

The modern version of the CWA was passed as the Federal Water Pollution Control Act Amendments (FWPCA) of 1972. The 1972 Amendments expanded the existing water quality standards in the 1948 Act, and included several new enforcement mechanisms. The Amendments' greatest achievement, however, was the creation of the ambitious national goal that all surface waters be fishable and swimmable by 1983, and that they receive zero pollution discharge by 1985.[56] The 1972 Amendments also required permits for every major surface water polluter or discharger, not just new pollution sources. These CWA permits were supposed to be based on technology performance standards, with increasing technology requirements over time. By 1977, major sources of water pollution were to utilize the best practicable technology, and they should implement the best available control technology by 1983. The Act's technology requirements only apply to point sources of water pollution, so nonpoint water pollution sources like farms, construction sites, urban stormwater runoff were exempt from regulation. For nonpoint sources, EPA was only authorized to develop non-enforceable voluntary guidelines for best management practices.

Several major regulatory permitting programs established under the CWA work in concert to curb pollution in U.S. waterways. The national pollution discharge elimination system (NPDES) is the permitting program for point source pollution discharges (from sources like pipes, ditches,

---

56. 33 U.S.C. §1251(a)(1).

and sewers) into waters of the United States.[57] The national and local pretreatment standards require industrial polluters to prevent pollution from being discharged to publicly owned treatment works (POTWs).[58] The Section 404 Dredge and Fill Discharge Permit Program, is administered by the U.S. Army Corps of Engineers (the Corps) to regulate the placement of dredge and fill materials into U.S. waters and wetlands.[59] Finally, the Sewage Sludge Use and Disposal Program establishes a permit system requiring creators, users, and disposers of sewage sludge to meet certain standards.[60]

The CWA also authorized EPA to set standards for toxic water pollutants similar to the CAA's standards for hazardous air pollutants. Additionally, because municipal wastewater plants are major water polluters, the 1972 Amendments provided $6 billion in federal subsidies for three years to improve those facilities. This grant was the largest federal public works program in history, but it was defunded by the Reagan Administration.[61] In 1987, however, Congress established the Clean Water State Revolving Fund to replace the Clean Water Act Construction Grants Program that President Reagan vetoed, and that revolving fund is still operational today.[62]

57. *Id.* §1342.
58. *Id.* §1345.
59. *Id.* §1345(f)(1).
60. *Id.* §1345.
61. Claudia Copeland & Mary Tiemann, Cong. Research Serv., Water Infrastructure Needs and Investment: Review and Analysis of Key Issues 5 (Dec. 21, 2010) (the Reagan Administration sought a phaseout of the Act's construction grants program by 1990).
62. U.S. EPA, *Learn About the Clean Water State Revolving Fund (CWSRF)*, https://www.epa.gov/cwsrf/learn-about-clean-water-state-revolving-fund-cwsrf (last visited Nov. 14, 2016).

Safe Drinking Water Act (SDWA)

Codification:
- 42 U.S.C. §§300f to 300j-26, ELR STAT. SDWA §§1401-1465

Regulations:
- 40 C.F.R. pt. 141 (2016)

Select History:
- Safe Drinking Water Act of 1974, Pub. L. No. 93-523, 88 Stat. 1660
- Safe Drinking Water Act Amendments of 1986, Pub. L. No. 99-339, 100 Stat. 642
- Lead Contamination Control Act of 1988, Pub. L. No. 100-572, 102 Stat. 2885
- Safe Drinking Water Act Amendments of 1996, Pub. L. No. 104-182, 110 Stat. 1613
- Public Health Security and Bioterrorism Preparedness and Response Act of 2002, Pub. L. No. 107-188, 116 Stat. 594

Agency:
- EPA
- U.S. States (though the Public Water System Supervision and Underground Injection Control (UIC) Programs)
- National Drinking Water Advisory Council

In the early 1970s, the Safe Drinking Water Act (SDWA)[63] directed EPA to set national standards for minimum contaminant levels in public water supplies and also required local water utilities to monitor water supplies and install EPA-approved technology to control their safety. Thus, EPA created primary drinking water standards to protect public health and secondary standards to assure qualities such as the taste, odor, and appearance of the nation's drinking water.[64] These drinking water standards considered the best available technology and the costs associated with using that technology. EPA sets maximum contaminant levels in national primary drinking water regulations. These standards include methods for cleaning the water. The standards also create testing requirements for drinking water sources in order to remain under the maximum contaminant level thresholds. SDWA programs are largely carried out by states and localities, as the SDWA allows EPA to grant primacy to states to implement the SDWA's standards, or even standards more stringent than those set out by the federal government.[65]

63. 42 U.S.C. §§300f to 300j-26, ELR STAT. SDWA §§1401-1465.
64. Press Release, U.S. EPA, EPA Voices Support for Safe Drinking Water Act (Mar. 8, 1973), https://archive.epa.gov/epa/aboutepa/epa-voices-support-safe-drinking-water-act.html.
65. *See infra* ch. 9 (discussing primacy—state and local).

The SDWA is the principal federal law that regulates hydraulic fracturing (commonly known as fracking) activities.[66] The law establishes a national regulatory program for underground injection control (UIC) to ensure the safety of drinking water where fluids (solids, liquids, or gases) are injected into the ground.[67] The UIC regulations promulgated by EPA must "contain minimum requirements for effective programs to prevent underground injection which endangers drinking water sources."[68] This standard is known as the endangerment standard, and it is the "major driving force" of EPA's fracking regulation scheme.[69]

C. *Chemical Tracking Laws*

In the mid-1970s, Congress created new types of environmental statutes. These chemical tracking laws regulated toxic materials used in the United States. Rather than focusing on particular natural resources or media, these "second generation laws" were specific to particular types of pollutants.[70] These statutes emerged with the growing awareness that industrial chemicals can cause cancer. Fears about carcinogens sparked the public's interest in cutting down on the use of toxic substances.[71] Several lawsuits, including the famous dichlorodiphenyltrichloroethane (DDT) litigation,[72] as well as the discovery of mercury and polychlorinated biphenyls (PCBs) in fish,[73] raised public concern about unwanted toxic contaminants in food and water. New health-focused statutes broadened EPA's oversight from water and air pollution emissions to "man-made chemical hazards as a whole." Prior pesticide laws[74] were mere "truth in labeling laws" designed to regulate food adulteration or product misbranding. These new laws introduced licensing schemes

66. Mary Tiemann & Adam Vann, Cong. Research Serv., Hydraulic Fracturing and Safe Drinking Water Act Regulatory Issues 7 (July 13, 2015), https://www.fas.org/sgp/crs/misc/R41760.pdf.
67. 42 U.S.C. §§300h-300h-5; 40 C.F.R. §§144-148 (2016).
68. 42 U.S.C. §300h(b)(1).
69. Tiemann & Vann, *supra* note 67, at 9.
70. Lazarus, *supra* note 17, at 83.
71. Richard Nixon, Remarks on Signing the National Cancer Act of 1971 (Dec. 23, 1971) (stating "we find that more people each year die of cancer in the United States than all the Americans who lost their lives in World War II").
72. U.S. EPA, *DDT Regulatory History: A Brief Survey (to 1975)*, https://www.epa.gov/aboutepa/ddt-regulatory-history-brief-survey-1975 (last updated Dec. 4, 2014) (excerpt from U.S. EPA, DDT, A Review of Scientific and Economic Aspects of the Decision to Ban Its Use as a Pesticide, Prepared for the Committee on Appropriations of the U.S. House of Representatives by EPA (1975) (EPA-540/1-75-022) (noting Environmental Defense Fund DDT lawsuit and Natural Resources Defense Council's lawsuits against EPA to force it to set standards for toxic water pollutants)).
73. PCBs were widely used as lubricants in electrical transformers.
74. Pure Food and Drug Act of 1906, Pub. L. No. 59-384, 34 Stat. 768; Federal Insecticide Act of 1910, Pub. L. No. 61-152, 36 Stat. 331.

requiring toxic substance manufacturers to test the effects of their product on human health and to track toxic substances sold in the United States.

These new programs, however, were difficult to implement because Congress did not follow the licensing model created for food and drugs. For instance, TSCA's pre-manufacture notice (PMN) required EPA to accept the data and safety information provided by the manufacturer. Thus, to get more information under TSCA, EPA must not only take the initiative to gather that information, but must also prove that there is a need to obtain the additional data.[75]

Moreover, efforts to register toxic chemicals were complicated because of the huge number of chemicals already on the market at the time of TSCA's passage and EPA's inability to collect sufficient data about those chemicals. Not only was information about chemicals hard to gather, the data was often limited and did not account for the chemicals' actual common usage practices that impact public health and safety, such as the actual use, misuse, and overuse of pesticides in day-to-day activities like farming, golf course maintenance, lawn care, etc.[76] Although TSCA's goals were ambitious, the implementation of the federal statute and the limitations placed on it by the judiciary made it one of the least effective environmental programs, requiring tons of research and analysis but producing regulations for only a handful of substances.[77] TSCA, as a model of the chemical tracking law schemes in the United States, demonstrates the disparities between the lofty goals of toxic substance control and oversight and the actual shortcomings of the regulatory programs developed to reach those goals.

75. 15 U.S.C. §2604(f) (2012).
76. ANDREWS, *supra* note 5, at 244.
77. Robert B. Haemer, *Reform of the Toxic Substances Control Act: Achieving Balance in the Regulation of Toxic Substances*, 6 ENVTL. LAW. 99 (1999).

Resource Conservation and Recovery Act (RCRA)/ Solid Waste Disposal Act (SWDA)

Codification:

- 42 U.S.C. §§6901-6992k, ELR STAT. RCRA §§1001-11011

Regulations:

- 40 C.F.R. §§239-273, 279-282 (2016)

Select History:

- Solid Waste Disposal Act, Pub. L. No. 89-272, Title II, 90 Stat. 2795
- Resource Recovery Act of 1970, Pub. L. No. 91-512, 84 Stat. 1228
- Resource Conservation and Recovery Act of 1976, Pub. L. 94-580, 90 Stat. 2798
- Used Oil Recycling Act of 1980, Pub. L. No. 96-463, 94 Stat. 2055
- Solid Waste Disposal Act Amendments of 1980, Pub. L. No. 96-482, 94 Stat. 2334
- Hazardous and Solid Waste Amendments of 1984, Pub. L. No. 98-616, 98 Stat. 3221
- Federal Facility Compliance Act of 1992, Pub. L. No. 102-386, 106 Stat. 1507
- Land Disposal Program Flexibility Act of 1996, Pub. L. No. 104-119, 110 Stat. 830
- Hazardous Waste Electronic Manifest Establishment Act, Pub. L. No. 112-195, 126 Stat. 1452

Agency:

- EPA
- EPA Office of Land and Emergency Management's (OLEM) Office of Resource Conservation and Recovery (ORCR)
- EPA-Approved State Delegated Programs

RCRA was passed in 1976 to be a waste management law, providing incentives for recycling and recovery of waste materials. RCRA amended the Solid Waste Disposal Act of 1965 (SWDA),[78] which focused on providing research and development funds to improve solid waste disposal technology and operations. The SWDA put EPA in charge of developing standards for waste disposal, and made states responsible for implementing and enforcing EPA's SWDA regulations.[79]

RCRA's amendments strengthened the SWDA's purpose and goals to ensure the proper disposal of solid waste on land. Industrial waste spills and deliberate waste dumping led Congress to require EPA to track hazardous wastes from cradle-to-grave, collecting manifests reporting the handling,

78. Pub. L. No. 89-272; *see* SENATE COMM. ON PUBLIC WORKS, 93D CONG., A LEGISLATIVE HISTORY OF THE SOLID WASTE DISPOSAL ACT 467 (Comm. Print 1974).
79. The Solid Waste Disposal Act of 1965, Pub. L. No. 89-272, 79 Stat. 997 (1965).

transportation, and ultimate disposal of the waste.[80] RCRA also ascribed civil and criminal penalties for noncompliance with RCRA and includes citizen suit provisions that empower members of the public to file lawsuits for noncompliance.[81]

Thus, under RCRA, EPA regulates hazardous industrial wastes and the industries that treat, store, or dispose of these wastes. RCRA requires facilities playing any role in the life cycle of toxic chemical substances to keep records about the chemicals, including cradle-to-grave manifests documenting all off-site shipments. This law effectively decreased dangerous waste disposal practices, including dumping toxic waste into oceans and other waterways and the open burning of wastes. RCRA also ended the commingling of industrial chemicals in municipal landfill waste.

80. Press Release, U.S. EPA, New Law to Control Hazardous Wastes, End Open Dumping, Promote Conservation of Resources (Dec. 13, 1976), https://archive.epa.gov/epa/aboutepa/new-law-control-hazardous-wastes-end-open-dumping-promote-conservation-resources.html.

81. *Id.*

Federal Insecticide, Fungicide, and Rodenticide Act (FIFRA)[82]

Codification:

- 7 U.S.C. §§136-136y, ELR STAT. FIFRA §§2-34

Regulations:

- 40 C.F.R. §§150-189 (2016)

Select History:

- Federal Insecticide, Fungicide, and Rodenticide Act of 1947, Pub. L. No. 80-104, 61 Stat. 163
- Federal Pesticide Act of 1978, Pub. L. No. 95-396, 92 Stat. 819
- Federal Insecticide, Fungicide, and Rodenticide Act Amendments, Pub. L. No. 96-539, 94 Stat. 3194
- Federal Insecticide, Fungicide, and Rodenticide Amendments of 1988, Pub. L. No. 100-532, 102 Stat. 2655
- Food Quality Protection Act (FQPA) of 1996, Pub. L. No. 104-170, 110 Stat. 1490
- Pesticide Registration Improvement Act of 2003, Pub. L. No. 108-199, Division G, Title V, 118 Stat. 419
- Pesticide Registration Improvement Renewal Act, Pub. L. No. 110-94, 121 Stat. 1000

Agency:

- EPA
- EPA Office of Pesticide Programs (OPP)
- EPA-Approved State Delegated Programs

FIFRA was originally enacted in 1947. The original version of the law required manufacturers to register all new pesticide products, and was merely a product labelling law until the 1972 Amendments, called the Federal Environmental Pesticide Control Act (FEPCA) created a new, more comprehensive pesticides governance scheme. FEPCA pretty much rewrote the pesticide law.[83] It reorganized pesticide regulation under EPA rather than the U.S. Department of Agriculture, which formerly regulated pesticides under the old FIFRA. FEPCA also required that pesticides only be used for specifically approved purposes that do not have unreasonable adverse effects on the environment.[84]

In its current form, FIFRA requires EPA to regulate the sale, distribution, and use of all pesticides in the United States. EPA must register every pesticide, requiring registrants to show that using the pesticide according to specifications will not cause unreasonable adverse effects on the environment.[85]

82. 7 U.S.C. §§136-136y, ELR STAT. FIFRA §§2-34.
83. ANDREWS, *supra* note 5, at 243.
84. 7 U.S.C. §136a(a).
85. *Id.* §136b(b) (defining the term).

FIFRA also requires pesticide, fungicide, and rodenticide producers to label their products. The labels contain information that instructs users about the proper usage of the chemical to best maximize the product's effectiveness and simultaneously protect the chemical applicator, consumers, and the environment. EPA's Office of Pesticide Programs (OPP) is in charge of regulating pesticides and carrying out FIFRA requirements.

Unlike the CWA and the CAA, FIFRA is primarily a risk-balancing statute. In order for EPA to register a pesticide, the Agency must weigh the environmental, health, and safety risks against the economic benefits of agricultural production. EPA must also repay the manufacturers and users for any stocks left unsold if a product is deregistered. Also, FEPCA grandfathered in all pesticides sold commercially before the 1970s, registering them until or unless EPA got around to reviewing them. Thus, the new law made it easier to keep using and selling old pesticides than to create and try to register new, possibly safer, ones.

Toxic Substances Control Act (TSCA)

Codification:
- 15 U.S.C. §§2601-2692, ELR Stat. TSCA §§2-412

Regulations:
- 40 C.F.R. pt. 700 to 799 (2016)

Select History:
- Toxic Substances Control Act of 1976, Pub. L. No. 94-469, 90 Stat. 2003
- Asbestos Hazard Emergency Response Act of 1986, Pub. L. No. 99-519, 100 Stat. 2970
- Radon Program Development Act of 1988, Pub. L. No. 100-551, 102 Stat. 2755
- Residential Lead-Based Paint Hazard Reduction Act of 1992, Pub. L. No. 102-550, 106 Stat. 3673
- Energy Independence and Security Act of 2007, Subtitle E, Pub. L. No. 110-140, 121 Stat. 1492
- Formaldehyde Standards for Composite Wood Products Act of 2010, Pub. L. No. 111-199, 124 Stat. 1367
- Frank R. Lautenberg Chemical Safety for the 21st Century Act of 2016, Pub. L. No. 114-182, 130 Stat. 448

Agency:
- EPA
- Centers for Disease Prevention and Control (CDC)
- U.S. Treasury Department
- U.S. Department of Housing and Urban Development

TSCA was originally passed to prevent new hazardous chemicals from entering the U.S. market. TSCA requires firms creating chemicals for commercial use to test their products and supply PMNs before chemicals are manufactured for use by the general public. Under TSCA, EPA can restrict the use of any chemical posing an "unreasonable risk." The statute allows a certain amount of risk, and the Agency must determine how much risk qualifies as "unreasonable." In addition to risk balancing, EPA can also take emergency action against imminent hazards under TSCA.

Like FIFRA, TSCA employs a screening strategy to conduct risk analyses on chemicals and prevent chemicals posing unreasonable risks to public health and the environment from entering the commercial market. EPA also regulates the sale of already marketed chemicals under TSCA. To that end, TSCA requires chemical manufacturers, importers, processors, and distributors to notify EPA of any substantial risks of injury to health or the environment as a result of the manufacture or sale of the products.

Despite its ambitious goals, TSCA has traditionally been considered one of the least effective environmental laws. The statute only addresses chemicals introduced after TSCA's creation. The law grandfathered in more than 60,000 chemicals that were already in commercial use prior to 1979—including some of the most toxic substances ever created by man. Although TSCA allowed EPA to review the already commercially available chemicals, EPA has only regulated a handful of the chemicals under their TSCA authority, largely because of the staggering amount of manpower required to review these chemicals. Some of the newer chemicals successfully reviewed and regulated under TSCA include asbestos, chlorofluorocarbons (CFCs), and PCBs.

Before 2016, TSCA had not been significantly updated in the decades since its creation. Environmental activists had largely written off the statute as being ineffective and outmoded.[86] President Barack Obama, however, signed the Frank R. Lautenberg Chemical Safety for the 21st Century Act into law on June 22, 2016.[87] This new law amends TSCA by creating new safety standards based on risks; creating clear and enforceable deadlines for reviewing chemicals; taking action on risks; and increasing the transparency of chemical information by banning unwarranted claims of confidentiality on chemical information.[88]

TSCA and RCRA have been compared as quite similar statutory schemes. Both laws include health-based and technology-based provisions, and both require some consideration of costs versus benefits. These toxic substance-focused statutes were created to supplement the media-based laws like the CWA and the CAA, because relying on media-focused statutes alone simply shifted the impacts of toxic pollutants from one media to the next. By considering the actual pollutants that affect the media, both RCRA and TSCA were meant to close the "last remaining loophole" in environmental law.[89]

D. *Cleanup Liability Laws*

CERCLA, as well as a combination of state laws, assign liability to polluters and property owners that contaminate their property.[90] CERCLA directs strict liability for the costs of responding to releases or threatened

86. James Hamblin, *Toxic Substances Will Now Be Somewhat Regulated*, The Atlantic, May 26, 2016.
87. Pub. L. No. 114-182.
88. *Id.*
89. Lazarus, *supra* note 17, at 83.
90. 26 U.S.C. §§4661-4662 (2012).

releases of hazardous substances.[91] Liable parties may also be compelled to clean up sites under the direction of government entities. Similar state laws also assign liability for pollution cleanup. Laws like the New Jersey Spill Condensation and Control Act[92] and the Pennsylvania Hazardous Sites Cleanup Act of 1988[93] ascribe liability for hazardous waste contamination in those states.

These environmental liability laws bring the financial impacts of environmental pollution directly to responsible parties, forcing them to pay for cleaning up their pollution. Some attorneys say that CERCLA has created an entire area of environmental practice as business owners all over the United States are sued to contribute to or perform cleanup at contaminated sites.[94]

91. Potentially liable parties are defined at 42 U.S.C. §9607(a).
92. N.J. Rev. Stat. §§58:10-23.11 to 58:10-23.11z (2016).
93. 35 Pa. Cons. Stat. §§6020.101 to 6020.1305 (2016).
94. Taft, Stettinius & Hollister, *CERCLA Then and Now: Looking Back Over the Last 30 Years* (Dec. 7, 2010), http://www.taftlaw.com/news/publications/detail/704-cercla-then-and-now-looking-back-over-the-last-30-years.

Comprehensive Environmental Response, Compensation, and Liability Act (CERCLA)

Codification:
- 42 U.S.C. §§9601-9675, ELR STAT. CERCLA §§101-405

Regulations:
- 40 C.F.R. pt. 279 (2016)
- 40 C.F.R. §§300.1 to 312.31 (2016)
- 40 C.F.R. pt. 451 (2016)

Select History:
- Comprehensive Environmental Response, Compensation, & Liability Act of 1980, Pub. L. No. 96-510, 94 Stat. 2767
- Superfund Amendments and Reauthorization Act (SARA) of 1986, Pub. L. No. 99-499, 100 Stat. 1615
- Omnibus Budget Reconciliation Act of 1990, Pub. L. No. 101-508, 104 Stat. 1388
- Community Environmental Response Facilitation Act of 1992, Pub. L. No. 102-426, 106 Stat. 2174
- National Defense Authorization Act for Fiscal Year 1997 (1996), Pub. L. No. 104-201, 110 Stat. 2422
- Asset Conservation, Lender Liability, and Deposit Insurance Protection Act of 1996, Pub. L. No. 104-208, 110 Stat. 3009
- Superfund Recycling Equity Act of 1999, Pub. L. No. 106-113, 113 Stat. 1501
- Small Business Liability Relief and Brownfields Revitalization Act of 2002, Pub. L. No. 107-118, 115 Stat. 2356

Agency:
- EPA
- All U.S. States, Territories, and Native American Tribes
- U.S. Department of Justice (DOJ)
- Agency for Toxic Substances and Disease Registry (ATSDR)

While RCRA prevented future toxic waste dumping, it failed to deal with the effects of past dumping, ignoring the places where hazardous waste was already leaching into soil and downstream wells and water supplies. In the late 1970s, the discovery of these past dumping sites, including the Love Canal landfill in New York State,[95] catalyzed advocacy for the cleanup of hazardous waste sites and against the creation of new ones. Outcry against hazardous waste sites in people's home communities became known as the not in my backyard (NIMBY) movement.

In 1980, Congress responded to the outcry over hazardous waste sites by enacting CERCLA, which is sometimes called the Superfund Act.[96] CER-

95. Phil Brown & Richard Clapp, *Looking Back on Love Canal*, 117 PUB. HEALTH REP. 95 (2002).

96. Comprehensive Environmental Response, Compensation, and Liability Act of 1980, Pub. L. No. 96-510, 94 Stat. 2767, amended by the Superfund Amendments and Reauthorization Act

CLA was created to guarantee that toxic waste dumping sites would be properly cleaned up. CERCLA charged EPA with identifying contaminated sites and developing the national priority list (NPL) of the most hazardous waste sites in America.[97] CERCLA also created the Agency for Toxic Substances and Disease Registry (ATSDR), which is charged with implementing CERCLA provisions. In addition, CERCLA allows EPA to take emergency actions to control immediate dumping hazards.

CERCLA creates joint and several liability for the costs of cleanup and damage for any entity that used contaminated sites. Anyone of the potentially responsible parties (PRPs)[98] can be held responsible for the entire cleanup cost without regard to their degree of fault. CERCLA also created a multibillion dollar federal trust fund (a Superfund) to pay for site cleanup. The original source for Superfund financing—industry taxation—expired in 1996, and the Superfund has since been filled with money appropriated by Congress and payouts won in federal cost recovery actions.

Like RCRA, CERCLA also proved effective, containing waste contamination risks at thousands of sites and creating a powerful incentive for safe waste management by threatening polluters with joint and several liability. By the 1990s, however, many cities wanted more flexible cleanup standards to permit brownfields redevelopment, or the industrial redevelopment of contaminated urban sites.

There are also limitations to the effectiveness of the Superfund as a source of money for contamination cleanup. The U.S. Government Accountability Office (GAO) notes that declining appropriations for the fund have delayed remedial cleanup activities at NPL sites.[99] The Superfund has been declared ineffective by many, with President William J. Clinton saying in 1993 that "we all know it doesn't work: The Superfund has been a disaster," because all of the money goes to attorney's fees, and none of the money funds the cleanup projects CERCLA was intended to compel.[100]

---

of 1986 (SARA), Pub. L. No. 99-499, 100 Stat. 1613. (The same act that created EPCRA also reauthorized CERCLA.)

97. 42 U.S.C. §9605(a)(8)(B) (requiring the creation and maintenance of this list).
98. *Id.* §9607(a).
99. U.S. GAO, Superfund: Trends in Federal Funding and Cleanup of EPA's Nonfederal National Priorities List Sites (2015) (GAO-15-812), http://www.gao.gov/products/GAO-15-812.
100. William J. Clinton, Remarks to Business Leaders, Pub. Papers, Feb. 11, 1993, at 88 book 1 (January 20 to July 31, 1993).

E. *Resource Management Laws*

Some environmental laws govern shared resources like public lands and natural flora and fauna within the United States. These laws manage the health, population, and habitats of living organisms, and create methods for limiting the amount of species and habitat destruction that can occur in these common resources as a result of human activities.

Endangered Species Act (ESA)

Codification:
- 16 U.S.C. §§1531-1544, ELR STAT. ESA §§2-18

Regulations:
- 50 C.F.R. pt. 402.03 (2016)

Select History:
- Endangered Species Act of 1973, Pub. L. No. 93-205, 87 Stat. 884
- Endangered Species Act Amendments of 1978, Pub. L. No. 95-632, 92 Stat. 3751
- Endangered Species Act Amendments of 1982, Pub. L. No. 97-304, 96 Stat. 1411
- Endangered Species Act Amendments of 1988, Pub. L. No. 100-478, 102 Stat. 2306
- National Defense Authorization Act for Fiscal Year 2004, Pub. L. No. 108-136, 117 Stat. 1392

Agency:
- EPA (Pesticides)
- National Oceanic and Atmospheric Administration (NOAA)
- U.S. Fish and Wildlife Services

Although it seems to focus only on protecting endangered flora and fauna, in reality, the ESA is one of the most powerful environmental protection laws. This law can stop both federal and private projects dead in their tracks if they threaten to impinge upon endangered species' habitats. The ESA was enacted in 1973, and it has since protected thousands of animals. Currently, more than 2,000 animals are considered threatened or endangered under the law.[101] The ESA has been called "the most comprehensive legislation for the preservation of endangered species ever enacted by any nation."[102] The ESA

---

101. NOAA Fisheries, *Endangered Species Act (ESA)*, http://www.nmfs.noaa.gov/pr/laws/esa/ (last updated Feb. 11, 2016).
102. Babbitt v. Sweet Home Chapter of Communities for a Great Or., 515 U.S. 687, 698, 25 ELR 21194 (1995) (quoting Tennessee Valley Auth. v. Hill, 437 U.S. 153, 180, 8 ELR 20513 (1978) (snail darter case)).

is more vital than ever as scientists report that the first confirmed species is extinct as a direct result of climate change.[103]

For environmental law purposes, however, the most effective aspect of the ESA is not necessarily the flora and fauna it protects, but the fact that the Act requires the protection of the habitat, or ecosystems in which that flora and fauna live. If a species is listed as endangered under the ESA, government and private activities on the land where that species resides are heavily restricted. The preservation of those lands for the purpose of preventing the extinction of the endangered flora and fauna has very few exceptions. Few actions and activities are allowed in places that serve as homes for endangered plants and animals.[104]

While the ESA goes a long way toward preventing the degradation of endangered species' natural habitats, it does not provide absolute protection. When the Act was amended in 1978, the legislature established an Endangered Species Committee[105] and gave that committee the authority to allow the extinction of animals by empowering the committee to exempt agencies from endangered species preservation requirements. The Endangered Species Committee is often called the "God Squad" because it decides which types of species live and die in U.S. habitats.[106] The committee is only allowed to use its exemption power under limited circumstances, when there are no other reasonable alternatives. There are also several ESA exemptions that allow certain activities in endangered species' habitats, even when those activities imperil the species.

ESA also creates "takings" prohibitions.[107] The statute explicitly prohibits anyone from "taking" (killing) endangered species. Initially, people got around the ESA takings provision by simply eradicating endangered species habitats before engaging in land use projects.[108] The U.S. Supreme Court case, *Babbitt v. Sweet Home Chapter of Communities for a Great Oregon*,[109] upheld U.S. Fish and Wildlife Service regulations labelling habitat modifications that hurt endangered species as takings,[110] making the "takings" provi-

---

103. Michelle Innis, *Australian Rodent Is First Mammal Made Extinct by Human-Driven Climate Change, Scientists Say*, N.Y. Times, June 15, 2016, at A6.
104. George Cameron Coggins & Robert L. Glicksman, 3 Public Natural Resources Law §29 (2d ed. 2013).
105. H.R. Rep. No. 95-1625 (1982), *reprinted in* 1982 U.S.C.C.A.N. 9463.
106. Jared des Rosier, Note, *The Exemption Process Under the Endangered Species Act: How the "God Squad" Works and Why*, 66 Notre Dame L. Rev. 825, 834 n.57 (1991).
107. 16 U.S.C. §§1538(a)(1)(B)-(C).
108. J.B. Ruhl, *Endangered Species Act Innovations in the Post-Babbittonian Era: Are There Any?*, 14 Duke Envtl. L. & Pol'y F. 419, 420 (2004).
109. 515 U.S. 687, 25 ELR 21194 (1995).
110. 50 C.F.R. §§17.3, 17.31 (2016).

sion of the statute a powerful tool in land use battles. Entire development projects can be halted by the very threat of harming an endangered species.

F. *Public Participation: Citizen Suits*

Citizen suit provisions are important parts of many environmental statutes discussed in this chapter. Citizen suits give U.S. citizens automatic standing to sue for implementation and enforcement of certain environmental laws, whether or not they suffer direct harm or have a direct stake in matters relating to an environmental statute or its implementation.[111] The ability to sue on behalf of environmental interests originates in the public trust doctrine and the idea that all people have a right to a clean environment. Citizen suits allow the public to participate more actively in the enforcement of environmental laws.[112] Citizens can sue agencies when agencies fail to enforce some environmental laws and sue EPA officials to compel them to fulfill duties like standard setting and promulgating regulations. Citizen suits also allow citizens to sue on behalf of communities at large and not just themselves.[113]

While many environmental laws, including the CWA,[114] the CAA,[115] the SDWA,[116] and RCRA[117] have these types of provisions, not all do. For example, under FIFRA, only "affected parties" are able to bring challenges in the courts, citizen public interest lawsuits are not allowed under that environmental statute.[118]

The presence of citizen suit provisions in environmental statutes increases the need for environmental information access in the United States. In order for citizens to file lawsuits to force standard setting or enforcement, people must first be able to assess the environmental dangers and needs of local communities and have sufficient information about issues of environmental concern. To participate in lawsuits, citizens must have access to the environmental data and information necessary to support their arguments in court.

---

111. Barry Boyer & Errol Meidinger, *Privatizing Regulatory Enforcement: A Preliminary Assessment of Citizen Suits Under Federal Environmental Laws*, 34 Buff. L. Rev. 833, 964 (1985).
112. Andrews, *supra* note 5, at 240.
113. *Id.* at 241.
114. 33 U.S.C. §1365.
115. 42 U.S.C. §7604.
116. *Id.* §300j-8.
117. *Id.* §6972.
118. 7 U.S.C. §136a(d)(1)(C)(ii).

V. The Spectrum of Federal Environmental Regulation

Environmental statutes work in concert with a vast number of other laws and regulatory programs run by a multiplicity of federal agencies from EPA to DOI, the Bureau of Land Management, the U.S. Food and Drug Administration, OSHA, and beyond. Environmental considerations have been infused into many aspects of American law, and there are few federal agencies that do not, in some way, participate in the larger environmental law scheme. The far reaches of environmental jurisprudence and its underlying regulatory projects result in an amorphous regulatory scheme that lacks clear scope and boundaries.[119]

Not only are the boundaries of environmental law somewhat nebulous, but environmental law continues to evolve and grow. Concepts like climate change and new technologies, such as state-of-the-art alternative energy sources continue to develop and become part of environmental law lexicon. In 1994, an executive order introduced the concept of environmental justice into federal law,[120] and thus every federal agency is now required to consider environmental justice, defined by EPA as "the fair treatment . . . of all people regardless of race, color, national origin, or income, with respect to the development, implementation, and enforcement of environmental laws, regulations, and policies,"[121] in all Agency goals and missions. Environmental justice, as a distinct legal concept, is still in its infancy as U.S. courts have yet to determine the extent to which agencies must pursue environmental justice.[122] As the evolution of environmental law continues, so to do governmental roles and ideas about environmental information, especially in light of the technological developments surrounding information availability online, as you will see throughout this book.[123]

VI. States' Roles in U.S. Environmental Regulation

Prior to the 1970s, most environmental problems were dealt with using common-law doctrines, like tort, nuisance, and trespass, and were often resolved in state courts or otherwise accounted for on a state level.[124] Since the cre-

119. David A. Westbrook, *Liberal Environmental Jurisprudence*, 27 U.C. Davis L. Rev. 619, 621 (1994).
120. Exec. Order No. 12898, 59 Fed. Reg. 7629 (1994).
121. U.S. EPA, *Environmental Justice*, https://www.epa.gov/environmentaljustice (last updated Aug. 8, 2016).
122. Andrew J. Doyle, *Environmental Justice on the Move*, Fed. Law., July 2015, at 3, 5.
123. Andrews, *supra* note 5, at 248.
124. Andrew T. Hawkins, *Federal Supremacy, Sovereign Immunity, and Cooperative Federalism: The Relationship Between Federal Facilities and State Environmental Laws*, 5 Tex. Tech. J. Tex. Admin. L. 63, 63-64 (2004).

ation of a more robust federal environmental law scheme, some of the states' environmental protection roles have been reallocated to the federal realm. Due to the local nature of environmental issues, however, implementation for most regulatory schemes requires close cooperation between EPA and states.[125] Thus, many U.S. environmental law programs are carried out in collaborations between federal and state governments. Sometimes, the federal government provides states with financial assistance to help the states achieve environmental standards. In other cases, states and the federal government participate in "cooperative federalism,"[126] where states participate in the implementation and administration of federal environmental programs. This participation helps state governments and the federal government meet or even exceed national standards by administering the federal regulations through state environmental agencies.[127]

There are several benefits to this EPA-approved delegation of power to the states. For one, EPA lacks the staff, funding, and manpower to run and monitor environmental protection efforts in every U.S. locale. Additionally, state citizens have the local knowledge to implement environmental programs in their municipalities, and they have greater monitoring capabilities due to their proximity to the natural resources surrounding their homes.

Some examples of federal and state environmental law partnerships and instances where the federal government provided primacy to states for implementation purposes include: EPA's creation of national ambient and emission standards under the CAA and the states' development of state implementation plans to achieve the CAA standards. Other statutes like the CWA and RCRA include similar standard-creating provisions, requiring EPA to set standards, and then allowing state environmental agencies to administer those standards. Thus, much federal environmental regulation is a matter of intergovernmental relations instead of purely federal regulation.

Some states create programs to meet the minimum federally set standards; other states create programs even more stringent than the baseline federal mandates.[128] No matter what kinds of partnerships individual states and federal governments engage in to enforce U.S. environmental law, it is impossible to deny that the nation's environmental governance scheme is deeply dependent upon the workings of state governments.

125. Andrews, *supra* note 5, at 248-49.

126. This is also discussed in Chapter 9.

127. Hawkins, *supra* note 125, at 75-76 (noting that the Commerce Clause is used as the authorizing mechanism for these partnerships between state and federal governments).

128. *See, e.g.*, California's air standards and emission trading program. CA Gov, *Cap-and-Trade Program*, https://www.arb.ca.gov/cc/capandtrade/capandtrade.htm (last visited Nov. 14, 2016).

VII. Conclusion

The U.S. federal government has, over the last 60 years, developed a complex environmental law regime. The current environmental regulatory scheme is sometimes conflicting and ineffective, but overall there is a unique body of law that attempts to preserve natural resources and combat the negative public health consequences of pollution. The various environmental laws and state-federal relationships at the core of their implementation create information access opportunities as well as complexities. This book exists to guide you to the information you need in order to formulate environmental decisions and effectively participate in environmental activities and processes.

Chapter 3: Researching Environmental Law

Law in Action

Marjorie Castillo lives in Wilmington, North Carolina. She has been reading about a new cement plant that Titan, a major U.S. cement manufacturer, is planning to build on the edge of her town. She is concerned because she has heard that cement plants cause a lot of air pollution, and her son suffers from asthma. She would like to find out more about the permit laws for cement plants, and she's curious to know whether anyone has ever filed a lawsuit challenging building permits for cement plants with any success.[1]

At first glance, the search for legal materials—cases, statutes, and agency regulations, etc.—does not seem much different than other types of research. After all, legal documents can be retrieved by a Google search, and there are all sorts of law blogs, law firm websites, and other online resources that talk about legal issues and cite statutes, cases, regulations, and even the U.S. Constitution. Anyone who has ever tried to figure out how to get a fishing license, fight a traffic ticket, or determine whether they have to mow the grass around their sidewalks has done some form of legal research.

A closer look at legal research, however, reveals that to find the most relevant information, researchers need to use a special process and look at certain types of materials created by government entities and legal specialists. Information may be easy to find on Google, but it is hard to verify. It is more difficult to find good explanations of the law without knowing where to look and how to distinguish the various resources. Legal research is not simple, and it cannot be accomplished with quick, superficial searches. To properly conduct legal research, information seekers must do more than simply find a law—they must also determine whether a document is a primary or secondary source; whether it is a mandatory legal authority or merely persuasive in

1. Press Release, Southern Environmental Law Center, Local Groups Appeal DENR and Titan America's Attempt to Eliminate Citizens' Right to Challenge Unlawful Air Pollution (Apr. 27, 2015), https://www.southernenvironment.org/news-and-press/press-releases/groups-appeal-denr-and-titans-attempt-to-eliminate-citizens-right-to-challe.

their jurisdiction; and whether the source contains quality, up-to-date information. Additionally, legal researchers must be able to identify what type of document they are looking at and its legal origins and force. Are they reading a court opinion, created by a judge, or a statute, created by a state or federal legislature? Finally, if the researcher wants to rely upon a particular primary source, they have to determine whether the authority is still good law that has not been overturned or superseded.

Since this book focuses on environmental information seeking and not general legal research, this chapter does not thoroughly cover every aspect of legal research. In this chapter, you will learn about basic principles of legal research, and how they can be applied in situations where you are seeking legal information about environmental law issues and topics.

I. Legal Research Basics

Before embarking on legal research tasks, you must understand the basic structure of the U.S. government and the types of documents it creates. In addition to the different branches of government producing different legal authorities, the U.S. government is based on a federalist system, so the various state governments' laws operate in concurrence with the laws of the federal government. As a result, researchers must often find authority from both state and federal levels of government to conduct effective legal research in a particular jurisdiction.

A. Primary v. Secondary Sources

There are two basic categories of legal information resources: primary sources and secondary sources. Primary sources are law, including constitutions, statutes, regulations, and cases. In comparison, secondary sources are sources that discuss and explain the law: things like law review articles, legal encyclopedias, and annotations. Both primary and secondary sources work together in the legal research process, because, while primary sources are the law, secondary sources are often necessary to help break down the law and understand how it works.

1. Types of Primary Sources

Primary sources include the binding, enforceable rules created by government. A brief overview of the U.S. government system is useful when discussing these rules, to understand where the rules come from and what forms they take. The U.S. government is divided into three branches established by the U.S. Constitution:

1. The legislative branch creates the nation's statutes. The federal legislature is bicameral, which means that it consists of two separate chambers. The chambers are called the U.S. House of Representatives and the U.S. Senate, respectively.

2. The judicial branch issues court opinions that interpret statutes and regulations, and also hears common-law cases where no statute or regulation is involved. The judicial branch has three levels of courts: trial courts, where cases first appear and issues of fact are resolved by judges, and sometimes juries; intermediate courts of appeals that hear cases being appealed from trial court decisions; and the high court, which hears select cases on final appeal. The high court in the federal judicial branch is called the U.S. Supreme Court, and it is the highest court in the nation.

3. The executive branch is headed by the president and consists of various federal agencies that create and enforce regulations. The executive branch implements and enforces the laws created by the legislative branch and interpreted by the judicial branch of government. Agencies, like the U.S. Environmental Protection Agency (EPA), create and enforce regulations to implement statutes. Administrative law, including agency regulations, is especially important in areas of law like environmental law that focus largely on regulatory and administrative activities.

State government systems are generally structured like the federal government system, with three branches of government and a bi-cameral legislature, though some of the terminology is different. For example, in New York, the trial courts are called Supreme Courts, and the state's high court is the Court of Appeals. Additionally, states' legislatures may call their separate chambers different names, like Assembly, instead of Senate or House of Representatives. State governments create, regulate, and hear cases about state laws, but state laws cannot conflict with federal law.

The four main sources of law in the U.S. legal system are: (1) constitutions; (2) statutes; (3) court opinions (also called case law or cases); and (4) regulations. Each of these is primary sources of law because they contain the law, and their words carry the force of law.[2] This chart explains the source and content of each type of primary source:

2. International law agreements including treaties may also impact U.S. environmental law. For instance, the Migratory Bird Treaty Act of 1918, 16 U.S.C. §§703 712, was created to implement the 1916 Migratory Bird Treaty between the United States and Great Britain (acting on behalf of Canada),

Primary Sources	Branches of Government	Contents
Constitutions	Government (foundational document)	Preeminent source of law establishing government structures and scope, and conferring particular rights and duties upon citizens.
Statutes	Legislature	Contain the laws emerging from the legislative process when a bill successfully passes through Congress and is signed by the president (often called legislation or laws).
Cases	Courts	The judge-drafted opinions (often called cases or case law) that interpret the rules created by the legislative and executive branches of government, or create legal rules where statutes are silent on the law (often called common law). Because the United States relies on the principle of stare decisis, case law creates precedents upon which future cases must rely.
Regulations	Executive Agencies	Rules created by government agencies when the legislature or president empower them to do so via enabling statute or executive order. Regulations often fill in the details of the broadly worded statutes.

■ Determining Whether a Primary Source Is Mandatory or Persuasive Authority. Once researchers locate relevant primary sources of law, they have to determine whether they are mandatory, or binding, authority. Mandatory authority is any primary source of law enacted by the jurisdiction controlling the outcome of the legal issue in question. Remember, primary sources include cases, statutes, regulations, and state and federal constitutions. Anything else is a secondary source and is never mandatory or binding authority. While mandatory authority is binding law for the issue at hand, persuasive authority is not binding; it is merely compelling, but has no legal force.

Determining jurisdiction is a key consideration when deciphering whether something is mandatory or persuasive authority. For example, the Michigan Natural Resources and Environmental Protection Act[3] is a primary source in Michigan, but not in Wisconsin, which is outside of the state law's jurisdictional scope. Similarly, *Shields v. Shell Oil Co.*,[4] a Michigan State Supreme Court case interpreting the Michigan law's statute-of-limitations provision,

which was created to protect the safe migration of birds in North America. Treaties and international agreements are discussed in further depth in Chapter 10.

3. Michigan Natural Resources and Environmental Protection Act (NREPA), 2015 Mich. Pub. Acts §§324.20101 et seq.
4. 621 N.W.2d 215 (Mich. 2000).

is a primary source in Michigan state court, but merely persuasive authority in Wisconsin, and other state courts.

2. Types of Secondary Sources

Unlike primary sources, secondary sources are not law. Instead, secondary sources include any resources that describe or discuss the law, from books to articles to encyclopedias and dictionaries. Anything that is not a primary source of law (a constitution, statute, case, or regulation) is, by default, a secondary source. Although secondary sources are not the law, they are still very useful to legal researchers because they are drafted by legal experts and help readers locate and understand relevant primary sources. They can also persuade a judiciary where there is no binding rule of law, or when one could argue for an exception to a rule or for a new rule of law. There are the ubiquitous types of secondary sources that exist for all topics, like blogs, news stories, and monographs, but there are also some special secondary sources that are specifically made for legal researchers, scholars, and practitioners:

- Legal Encyclopedias. Similar to other encyclopedias, legal encyclopedias provide general, brief commentary and explanations of the full range of law. Legal encyclopedia sets are usually specific to particular jurisdictions. The major general legal encyclopedias, providing coverage of all U.S. jurisdictions, are *American Jurisprudence* (Am. Jur.), now in its second edition, and *Corpus Juris Secundum* (C.J.S.). State encyclopedias are available for some, but not every state. There are also several encyclopedia sets for more specific legal focuses, like *American Jurisprudence Trials.*
- Legal Treatises, Hornbooks, and Nutshells. In contrast to legal encyclopedias, which provide broad, general overviews of legal topics, legal treatises and hornbooks present in-depth, detailed explanations of specific areas of law. While hornbooks are a single monographs providing information about subjects covered in law school courses, treatises can span many volumes, providing more practical and tangential information on the legal topics, they are similar in that they hone in on a specific topic of law (like contracts law or evidence law). Hornbooks provide concise, organized overview discussions of legal topics. Some particularly useful treatises and texts for environmental law are listed later in this chapter.

- Legal Dictionaries. Like legal encyclopedias, legal dictionaries are similar to general dictionaries, but they define legal words in legal contexts. They also contain relevant Latin and foreign phrases that are used in legal practice and give examples of words and phrases being used in legal situations. The most well-known legal dictionary is *Black's Law Dictionary*, now in its 10th edition.[5]

- Restatements of the Law. Restatements are written by the American Law Institute (ALI), which is an organization of professors, judges, and lawyers with expertise in various areas of law. Restatements organize, or restate, U.S. common law into distinct legal rules, and provide hypothetical illustrations of the rules. Although restatements are not primary sources of law, they are highly persuasive and sometimes influence the court decisions.

- American Law Reports (ALRs). ALRs are annotated law reports that provide analysis on a variety of legal topics from all jurisdictions. They report selected cases and analyze those cases with accompanying articles. The articles provide analysis, relevant citations to associated statutes and secondary sources, and cite and disccuss other cases from different jurisdictions that have also decided cases on the topic. ALRs are kept up to date and provide comprehensive reviews of specific legal issues.

- Law Review Articles. Law reviews are scholarly legal periodicals usually curated and edited by law students or organizations affiliated with law schools. They contain long articles and shorter essays and notes written by law professors, legal practitioners, and law students. The articles are on all sorts of topics, but they are usually far more specific and sometimes a bit esoteric. Law reviews, however, often contain useful overviews of legal topics, legislative, judicial and regulatory histories, as well as plentiful footnotes citing primary authorities of law.

- Bar Journals. These periodicals published by local, regional, and national bar associations are trade magazines that feature news about the bar, legal analysis from members of the bar, and in-depth stories relevant to the particular bar and jurisdiction. Bar journals are especially useful as current awareness tools for members of a particular bar to stay informed about their jurisdiction and areas of specialty and interest.

5. Black's Law Dictionary (10th ed. 2014).

- CLE Materials. Materials created during Continuing Legal Education (CLE) seminars are sometimes published as monographs and multi-volume sets of books. The information shared during CLE sessions is usually very practical, focusing on topics such as how to file and litigate certain types of cases, and how to stay within ethical boundaries while engaging in certain legal practices. Major CLE providers, like the Practicing Law Institute (PLI) publish their CLE proceedings, and they are available in commercial legal databases.

- Study Aids. Law school study aids, like West's Nutshell Series, Gilbert Law Summaries (Gilbert's), and Emanuel Law Outlines (Emanuel's), help law students break down the complicated legal concepts from their law school classes by providing outlines, supplemental text, and sample essay questions and answers. These books are designed to help law students prepare for their exams, but they are also helpful to legal researchers unfamiliar with doctrinal areas of law commonly taught in law school.

- Law Firm Newsletters. Law firms sometimes publish and send newsletters on legal topics drafted by lawyers in their specialized practice groups. Attorneys write letters informing potential clients of their rights, of the inner workings of particular types of transactions, and on updates in the legal field. These legal information resources are often available online, and they may contain summaries of current legal issues and up-to-date information on statutes, case law, and regulations.

Tip: *Although a legal researcher's end goal is ordinarily to find the best primary sources to support their legal issue, secondary sources are usually the best place to start when looking for legal information. This is because secondary sources provide explanations of legal topics, as well as references to useful and relevant primary sources, and also provide commentary and critical background and context for those primary sources of law.*

B. *Finding Specific Primary and Secondary Sources*

As previously mentioned, legal researchers may find many primary and secondary legal sources through Google and other general Internet search engines. While it is true that "Googling" will retrieve a significant amount of legal information, Internet search results can be a bit overwhelming and full of superficial and irrelevant items. It is easier to find substantive, deeper, and more thorough examinations of legal issues and topics using legal specialty websites, namely Westlaw and Lexis Advance.

Lexis Advance and Westlaw, the two major online legal information providers, supply premier legal research services to lawyers, courts, and agencies. These two competing services each provide a wealth of legal information, with thorough secondary texts and annotated and updated primary sources containing all sorts of helpful research tools and search functions.

There are also several other online legal research services, including Bloomberg Law, which provides corporate research tools in addition to legal research databases, and lower cost alternatives like Fastcase (which recently acquired LoisLaw, the competing legal research platform), which provide databases containing primary sources and limited secondary sources, but lack a sophisticated citator and the digest features that Lexis, Westlaw, and Bloomberg provide. Another legal database, HeinOnline, contains digital portable document format (PDF) archives of law reviews and other legal periodicals, as well as deep historical and current database archives of constitutional, legislative, regulatory, and international law documents. There are also specialized online databases and resources that contain materials for certain legal research areas, including ProQuest Legislative Insight, which offers extensive U.S. legislative history resources; CCH Intelliconnect, which contains specialized tax and accounting resources; and several other specialized commercial databases.

I. Finding Secondary Sources

In legal research, anything that is not a primary source of law (a constitution, statute, case, or regulation) is, by default, a secondary source. While a large portion of legal information materials are secondary sources, the specific secondary sources discussed in this chapter are the best starting points for the legal research process.[6] These specialized legal secondary sources provide topical overviews and excellent footnotes and references containing the precise citations to the relevant primary sources—cases, statutes, regulations, and constitution sections. Unfortunately, these comprehensive secondary sources are often the hardest things to find for free because they are published by for-profit enterprises whose owners are not required to make them available to the general public.

6. Many resources contain useful information about the law, however, certain resources tend to be better for making progress in legal research tasks. For example, magazine articles are secondary sources, but legal encyclopedias are special secondary sources designed to help people find useful primary authority.

For example, it is unlikely that researchers will be able to access sources like multi-volume treatises or legal encyclopedias online without access to a major subscription database like Westlaw or Lexis Advance. Nevertheless, if you have access to a nearby academic, court, or government-run law library, it is likely that those treatises and encyclopedias will be available to you on-site in those libraries.[7] Additionally, many law libraries create online research guides that will provide you with links and information about valuable environmental law secondary sources. Moreover, many law review articles and other scholarly writings about a vast array of legal topics can be accessed using the free website Google Scholar.

If you have Lexis Advance or Westlaw access, you can search environmental law secondary sources by topic—both services contain a special environmental law content type to narrow your search focus to environmental law-related results. You can also search generally, using the phrase environmental law as a search term to narrow your results to that area of law to ensure that the phrase will be contained in your search results.

If you do not have access to one of these legal search providers, you can visit the Google Scholar search page online and type in a legal search query, making sure you have selected articles and not case law as your content type. The search will retrieve law review articles that contain useful citations and footnotes. For instance, a search for national pollutant discharge elimination system (NPDES) permits on Google Scholar retrieves a *Harvard Environmental Law Review* article by environmental law scholar Jeffrey M. Gaba called "Generally Illegal: NPDES General Permits Under the Clean Water Act" available free of charge online.[8] This article, while about a specific aspect of the NPDES program, provides background and citations related to NPDES permitting practices. The citations to primary authority within the law review article can be used to easily locate and retrieve the relevant statutes, regulations, and cases relating to the NPDES program.

■ Evaluating Online Sources. When searching online for secondary sources, you must evaluate the information you find. The Internet democratizes information, giving all people and entities the opportunity to publish information online. The wealth of information available using the Internet is wonderful, allowing researchers to find every environmental law in their state with a push of a button, and offering up hundreds of responses to queries like "is climate change real?" Since anyone can publish anything online, it is impor-

7. Additionally, many libraries participate in the Federal Depository Library Program (FDLP), and receive government documents provided by the Government Printing Office (GPO) to add to their library collections.
8. 31 Harv. Envtl. L. Rev. 409 (2007).

tant, however, to review the information critically before accepting it as part of your legal research. There are four major factors to consider when you find an online resource you would like to include in your information seeking:

1. *Who is the author?* Was the data about deep sea oil wells published or gathered by an environmental advocacy organization, like the National Resources Defense Council (NRDC), which advocates for environmental protection, or was is it from the British Petroleum (BP) website?
2. *Is the information objective?* Is the blogger's description of the water quality in the Hudson River from a biased perspective or a personal account, or is it unbiased and detached from judgment? Is the online information defending a certain perspective or persuading readers toward a particular viewpoint, or is it neutral?
3. *Is the information accurate?* Can the data online about organic pesticides be scientifically verified? Does the report cite any scholarly studies or scientific or government websites? Or does the author not provide a source for the information he uses in his writing? Wikipedia is an unreliable online source for information because its content can be added and changed by anybody, so it is sometimes unverifiable and inaccurate. If you find information through Wikipedia, you must follow its citation links to verify its accuracy.
4. *Is the information current?* Statutes and regulations often undergo amendments, and case law is often overturned, changing prior precedent, so legal information goes out of date quickly. Thus, when searching for legal information, it is important to know when information sources were last updated. For instance, a webpage about the definition of "waters of the United States"[9] under the Clean Water Act that was published online before 2014 does not take into account the 2014 Amendments to that definition.[10]

A good secondary source leads researchers to primary sources by citing or linking to statutes, cases, and regulations. The next section will offer advice on where to find these primary sources when they are not directly linked from an online secondary source.

9. 33 C.F.R. pt. 328.
10. Clean Water Rule: Definition of "Waters of the United States," 80 Fed. Reg. 37053-127 (June 29, 2015).

2. Finding Primary Sources

Finding primary sources online is sometimes easier than accessing relevant and authoritative secondary sources, as federal and state governments usually recognize the public utility of providing these official government materials online. While availability varies, depending on the technology and resources allotted to online resources by each government, generally, statutes, cases, regulations, and constitutions from the federal government and the 50 states can be found somewhere on the Internet.

When looking for primary resources online using general search engines rather than specialized legal databases, remember to vet your source using the evaluation questions in the previous section. Google Scholar,[11] Justia,[12] and the Library of Congress[13] tend to be good case law resources. State and federal websites, as well as websites like Cornell's Legal Information Institute (LII)[14] provide statutes and regulations. The primary sources you find online free of charge may be outdated or contain errors, so it is best to check when the page was last updated and to double check the text against another online version for accuracy. Free sources of online primary sources are unofficial and can rarely be relied on in a court of law.[15] Also note that online sources of primary authorities rarely provide or maintain older versions of the laws. To find prior versions of primary sources, you may have to access a major online legal database, an archive like HeinOnline, or a library's microfilm collection.

Additionally, while primary sources may be available for free online, the free versions lack the editorial upgrades associated with the major online legal databases. Commercial legal databases cost money in part because of the value they add to primary sources. They offer annotations for statutes, headnotes for cases, cross-references and hyperlinks to secondary and primary sources dealing with the law you are looking at, and most importantly, information about whether the primary sources are still good law.

---

11. Google Scholar, https://scholar.google.com/.
12. Justia, https://www.justia.com/.
13. Library of Congress, https://www.loc.gov/.
14. Legal Information Institute, https://www.law.cornell.edu/.
15. One exception to online resources as generally not official resources are those items online that are marked with the GPO Seal of Authenticity, see https://www.gpo.gov/authentication/faq/ (last visited Nov. 28, 2016).

Tip: *A note about citators: Lexis Advance and Westlaw provide citator services as part of their legal research features. Lexis Advance's citator is called Shepard's, and Westlaw's is KeyCite.*[16] *Citators are tools that let you know whether a primary source—case, statute, regulation, or constitution—is still good law. By tracking the history of the primary sources and their treatment in the courts, legislature, and executive, citators can tell you whether your case has been overturned or whether your statute or regulation is about to be or has recently been amended. Through colorful icons like red and yellow flags, stop signs, and other notations, citators mark each primary source with warning signs if they are no longer good law. Citators also link to primary and secondary sources that have cited the source, providing quick access to those authorities.*

■ Judicial Materials (Court Opinions). The main type of information legal researchers seekers normally seek from the judiciary is a court's final published opinion.[17]

Depending on which court's opinions you seek, case law can be found in various places on the Internet. One of the best free sources for case law is Google Scholar.[18] If you select "Case law" instead of "Articles," you can run a Google search in a huge database filled with published court opinions from U.S. federal and state appellate courts. While Google's search results lack the features found in Westlaw, Lexis, and Bloomberg Law versions of cases (including headnotes, hyperlinks to statutes and cases, and the all-important citator feature that helps researchers understand whether a case is still good law), Google does have some useful tools for analyzing court opinions beyond their text.

For instance, when looking at *Sierra Club v. Morton*[19] (a landmark case about standing to sue in environmental law cases), using Google Scholar you can see how many cases have cited the *Morton* case, and look at those cases. You can also see law review and other scholarly articles related to the case. Finally, the Google Scholar "How Cited" feature provides an overview

---

16. Several other online legal research products have also begun creating citator features. Bloomberg Law's Bcite is currently under development and works similarly to Shepard's and KeyCite (http://www.bna.com/bloomberglaw-case-law-bcite/). Other products like Casetext (https://casetext.com/) are thinking of new ways to verify whether cases are still good law, like crowdsourcing case information to create a community-built legal research citator.
17. Note that deliberation about opinions is confidential to the chambers of the court (and also, such materials are not primary sources), and unavailable to the public, and because unpublished opinions are generally not considered to have precedential value, and thus do not cleanly fit into the primary sources cannon.
18. *See* Google Scholar, *supra* note 11.
19. 405 U.S. 727, 2 ELR 20192 (1972).

Figure 3.1 *Sierra v. Morton* on Google Scholar

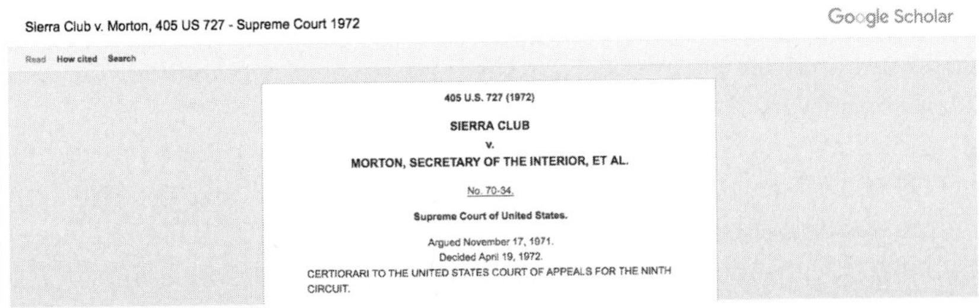

If you retrieve *Sierra Club v. Morton*, 405 U.S. 727 (1972) on Google Scholar, the image will contain the text of the case, as well as a "How Cited" link in the upper-left hand side of the page, which pulls up other cases and related documents that cite the case.

of how the case has been treated by later decisions, listing cases that have cited *Sierra Club v. Morton* with excerpted sentences from the citing cases that reflect the later courts' holdings. While the How Cited feature lacks the editorial benefits of commercial citators like cautionary flags and warning signs, the How Cited excerpts provide researchers with citations and case quotes. These case citations indicate whether the case has received positive or negative treatment since its publication, albeit with some additional case law reading and analysis. Other online services providing free access to court opinions lack Google Scholar features. Services like Justia, caselaw.com, and findlaw.com provide the text of published court opinions with nothing more.

If you are looking for information about a case beyond its final judicial opinion, court dockets for many cases are available online, as well. A docket is a record of court case proceedings containing the play-by-play information about a case. The types of information and the amount of detail provided by dockets vary by jurisdiction. Generally, dockets are the best place to find out about unreported cases, and to get case information that is not published in a court reporter.[20]

The docket lists the documents filed by the parties in the case, from the initial complaint through the various motions and orders, to the final order or opinion. Dockets also provide detail about party and attorney involvement in a case, and the events that occurred during the lifecycle of a case (including discovery, orders, settlement details, and other ancillary litigation activities). Dockets are especially useful for researching trials, which

20. Case reporters are books that contain court opinions. Not all court opinions are published in a reporter; only cases with precedential value are usually selected for publication in a reporter. These unreported cases are sometimes called unpublished opinions.

can last many years and involve many different events throughout the lifespan of the litigation.

Some jurisdictions have free online docket systems, some offer subscription-based online docket systems, and some courts require you (or a court runner) to visit the court in-person to obtain court dockets. There is no uniform docket coverage in the United States, so you must check the docket retrieval procedures for the specific court that generated the docket. Federal dockets can usually be found in the PACER (Public Access to Court Electronic Records) system. PACER provides case and docket information from federal appellate, district, and bankruptcy courts as part of its duty to provide public access to court information from a centralized service. PACER, however, is not free. PACER access requires registration and the use of a password, and prints of docket materials are 8 cents per page. Bloomberg Law is a subscription legal research service that excels at docket services. If you regularly seek dockets, it may be worth getting a subscription to their services.[21]

Something to keep in mind when it comes to searching for documents from environmental litigation is that because parties to environmental litigation are often industrial corporations and environmental nonprofit organizations, you can often find information related to the litigation, including press releases and complaints, on those organizations' websites. Also, litigation involving federal agency enforcement of an environmental law likely has press releases and related materials available on agency websites, especially the U.S. Department of Justice (DOJ) and EPA websites, because those agencies are charged with enforcing many of the environmental laws. The proceedings of administrative law judges are sometimes available online, as well.[22]

■ Legislative Materials (Statutes). Statutes are the primary sources created by the legislative branch of government. Statutes are often referred to simply as "laws," even though they are not the only form of binding primary authority. While many other documents are generated during the lawmaking process, the only legally binding documents from the legislative process are the final laws passed by the U.S. Congress and then signed into law by the president. State governments and the federal government each create their

21. Note that online docket searching has its limitations. Online docket availability only goes back to the late 1990s, thus older dockets are usually stored offsite and require a courier service to obtain. Federal dockets may be archived by the National Archives and Records Administration.

22. For instance, the Filings, Procedures, Orders, and Decisions of EPA's Administrative Law Judges, https://www.epa.gov/alj and EPA's Environmental Appeals Board, https://yosemite.epa.gov/oa/EAB_Web_Docket.nsf.

Figure 3.2 DOJ Environmental and Natural Resources Division Press Room Website

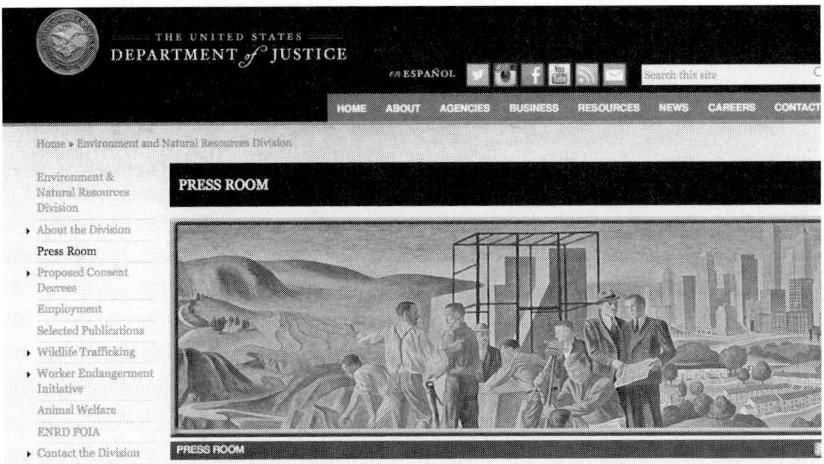

The DOJ maintains online links to their press releases and select consent decrees. Some larger cases, like the *Deepwater Horizon* litigation, have their own DOJ website containing links to various case-related documents. *See* https://www.justice.gov/enrd/deepwater-horizon.

own set of statutes. While the online locations of state governments' statutes vary, they are typically available on the states' legislative websites.

Federal statutes signed into law are published as slip laws, and assigned public law numbers (Pub. L. No.). Then, at the end of the legislative session, all of the slip laws are collected and published chronologically (by date of passage) in the Statutes at Large (Stat.).[23] These laws are then codified in the U.S. Code (U.S.C.). To find the slip law or public law versions of federal statutes, you can search in "Public and Private Laws" on the FDsys.gov website or look them up by Pub. L. No. on Congress.gov. Stat. sections can be viewed on FDsys.gov and also on the Law Library of Congress website.[24]

The U.S.C. is available on several government websites. FDsys.gov, the Government Printing Office (GPO) website, provides the U.S.C., as does Congress.gov. The House's Office of Law Revision Counsel (OLRC) provides an online version of the U.S.C. as well.[25] The OLRC also provides a helpful

23. These two types of citation formats: Pub. L. No. and Stat. references are important when legal information seekers are looking to conduct legislative history research.
24. Library of Congress, *United States Law*, http://www.loc.gov/law/help/guide/federal/uscode.php#statutes (last visited Nov. 28, 2016).
25. Office of the Law Revision Counsel, *United States Code*, http://uscode.house.gov/ (last visited Nov. 28, 2016).

Figure 3.3
Statutes Publication Process

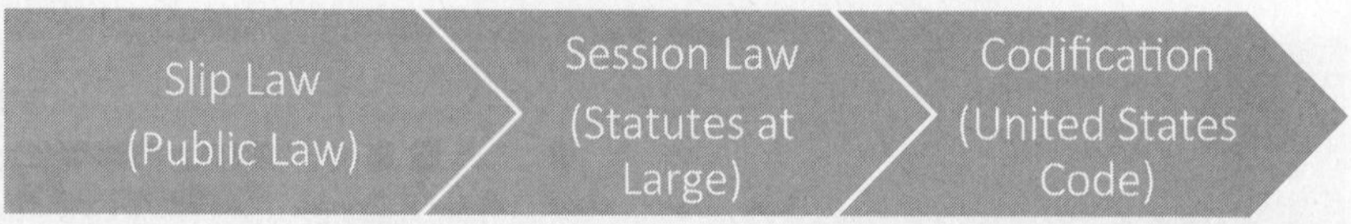

When a President signs a bill into law, the new law is published as a singular document, or slip law. At the end of each legislative session, all of the slip laws for the Congress are published in the Statutes at Large publication, in chronological order based on their passage. Statutes are then codified, and integrated into the United States Code by topic and statutory scheme.

popular name tool that lets researchers find sections of the U.S.C. by their popular names.[26] For example, looking up the Comprehensive Environmental Response, Compensation, and Liability Act (CERCLA) leads researchers to the Pub. L. No.,[27] the Stat. citation,[28] and the U.S.C.,[29] all hyperlinked to the underlying text of those documents. If you need more information about how to find statutes, the Library of Congress has created a great overview and guide for finding federal statutes online.[30]

The major online legal databases also have their own annotated versions of state and federal statutes. While these versions are not official, as they are not published by the U.S. government, the annotations, cross references, and historical notes are critical for researchers. Westlaw's version of the federal annotated code is called the *United States Code Annotated* or U.S.C.A., Lexis' is the *United States Code Service* or U.S.C.S. Each annotated federal code includes annotations to provide commentary, analysis, background, and notes of decisions that describe major cases interpreting the statutes. Both services' annotated codes also provide some legislative history material on the federal statutes.

While legislative history materials—the documents drafted and gathered by the legislature during the legislative process—are not primary sources, they can help determine the legislature's intent for passing a particular statute and provide context to statutory language. For instance, looking into legislative intent is necessary to resolve the differing interpretations by the U.S.

26. Office of the Law Revision Counsel, *Popular Name Tool*, http://uscode.house.gov/popularnames/popularnames.htm (last visited Nov. 28, 2016).
27. Pub. L. No. 96-510.
28. 94 Stat. 2767.
29. 42 U.S.C. §§9601 et seq.
30. Library of Congress, *Researching Federal Statutes*, https://www.loc.gov/law/help/statutes.php (last visited Nov. 28, 2016).

Court of Appeals for the Second Circuit[31] and the U.S. Court of Appeals for the Seventh Circuit[32] on whether certain types of municipal waste are considered hazardous waste under the Resource Conservation and Recovery Act (RCRA).[33] The Law Librarians' Society of Washington, D.C., created a guide to help legal researchers who want to look beyond the words of a final, codified statute to its legislative history.[34] It contains an overview of the legislative process and its associated documents, including a comprehensive list of all of the types of documents that comprise legislative history. There are many nuances to legislative history research, and many ways to obtain the documents created during the legislative process. Although there is no one correct source or process for conducting legislative history research, there are several tips for finding and compiling legislative histories:

- Start your legislative history research by locating the Pub. L. No. and Stat. citations. Legislative history documents will likely be grouped according to those citations, so knowing them is essential to finding the statute's legislative history. The Pub. L. No. and Stat. citations can be found at the bottom of the statute publication, whether it is the U.S.C, U.S.C.S., or U.S.C.A. There are also Pub. L. No. and Stat. citations for every amendment to the statute, which may also be useful if you are researching a particular amendment or learning how the law has developed over time. Appendix 1 lists the Pub. L. No. and Stat. citations for many of the major environmental laws.
- Understand the legislative process and what types of documents are generally created during each step. For example, testimony is often given at hearings, and floor debates often occur when the bill leaves a committee and is read in the legislative chamber. ProQuest's legislative Insight, Congress.gov, and Library of Congress, as well as other online resources with legislative focuses, provide more details about the legislative process, and also may provide details about how the statute or bill you are researching progressed through the legislative process. These online resources are also the best places to find the

31. Environmental Defense Fund v. Wheelabrator Techs., Inc., 931 F.2d 211, 21 ELR 20845 (2d Cir. 1991).
32. Environmental Defense Fund v. City of Chicago, 948 F.2d 345, 22 ELR 20125 (7th Cir. 1991), *cert. granted, judgment vacated sub nom.* City of Chicago v. Envtl. Defense Fund, 506 U.S. 982 (1992).
33. Jane Ellen Warner, *Environmental Law—the Household Waste Exclusion Clarification; 42 U.S.C. Section 6921(i): Did Congress Intend to Exclude Municipal Solid Waste Ash From Regulation as Hazardous Waste Under Subtitle C?*, 16 W. New Eng. L. Rev. 149, 151-52 (1994).
34. Law Librarians' Society of Washington, DC, *Federal Legislative History Research: A Practitioner's Guide to Compiling the Documents and Sifting for Legislative Intent*, http://www.llsdc.org/federal legislative-history-guide (last visited Nov. 28, 2016).

Figure 3.4

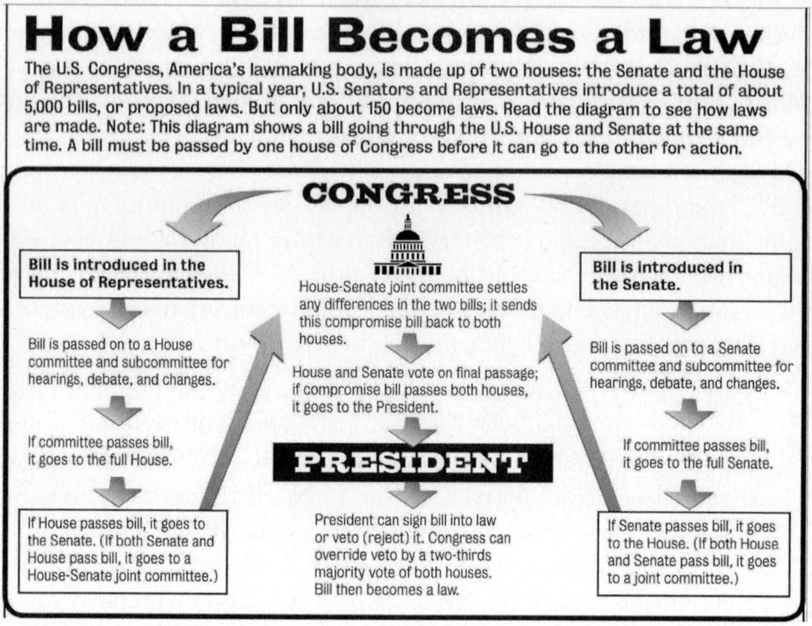

legislative history documents themselves. While the chart of the legislative process may look straightforward, it is rare that a piece of legislation smoothly travels from beginning to end in a linear fashion. Many bills die during the legislative session, not even making it to a final vote on the floor of the legislative chamber. Studying a statute or one of its amendments as it moved through the legislative process often involves the forensic work of trying to piece together what happened to the law during its time in the legislature, reviewing floor debates, testimony, legislative news, draft versions of the bills, and the actions of the legislators involved.

- Know that because the legislative process for each statute is unique, the associated legislative history documents will vary from statute to statute. Some laws have extensive legislative history materials, and others have very small bodies of legislative history. The number of hearings, debates, markups, and other actions that were taken on a bill during the legislative process will determine whether there are volumes of legislative history or very little information about the history of the statute.

Aside from searching for legislative history materials in order to determine legislative intent, legal researchers may wish to track a bill or monitor it for pending legislation that could address a particular environmental issue. Many environmental interest groups (both pro-industry and pro-environment) follow, track, and report on legislative news relevant to their interests on their websites, through their social media accounts, or in special legislative update blogs. Also, some websites such as the nonprofit website Govtrack[35] allow users to log in and set alerts for legislation, members of Congress, and committees. Congress.gov also provides overviews of legislation and allows visitors to the site to set alerts for particular bills or to the Congressional Record, which lists the daily events in Congress.[36]

■ Executive Materials (Regulations). The agencies within the executive branch of the government create and enforce regulations to fulfill the purpose of federal statutes or carry out Executive Orders. Generally, statutes passed by the legislature are broad and overarching, lacking specificity about how the government will carry out the statutory decrees and what will constitute statutory compliance by the public. Congress deliberately creates broad legislation, as the legislature is not composed of specialists, and legislators often lack the time to examine the issues underlying any particular statute because they consider many bills in a legislative session. Agencies, on the other hand, are staffed with scientists, engineers, and other specialists dedicated to enacting and overseeing regulatory programs. Agencies conduct most of the day-to-day governance of environmental issues and programs in the United States, so it is often necessary to seek agency information when doing environmental work. The regulations created by agencies subject to enabling statutes have the force and effect of law,[37] and courts defer to agencies' interpretations of statutes.[38]

Codified regulations are published in the *Code of Federal Regulations* (C.F.R.). Like statutes, this code is available in commercial databases. Although Westlaw and Lexis Advance do not create distinct annotated print C.F.R. publications, their online versions of the C.F.R. do contain tools like links to related cases, statute citations, and citation information to help determine whether the regulation is still good law. The C.F.R. is also available online at FDsys.gov, free of charge. To find regulations related to a particular agency, check the agency's website. Agency websites usually have regulatory information links where the agency provides the regulations it

35. Govtrak, https://www.govtrack.us/ (last visited July 28, 2016).
36. Congress.gov, *Alerts Help*, https://www.congress.gov/help/alerts (last visited July 28, 2016).
37. United States v. Mead Corp., 533 U.S. 218, 230 (2001).
38. Chevron, U.S.A., Inc. v. Natural Resources Defense Council, 467 U.S. 837, 14 ELR 20507 (1984).

promulgates and other information about the oversight and programs associated with the regulations.

Regulations, like statutes, undergo a process to become a final, codified, legally binding rule. The Administrative Procedure Act (APA) sets forth the rulemaking process that federal agencies use to create regulations.[39] There are two types of rulemaking processes under the APA, formal and informal rulemaking. Formal rulemaking is rarely undertaken by federal agencies, as it involves in-depth hearing-like processes. Informal rulemaking is far more common. Informal rulemaking is also called notice-and-comment rulemaking because the process requires agencies to notify the public about proposed rules and to give interested parties the opportunity to comment on them. For more information about the administrative rule promulgation process, see Chapter 7 and Figure 7.1, which depicts the steps of the rulemaking process.

Like statutes, regulations also have histories known as the rulemaking record. Regulatory records are a blend of statutes (and sometimes executive orders) and various documents, depending on the type of rulemaking. Generally, each rule has an associated:

- Enabling statute or executive order: Agencies can only create regulations when given the authority to do so by Congress or by the president. Congressional authority is given through enabling statutes, delegating authority to an agency to act on an issue. Presidential authority is granted by executive order, although enabling statutes are more common forms of agency rulemaking authority.
- Proposed Rule: In response to an enabling statute or executive order, an agency will often create a proposed rule and publish it in the *Federal Register* to provide the public with notice about the proposal, and also to solicit public comments. The proposed rule provides extra information about the rule, like the rule's purpose—and the reason for proposing the rule.
- Final Rule: If the proposed rule is finalized, and the agency issues a final rule, it must be published in the *Federal Register* prior to its codification. The final rule will likely provide a summary of the comments made about the rule and responses to those comments. After a final rule is issued, it is codified in the C.F.R.

39. 5 U.S.C. §553, *available in* ELR STAT. ADMIN. PROC. States each have their own administrative procedure laws that set forth rulemaking processes for state agencies.

You will notice that notice of and information about most steps of the rulemaking process appear in the *Federal Register*. The *Federal Register* is the "Daily Journal of the United States Government."[40] The APA requires agencies to use *Federal Register* publication to notify the public about rulemaking activities. See Chapter 7 for more information about rulemaking and the *Federal Register*.

Like the legislative process, the rulemaking process is not always simple and straightforward. Some proposed rules never become finalized rules, and there may be multiple notice-and-comment periods and revisions for a single rule or concept. To view the rulemaking process of a particular regulation, you can use the Regulation Identifier Number (RIN) assigned to the rulemaking to locate the associated *Federal Register* entries about the regulation. The RIN is noted at the top of each *Federal Register* entry concerning the particular rulemaking.

In addition to issuing regulations, agencies also create other types of documents supporting their rulemaking processes and regulations. For example, guidance manuals describe how to comply with their regulations and adjudication materials from enforcement actions related to upholding and properly executing regulatory requirements. Agency guidance and adjudication records can often be found on the websites of the agencies that create them. For example, the Bureau of Land Management (BLM) has a webpage called Laws, Regulations, and Policies, which contains links to BLM enabling statutes, BLM regulations, Interior Board of Land Appeals decisions and orders, and also BLM manuals and handbooks.[41] Generally, the best starting point for agency information is on the agency's webpage, especially for federal agencies that have online reading room requirements through the electronic Freedom of Information Act (E-FOIA).[42] Remember though, that agency guidance does not carry the force of law.

The federal government has recently created and improved their regulatory websites, and there are now more online tools for finding, looking at, and participating in rulemaking processes. Until recently, the rulemaking process was a bit opaque as it was happening, but now, three major websites have created far more transparency in agency regulation work:

- Regulations.gov. This website was created to increase public participation in federal rulemaking and to carry out the eRulemaking Program.

40. *See* https://www.federalregister.gov/.
41. Bureau of Land Mgmt., *Laws, Regulations, Policies, Court Decisions*, http://www.blm.gov/wo/st/en/info/regulations.html (last updated May 20, 2016).
42. U.S. Department of Justice, *FOIA Guide, 2004 Edition, FOIA Reading Rooms*, https://www.justice.gov/oip/foia-guide-2004-edition-foia-reading-rooms (last visited Nov. 28, 2016).

It is managed by EPA. This website is special because it not only allows you to see each proposed rule's docket (a list of all of the actions that have happened in the rulemaking with links to the associated documents and public comments); it also allows you to provide online comments on proposed rules.

- Federalregister.gov. This website is run by the Office of the *Federal Register*, the National Archives and Records Administration, and the GPO. It is an online version of the publication, and contains excellent navigation and search tools so you can search the *Federal Register* by keyword, RIN, topic, and entry type (aka, proposed rulemaking, final rule). Federalregister.gov entries also contain links to the related C.F.R. and U.S.C. sections, and provides a public inspection desk where you can see previews of *Federal Register* entries.
- Reginfo.gov. This website is run by the Office of Information and Regulatory Affairs (OIRA), and provides an overview of all of the regulatory activity going on in the federal government. OIRA oversees the rulemaking process, and their regulatory dashboard provides public access to information currently undergoing OIRA review. The dashboard contains material like graphical representations of regulatory actions currently under review, and the agencies' bi-annual Unified Agendas containing their upcoming rulemaking plans.

State agency rulemaking coverage varies, but is generally not as thorough or accessible as the information available on federal websites, although some states have established excellent online systems. Generally, for state rulemaking materials start your research on the particular state agency's webpage, or on the main webpage created by the state government containing the state's regulatory code and regulatory oversight information. Chapter 7 discusses information related to the federal rulemaking process in greater depth, as it is often very useful for environmental information seekers.

II. Legal Research Process: How to Conduct Legal Research

While legal researchers are occasionally looking for a discrete item, like a particular rule or case, legal research is more often a far broader task, involving the application of a set of facts to a particular law or legal doctrine, or even determining whether something is a legal issue in the first place. Different legal research experts suggest different legal research plans and processes for conducting legal research. While the plans vary, their central concepts are the same: to conduct effective legal searching, one must organize their thoughts and create an issue statement, start their research in secondary sources to find primary sources that apply to their issue, and then verify that the primary sources they find are still good law. This book will adopt the legal research process used by the CUNY School of Law Legal Research faculty, as described in their book, *Progressive Legal Research*, which lays out the process in six steps.[43]

The six steps for conducting legal research are[44]:

Step 1: Organize your facts and frame the legal issue	Organize your facts in the context of the "5 Ws" (Who, What, When, Where, and Why) and create a preliminary issue statement.
Step 2: Determine the jurisdiction	Think about factors that impact jurisdiction. Consider where the events took place and where the parties are located. Consider whether your issue will be governed by federal or state laws. (Be aware that sometimes you cannot figure these things out without further research.)
Step 3: Find and read relevant secondary sources	Read secondary sources to improve your understanding of the legal issue through analysis and topic overviews, and note relevant primary sources discussed in the secondary sources.
Step 4: Find and read primary sources	Generate a list of statutes/cases/regulations. Determine whether you need to redefine your understanding of the legal issue or if you need additional information.
Step 5: Update primary sources	Determine whether the primary source is current and whether it is still "good law."
Step 6: Examine additional primary and secondary sources	As needed, revisit primary and secondary sources based on what you have learned.

Like most processes, these steps do not always play out perfectly and go in this exact order. Sometimes, the legal issue or jurisdiction is initially unclear, or you must search secondary sources to clarify an issue statement. Sometimes, after conducting steps 1-6, you may realize that you are off course or need to get new information, and start all over again. Generally,

43. *See* Jonathan Saxon et al., Progressive Legal Research 21-22 (forthcoming 2017).
44. *Id.*

however, following this legal research process is an ideal way to find and make sense of legal information.

For instance, Marjorie Castillo, from the Law in Action example at the beginning of this chapter, could follow this legal research process to learn more about permit laws for cement plants and whether anyone has ever filed a successful lawsuit challenging building permits for cement plants. Marjorie's search strategy may look something like this:

Step 1: She would start by sorting out the facts in her issue (who: citizens, Titan, cement manufacturers; what: cement plants, permit laws for cement plants, air pollution; when: before the plant gets built, after a building permit is issued, as soon as possible; where: North Carolina, or elsewhere in the United States; why: to protect public health) and developing an issue statement ("aside from learning more, generally, about cement plant permit laws, the issue is whether anyone has ever filed a successful lawsuit to challenge building permits for cement plants"). Her jurisdictional scope is unclear, so for **Step 2** of the process, it seems as if any jurisdiction in the United States would be useful to Marjorie, but especially law enforceable in North Carolina, where she lives.

Then Marjorie would proceed to **Step 3** and begin looking for secondary sources discussing air pollution-related permits for cement plants and lawsuits associated with those permits. A quick Google search for the terms "air pollution cement plants permit" did not retrieve very useful information, so she added the term "EPA" to the search to see whether the federal agency had addressed the topic. That search led to EPA's Air Enforcement Program webpage, which contained a link to information about pre-construction permits for cement manufacturing plants, noting that cement plants are the third largest industrial source of air pollution.[45] Following that link leads Marjorie to a list of major cement plant lawsuits and settlements, with attachments to some of the complaints containing the relevant Clean Air Act (CAA) statutes and cases interpreting those provisions.[46]

Armed with the new information she's gathered from secondary sources, as well as several primary source citations, Marjorie proceeds to **Step 4** of the legal research process, going to Google Scholar's case law database to search for cases interpreting the primary sources in her jurisdiction. Finding that she was not retrieving helpful cases, she returned to **Step 3** to gather more secondary sources, running searches in Google Scholar to read law

45. U.S. EPA, *Air Enforcement*, https://www.epa.gov/enforcement/air-enforcement (last updated Aug. 4, 2016).

46. U.S. EPA, *Cement Manufacturing Enforcement Initiative*, https://www.epa.gov/enforcement/cement-manufacturing-enforcement-initiative (last updated Oct. 19, 2015).

review articles about permit injunctions in environmental law. She found a few articles on point that cited relevant cases.

Marjorie learned that the federal government delegates its air quality permitting duties to the state, and found that the relevant permitting statute in North Carolina is N.C. Gen. Stat. Ann. §143-215.108(a)(2) and the associated construction and operation permitting regulations are 15A N.C. Admin. Code 2Q.0301-.0307. Moving back to **Step 4** to examine the statute, regulations, and related cases, she learned that courts rarely grant injunctions like the one she would like to obtain, and especially in North Carolina.[47] She went to her local law library to utilize Westlaw or Lexis Advance to look at the statutes, regulations, and court opinions, making sure the primary authorities of law are still current and have not been overturned as described in **Step 5** of the legal research process. At this point, Marjorie felt that her legal research process was complete, and she now had the information she needed to make a decision about her next steps.

As you can see from Marjorie's example, carrying out the legal research process is not always a straight path, and the answers are not always immediately clear. Completing each successive step, however, is useful for finding important and helpful legal information.

IV. Researching Environmental Law

Beyond the basic legal research process, environmental law research focuses on specific resources, agencies, statutes, cases, and concepts. To conduct environmental law research, you must also familiarize yourself with the particulars of environmental law, including: (1) the relevant law; (2) the actors and agencies involved; (3) the documents produced; (4) the vocabulary used; and (5) a foundational understanding of some basic scientific concepts. Environmental law researchers are dealing with a statutory scheme that is unique—focused on the management of living breathing ecosystems. Chapter 11 can help you make sense of the scientific concepts involved in environmental law. Additionally, the following primary and secondary sources provide excellent starting points for environmental law-focused research:

47. John Murwaski & Anne Blythe, *EPA: NC Risking Federal Takeover of Environmental Regulation*, News & Observer, Nov. 14, 2015, http://www.newsobserver.com/news/business/article44892177.html (discussing North Carolina's tough citizen suit standing issues).

A. *Statutes*

Environmental law is not found in just one section of the U.S. Code. Instead, statutes that deal with environmental issues can be found throughout the code with large concentrations located in the following[48]:

Title 7	Agriculture
Title 16	Conservation
Title 21	Food and Drugs
Title 30	Mineral Lands and Mining
Title 33	Navigation and Navigable Waters
Title 42	The Public Health and Welfare
Title 43	Public Lands
Title 54	National Park Service and Related Programs

B. *Regulations*

Titles and parts of the C.F.R. do not always correspond with the same numeric titles and sections in the U.S.C. Many environmental statutory provisions will be in one title of the U.S.C. and in a different title in the C.F.R. This is most apparent with the Clean Water Act (CWA), where the statute can be found at 33 U.S.C. §§1251 et seq., but the regulations can be (mostly) found in 40 C.F.R.

Regulations that deal with environmental law can be found under:

Title 7	Agriculture
Title 9	Animals and Animal Products
Title 10	Energy
Title 18	Conservation of Power and Water Resources
Title 21	Food and Drugs
Title 30	Mineral Resources
Title 33	Navigation and Navigable Waters
Title 36	Parks, Forests, and Public Property
Title 40	Protection of Environment
Title 42	Public Health
Title 43	Public Lands: Interior
Title 50	Wildlife and Fisheries

48. Additionally, Appendix 1 contains statutory citation information for most of the major environmental laws.

C. Cases

Many commercial legal databases can filter cases into discrete topics using an online digest. Cases are not published by topic, but chronologically in order of the date they were released. Digests create topical categories for reported cases. On Westlaw, the digest is known as the Key Number System. Lexis has a similar Headnote System. The Key Number for environmental law in Westlaw is 149E. If you select the "West Key Number System" in Westlaw's "Tools" tab, you can click through various environmental law topics to find related cases dealing with those aspects of environmental law.

Westlaw and Lexis Advance's digests are useful for limiting searches to specific environmental legal issues.

Figure 3.5 Westlaw's Key Number System

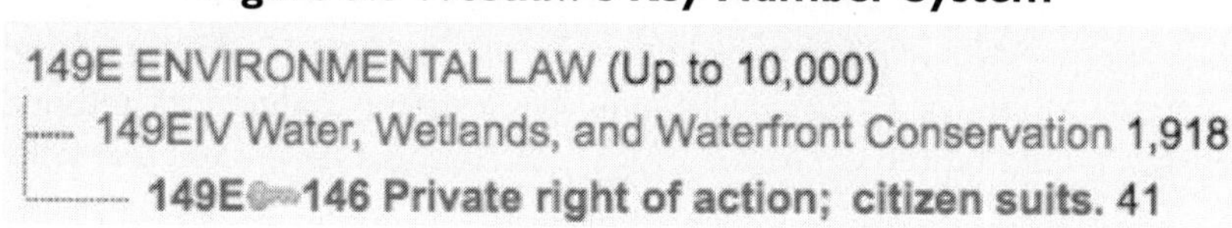

The key number system begins with broad topic numbers containing smaller "Keynote" subcategories. Here is an example from topic 149E—Environmental Law with the subtopics of Water, Wetlands, and Waterfront Conservation, and further, the topic of private rights of action and citizen lawsuits.

D. Secondary Sources

There are specific secondary sources written about environmental law topics. These secondary sources inform readers about various environmental law topics and schemes, and provide plentiful footnotes to landmark cases, key statutes, and regulations. Some of the most well-known environmental law secondary sources include:

- Daniel A. Farber, *Environmental Law in a Nutshell* (9th ed. 2014). This work provides a shorter overview of environmental law and is a great starting point for understanding the basic U.S. environmental law scheme. For the international environmental information seeker there is: Lakshman Guruswamy, *International Environmental Law in a Nutshell* (4th ed. 2011).
- James Salzman & Barton H. Thompson Jr., *Environmental Law and Policy* (4th ed. 2014). This text frames the environmental policy debate

and examines the overarching environmental regulatory scheme in the United States.

- Richard Lazarus & Oliver Houck, *Environmental Law Stories* (2005). This reader gives researchers the behind-the-scenes stories for some of the most important environmental cases in the history of the United States.
- The American Bar Association's (ABA's) Section of Environment, Energy, and Resources (SEER) *Basic Practice Series.*[49] For the last several years the SEER has published primers to many of the major federal environmental statutes including: the CAA, CERCLA, the CWA, the Emergency Planning and Community Right-To-Know Act, the Endangered Species Act (ESA), the Federal Energy Regulatory Commission, RCRA, and the Toxic Substances Control Act (TSCA). These texts (mostly now in their second edition) are easy to understand and provide researchers with the basic provisions of each law.[50] For the slightly more sophisticated legal information seeker, SEER also publishes a series of handbooks and practice guides on the CAA, the CWA, the National Environmental Policy Act (NEPA), and global chemical control.[51]
- The Environmental Law Institute also publishes a series of deskbooks on both statutes and environmental topics. The books provide expert commentary and text excerpts from key statues, regulations and guidance documents. Current statutes covered include: the ESA, NEPA, the Oil Pollution Act, RCRA, and TSCA.[52] Current topics covered

49. To purchase, search the ABA's Store for "Basic Practice Series" under the category of environmental law: http://shop.americanbar.org/ebus/Store.aspx?term=Basic+Practice+Series&sortby=Relevance&perpage=24 (last visited Aug. 8, 2016).
50. *See* Roy S. Belden, Basic Practice Series: Clean Air Act (2d ed. 2011); Carole Stern Switzer & Peter Lawrence Gray, Basic Practice Series: CERCLA, Comprehensive Environmental Response, Compensation, and Liability Act of 1980 (Superfund) (2d ed. 2008); Joel Gross & Kerri Stelcen, Basic Practice Series: Clean Water Act (2d ed. 2012); Peter L. Gray, Basic Practice Series: EPCRA (Emergency Planning and Community Right-To-Know Act) (2002); Sam Kalen & Murray Feldman, Basic Practice Series: ESA (Endangered Species Act) (2d ed. 2012); James Hall McGrew, Basic Practice Series: FERC (Federal Energy Regulatory Commission) (2d ed. 2009); John W. Teets et al., Basic Practice Series: RCRA (Resource Conservation and Recovery Act) (2002); Lynn L. Bergeson, Basic Practice Series: TSCA (Toxic Substances Control Act) (2000).
51. *See* The Clean Air Act Handbook (Julie R Domike & Alec Chatham Zacaroli eds., 4th ed. 2016); The Clean Water Act Handbook (Mark A. Ryan ed., 3d ed. 2011); Albert M. Ferlo et al., The NEPA Litigation Guide (2d ed. 2012); Global Chemical Control Handbook: A Guide to Chemical Management Programs (Lynn L. Bergeson ed., 2014).
52. *See* Lawrence R. Liebesman & Rafe Petersen, ESA Deskbook (2d ed. 2010); Nicholas C. Yost, NEPA Deskbook (4th ed. 2014); Russell V. Randle, Oil Pollution Deskbook (2d ed. 2012); Susan M. McMichael, RCRA Permitting Deskbook (2011); Carolyne R. Hathaway et al., TSCA Deskbook (2d ed. 2012).

include: climate change, environmental crimes, natural resource damage, Superfund, wetlands, and more.[53]

- Thomson Reuters *Environmental Law Series*. This multi-volume series available both in print and on Westlaw contains a master index, as well as more than 20 distinct environmental topics and regulatory schemes. Titles include the *Clean Air Act Handbook*; *Defending Pesticides in Litigation*; *Environmental Regulation of Land Use*; *Environmental Science Deskbook*; *Environmental Spill Reporting Handbook*; *Hazard Communication Handbook*; *Law of Chemical Regulation and Hazardous Waste*; *Law of Environmental Protection*; *Law of Independent Power*; *Law of Solid Waste*; *Law of Toxic Torts*; *Law of Water Rights*; *Law of Wetlands Regulation*; *Managing Environmental Risk*; *NEPA Law and Litigation*; *OSHA Environmental Compliance Handbook*; *RCRA and Superfund: A Practice Guide*; *Public Natural Resources Law*; and *State Environmental Law*.
- William H. Rodgers Jr., *Environmental Law* is another multi-volume set. It has been published since 1986, and it is broken out into elemental topics including: air & water; pesticides & toxic substances; and hazardous wastes. Rodgers' work is known for its excellent case-related footnotes.
- Matthew Bender, a major legal publisher source both in print and on the online legal research platform Lexis, also has a number of relevant environmental titles: *Agriculture Law*; *Environmental Law Practice Guide: State and Federal Law*; *Environmental Law in Real Estate and Business Transactions*; *Grad's Treatise on Environmental Law*; *Insurance Coverage for Environmental Claims*; *The Law of Hazardous Waste: Management, Cleanup, Liability, and Litigation*; and *The Law of Oil and Gas Leases*.
- The Law Journal Press (LJP) also publishes a number of niche-related environmental treatises that are available in print, online, and through Lexis: *Environmental Enforcement: Civil and Criminal*; *Environmental Law Lexicon*; *Environmental Regulation of Real Property*; and *Green Buildings: Law, Contract, and Regulation*.

53. *See* Tom Mounteer, Climate Change Deskbook (2009); Judson W. Starr et al., Environmental Crimes Deskbook (2d ed. 2014); Valerie Ann Lee & P.J. Bridgen, Natural Resource Damage Assessment Deskbook (2d ed. 2014); Allison Rumsey & Michael Daneker, Superfund Deskbook (2d ed. 2014); Margaret Strand & Lowell M. Rothschild, Wetlands Deskbook (4th ed. 2014).

In addition to treatises and texts that provide overviews of various environmental law concepts and schemes, there are also books that focus on environmental advocacy, providing overviews of environmental law and instructions on participating in environmental advocacy in policy and litigation. These titles include:

- Elizabeth Mullin, *The Art of Commenting: How to Influence Environmental Decisionmaking With Effective Comments* (ELI Press 2d ed. 2013).
- Richard G. Stoll, *Effective EPA Advocacy: Advancing and Protecting Your Client's Interests in the Decision-Making Process* (2011).
- Jason J. Czarnezki, *Everyday Environmentalism: Law, Nature, and Individual Behavior* (ELI Press 2013).
- Christopher D. Stone, *Should Trees Have Standing? Law, Morality, and the Environment* (3d ed. 2010).

Chapter 4: Environmental Laws and Information Access

Law in Action

Bert Sampson, a microbiologist at Albany College in New York State, is studying the Hoosick Falls public water supply as part of his research. During routine testing, he discovered elevated levels of perfluorooctanoic acid (PFOA), a toxic carcinogen, in the water. Suspecting the levels could be connected to a nearby plastic manufacturing plant, Sampson wants to know what type of chemical monitoring is done for local drinking water systems. What types of state and federal laws require the reporting of the amounts of PFOA in the water? Where can he locate further data and information as well as report his findings?

I. Overview: Environmental Laws' Access Provisions

As we have seen, the U.S. environmental statutory scheme is uniquely diverse and presents the public with critical access to a wide assortment of environmental information. The U.S. Environmental Protection Agency (EPA) is the agency responsible for managing the majority of these information resources. "Access to accurate information sufficient to effectively participate in managing human health and environmental risks" is one of the primary tenets of the Agency's mission to protect human health and the environment.[1]

While U.S. environmental law provides an array of information access opportunities, there are a number of transparency obstacles in the current environmental law regime. First, the amount of environmental information that is available is staggering. As an illustration, for the last 45 years EPA has been responsible for the administration of at least 29 major federal laws and several executive orders. Moreover, these laws apply at both the national and the state level and require the "public[ation of] regulatory, scientific, administrative, educational, and enforcement materials on subjects ranging from air to water to climate change."[2] The Agency both creates and takes in

1. U.S. EPA, *Our Mission and What We Do*, https://www.epa.gov/aboutepa/our-mission-and-what-we-do (last visited Nov. 28, 2016).
2. Taryn L. Rucinski, *An Environmental Legal Practitioner's Guide to EPA's Website*, 41 ELR 10416, 10416 (May 2012).

billions of pieces of information under these projects and requirements. The volume and variety of programs managed by EPA under various mandates makes it almost impossible to dedicate the time and resources needed to streamline and improve Agency information access practices. In short, it is easy to get lost in the "green haze" of EPA.gov and other government websites that publish information collected pursuant to environmental laws.[3]

Second, there is no uniformity to the information access provisions that appear in U.S. environmental laws. Different laws require different types of information collections, which are processed and then presented in various ways depending upon statutory and regulatory requirements. Some laws require in-depth reports while others require only minimal numeric data inputs. EPA, under the U.S. environmental law scheme, creates a stream of information and data that is virtually impossible to reconcile into a holistic snapshot for any given locality or natural resource. In addition, because EPA's information resources are derived from specific, disparate environmental laws, the information tends to be siloed into different statutory repositories. Environmental information in EPA databases is usually statute-specific with very little cross-applicability. In essence, U.S. federal environmental law does not prescribe uniformity in the type of information collected nor in the delivery method for disseminating that information.

Last, the type of data made available to the public varies greatly depending upon the law. Some environmental laws focus on the public's right to know, explicitly creating information access to municipal officials and members of the general public. Other laws mandate pollution disclosures be made to federal and state agencies, but not necessarily to the public. It is not always the case that information collected by the government under environmental statutes can be handed over to private citizens for environmental decisionmaking purposes. Much of the information collected pursuant to U.S. federal environmental law is subject to privacy measures exempting certain types of information from public access. Research data and other trade secret materials may be unavailable despite the statutory information mandates. In fact, private entities required to submit data under U.S. environmental law can preemptively mark their documents as confidential business information (CBI) to avoid Freedom of Information Act (FOIA) disclosure to the public.[4]

Environmental laws do not generally preclude information seekers from using FOIA, information collected under the various environmental laws, from permitting information to environmental impact statements to pol-

---

3. *Id.*
4. For a further discussion of the limits of CBI disclosure, *see infra* Chapter 8.

lution reports, can be requested through FOIA. Best practice advises a search of EPA.gov, state environmental websites, and other online locations suggested by this chapter prior to using FOIA, but if the information is unavailable or very difficult to find, you are entitled to make a FOIA request to a federal agency, or to state agencies through state freedom of information laws.[5]

II. Searching EPA.gov

EPA is by far the largest collector of environmental information in the United States. Public access to much of that information is guaranteed by statute, regulation, and through the Agency's mission statement.[6] Ironically, these factors do not make the process of searching for information on EPA.gov any easier. It is best to approach EPA.gov methodically in order to find what you need. An effective search strategy for accessing information through EPA.gov is to start with a comprehensive search plan focused on the type of information you are looking for. Because EPA is responsible for administering many major federal environmental programs,[7] has multiple regional and internal offices, and is charged with publishing regulatory, scientific, administrative, educational, and enforcement materials on a vast array of topics,[8] there are many layers of information to sift through in the mammoth EPA.gov system. Environmental information seekers using EPA.gov should thus engage in this multi-step process:

> Determine what type of information you need, considering the intended audience. Ask: "Who is this information for?" This question helps the information seeker, allowing them to cut through the layers of EPA.gov, as the most easily accessible webpages are also the most general, and least sophisticated. For example, if you are looking for specific data, like complex information on a particular waterway's total maximum daily load (TMDL) allowances under the Clean Water Act (CWA), you should forego EPA.gov's introductory homepage in favor of an underlying CWA database. A good starting point is often one of EPA's law summary pages. For each environmental law, EPA has created a webpage called Summary of the [law name]. These summary pages contain a useful overview of the statute, statute citation information,

5. *See infra* Chapter 5 for a more in-depth discussion of FOIA.
6. See U.S. EPA, *Radiation Protection at EPA: The First 30 Years* https://www.epa.gov/sites/production/files/2015-05/documents/402-b-00-001.pdf (last visited Nov. 28, 2016), discussing the Reorganization Plan No. 3 of 1970 and stating that the sole purpose of EPA is to "effectively ensure the protection, development, and enhancement of the total environment").
7. Rucinski, *supra* note 2, at 10417.
8. *Id.*

and links to other webpages dealing, in more depth, with particular aspects of the law.

Evaluate the subject matter of your search. EPA.gov divides its information collections into broad topics like air, climate change, and toxic substances. Identifying the natural resource at issue or a particular environmental program overseeing the information you need will help you determine which set of EPA.gov webpages will be useful to you. For example, searching the "Emergencies" category on EPA's website will not retrieve information about national wastewater discharge standards, which are regulated under the CWA. Although the broad category of "Emergencies" may seem inclusive, the associated links only relate to specific EPA programs overseeing certain types of emergency preparation.

Understand that EPA.gov is not always logically organized. Much of the interrelated information on the website is not rationally or conveniently linked, which forces the website's users to repeat searches for information in several ways to ensure that they get all of the relevant information.[9] Searching on EPA.gov can also be frustrating because of the number of broken links and error messages that occur during searching and navigation.

To get the most out of EPA.gov, review which programs oversee your area of interest, and which search terms (and often acronyms) are related to those programs. Be prepared to click through many seemingly vague links, and to backtrack often as you navigate EPA.gov. Sometimes it is even more efficient to rely on a general search engine, like Google, to lead you to your information, rather than to navigate the general, entry-level EPA.gov webpages.[10] Also, note that there is a "Search EPA.gov" search box in the upper right-hand corner of every EPA.gov webpage. The "Search EPA.gov" search box allows you to run a key word search within EPA.gov rather than clicking through links, if the links are not clear or fail to get you to the information you need.[11]

III. Envirofacts: What Is It, What Does It Cover?[12]

Envirofacts, EPA.gov's major online information access portal, was created in the mid-1990s to function as the "one-stop source for environmental information."[13] At its inception, Envirofacts was a big step forward for pub-

---

9. *Id.*
10. *Id. at* 10418.
11. *Id.*
12. *Id.*
13. U.S. EPA, *Envirofacts*, http://www3.epa.gov/enviro/ (last visited Nov. 28, 2016); Michael L. Ford, *Database Review: EPA's Envirofacts Warehouse*, 4 Envtl. Law. 713-19 (1998).

lic access to environmental information in the United States. Historically, EPA's information was much more limited; prior to Envirofacts' creation, the information available on Envirofacts was only available through in-house inspection at EPA libraries,[14] or through FOIA requests. The EPA's stated goal in creating Envirofacts was to "make all EPA information that is subject to FOIA publicly accessible."[15]

In substance, Envirofacts provides access to databases that store toxic chemical release data, water discharge permit compliance records, hazardous waste handling processes, Superfund statutes, and air emission estimates. On its surface, the portal has a relatively clean design; however, in practice accessing the underlying data can be difficult and confusing.[16] Many Envirofacts searches lead to disorganized and poorly explained results that leave Envirofacts users unsure of their search results' meaning and relevance. Using the advanced search functions in Envirofacts leads to better search results, so it is recommended to stick to the advanced search functions on the website when possible.

Envirofacts has an advanced search feature that allows you to enter a variety of fields and keywords. Entering a ZIP Code, for example 11101 (the ZIP Code for Long Island City, New York, provides a map with data points and a list of facilities at the bottom of the page. Searching for a particular industrial site, for instance, Carbona Products, in the advanced search feature retrieves a map marking the location of the industrial site as well as summary report, facility report, and compliance report for the company. The advanced search feature is the best way to search Envirofacts, as a more general search does not provide such precise results.

IV. Accessing Environmental Statute Information

In response to the variety of information-collecting provisions that appear in U.S. federal environmental laws, this book attempts to group environmental laws' information access provisions into four categories that reflect the general requirements for information reports and data points:

1. Prospective emergency plans and overviews (generating safety plans and studies about potential environmental impacts);
2. Manifests for hazardous materials (generating information about the creation, transport, sale and destruction of hazardous materials);

---

14. *See infra* Appendix 2.
15. *See* Pat Garvey, *Web-Enabled Warehouse Brings Power to the People*, 4 Envtl. L. Rev. 713 (1997).
16. *Envirofacts*, *supra* note 13.

3. Pollution reports (generating pollution data after it has occurred); and
4. Public notice and mandatory labeling for potentially dangerous materials and events (generating information about safe handling of toxic materials).

One benefit of these environmental law provisions is that much of the information generated under the environmental law programs is published online, and can be accessed without filing a FOIA request. A down side, however, to having a variety of information collection provisions is that each style of law creates unique types of environmental information, and each law has a different database or information access point created by either EPA, another government agency, or even a private entity. Although EPA has attempted to streamline this information, there is currently no single point of entry from which to access all of the types of information generated by environmental law requirements.

Although this section focuses on federal environmental laws, researchers should be mindful that state and local statutes and regulations provide additional layers of environmental information access depending on the geographical location. Information regarding state access to environmental information is addressed further in Chapter 9.

A. *Access to Planning Information*

Among the environmental statutes administered by EPA are the National Environmental Policy Act (NEPA) and the Emergency Planning and Community Right-To-Know Act (EPCRA), two major environmental laws that require the preparation of comprehensive documents about prospective environmental harm. These future-focused statutes provide information about possible environmental risks resulting from human and industrial activities before an activity or construction can commence. Both NEPA and EPCRA require the collection and dissemination of predictive information about the potential consequences of certain activities. These laws are the most overt transparency provisions in the environmental law statutory lexicon, as they warn the public about environmental risks and elicit public feedback.

1. National Environmental Policy Act

The National Environmental Policy Act (NEPA)[17] is an information forcing statute that prompts federal agencies to consider the environmental impacts of federal and federally funded activities. While not demanding any particular result,[18] the statute requires that agencies take a hard look at the potential environmental ramifications of government activities. These studies are drafted into environmental impact statements (EIS) that are available to the public.[19] NEPA also requires the agencies involved in the EIS drafting to give the public time to comment,[20] so that the public can communicate any concerns with proposed projects to the agencies involved prior to commencing the projects.[21]

a. How to Find EIS Documents

EPA maintains a public database for drafted and completed EIS called the "Environmental Impact Statement Database" or "EIS Database." Currently, the database has records of all EIS received by EPA since 1987, EPA comment letters on EIS since 2001, and portable document format (PDF) versions of all EIS that EPA has received since October 2012.[22] Additionally, EIS and their supplemental studies, annexes, and draft materials can be found in agency reading rooms, and are sometimes posted in agencies' online reading rooms for wider public access. ProQuest, an online subscription service available in many public and academic libraries, also contains an Agricultural and Environmental Sciences Database that catalogs EIS.[23]

Prior to final publication, all EIS drafts are filed with EPA. Each week, EPA publishes a Notice of Availability for new EIS documents in the *Federal Register.*

17. National Environmental Policy Act of 1969, Pub. L. No. 91-190, 83 Stat. 852 (1970) (codified at 42 U.S.C. §§4321-4370d, ELR STAT. NEPA §§2-209).
18. *See* Marsh v. Oregon Natural Resources Council, 490 U.S. 360, 361, 373, 374 19 ELR 20749 (1989) (requiring agencies take a "hard look"); Robertson v. Methow Valley Citizens Council, 490 U.S. 332, 19 ELR 20743 (1989) (noting NEPA does not mandate pro-environmental results—just informed results). For authoritative lists of cases interpreting NEPA, see NEPA.Gov., *NEPA Case Law Review*, https://ceq.doe.gov/legal_corner/case_law_review.html (last visited Nov. 29, 2016) (the most recent cases listed are from 2005).
19. 42 U.S.C. §4332.
20. 40 C.F.R. §1506.6 (2016).
21. This citizen's guide to NEPA is useful for engaging with the EIS and EA drafting process as a member of the public. *See* COUNCIL ON ENVTL. QUALITY, A CITIZEN'S GUIDE TO THE NEPA 11, 13 (2007), http://www.transwestexpress.net/about/docs/A_Citizens_Guide_to_NEPA.pdf.
22. U.S. EPA, *Environmental Impact Statement (EIS) Database*, https://cdxnodengn.epa.gov/cdx-enepa-public/action/eis/search (last visited Nov. 29, 2016).
23. ProQuest, *Natural Science Collection*, http://www.proquest.com/products-services/natural_science.html (last visited Nov. 29, 2016).

Figure 4.1 EIS Notice of Availability

Federal Register / Vol. 81, No. 199 / Friday, October 14, 2016 / Notices 71041

DEPARTMENT OF AGRICULTURE

Forest Service

DEPARTMENT OF THE INTERIOR

Bureau of Land Management

Notice of Availability of the Mountain Valley Pipeline Project and Equitrans Expansion Project Draft Environmental Impact Statement and the USFS Draft Associated Land and Resource Management Plan Amendments

AGENCY: Forest Service, USDA; Bureau of Land Management, Interior.

ACTION: Notice of availability.

SUMMARY: In accordance with the National Environmental Policy Act of 1969 (NEPA), as amended; the Federal Land Policy and Management Act of 1976, as amended; and the National Forest Management Act of 1976, as amended (NFMA), the Bureau of Land Management (BLM), the U.S. Forest Service (USFS) and the U.S. Army Corps of Engineers (USACE) have participated as cooperating agencies with the Federal Energy Regulatory USFS consideration of LRMP amendments, and/or the USFS consideration of submitting a concurrence to BLM, to the FERC by any of the four methods listed below. The FERC encourages electronic filing of comments and has expert staff available to assist you at (202) 502–8258 or *efiling@ferc.gov*. Please carefully follow these instructions so that your comments are properly recorded.

(1) You can file your comments electronically using the eComment feature on the FERC's Web site (*www.ferc.gov*) under the link to Documents and Filings. This is an easy method for submitting brief, text-only comments on a project;

(2) You can file your comments electronically by using the eFiling feature on the FERC's Web site (*www.ferc.gov*) under the link to Documents and Filings. With eFiling, you can provide comments in a variety of formats by attaching them as a file with your submission. New eFiling users must first create an account by clicking on "eRegister." If you are filing a comment on a particular project, please select "Comment on a Filing" as at 866–208–3676, or for TTY, contact 202–502–8659. The eLibrary link also provides access to the texts of formal documents issues by the FERC such as orders, notices, and rulemakings.

SUPPLEMENTARY INFORMATION: This NOA is specific to the BLM and the USFS and provides notice that these agencies have participated as cooperating agencies with FERC in the preparation of the MVP Project and Equitrans Project Draft EIS. The Mountain Valley Pipeline route would cross about 3.4 miles of lands managed by the USFS, the Jefferson National Forest, in Monroe County, West Virginia and Giles and Montgomery counties, Virginia. The Equitrans Expansion Project would not cross the Jefferson National Forest.

The FERC is the NEPA Lead Federal Agency for the environmental analysis of the construction and operation of the proposed MVP and Equitrans Projects. Under the Mineral Leasing Act (30 U.S.C. 185 *et seq.*), the BLM is the federal agency responsible for issuing right-of-way grants for natural gas pipelines across federal lands under the jurisdiction of two or more federal agencies. Therefore, the BLM is

This EIS Notice of Availability for an EIS on the Mountain Valley Pipeline Project and Equitrans Expansion Project comes from the Bureau of Land Management. The notice describes the contents of the EIS, and also contains information about how to comment on the EIS.

This notice starts a 45-day public comment period for each draft EIS.[24] There is also a required 30-day waiting period for agencies after a final EIS is issued before they can implement the EIS' underlying plan.[25]

The EIS records in EPA's EIS Database include information like: the title, number, document type, *Federal Register* date, any comment or review period dates, amended notice dates, websites, EPA comment letter dates, the states or territories, lead agency, and a contact's name and phone number. The pages also include links to PDFs of the EIS and related documents, if available. From October 2012 onward, PDFs are available, but prior to that month, PDF availability is patchy, and sometimes only a comment letter (or nothing at all) is available from the actual EIS. Some libraries also maintain collections of certain EIS documents.[26]

The public's EIS comment letters and EPA's responses are also available through EPA's EIS Database, as are the EPA's comment letters to the agencies drafting the EIS. These comment letters can be used for informational and

24. 40 C.F.R. §1506.10(c) (2016).
25. *Id.* §1506.10(b)(2).
26. *See also* Northwestern University, *Transportation Library*, http://www.library.northwestern.edu/libraries-collections/evanston-campus/transportation-library/services/reference-services/search-eis (last visited Nov. 28, 2016).

advocacy purposes. For example, a Minnesota environmental organization referenced an EPA EIS comment letter to several state and federal agencies to highlight concerns about a proposed mining project.[27] In their article, the environmental group pointed out that EPA does not think the EIS draft adequately addresses the potential pollution that the mining project could create.[28] This type of environmental critique would be impossible without access to EIS records.

Figure 4.2 Draft EIS

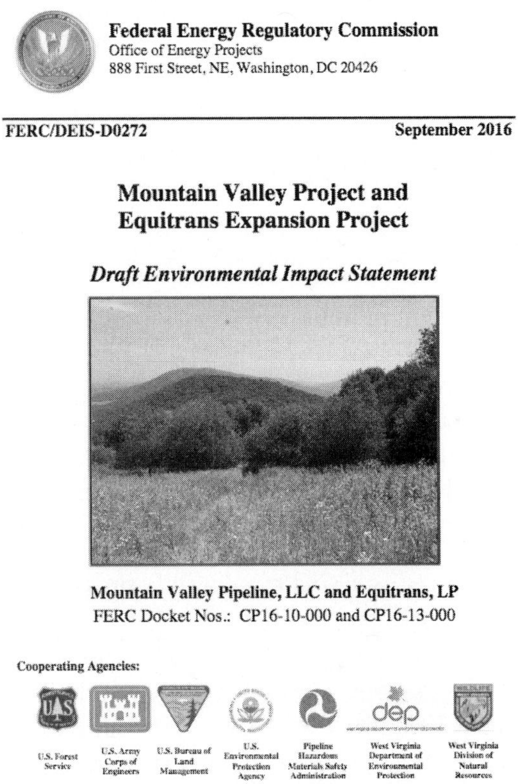

Federal Energy Regulatory Commission
Office of Energy Projects
888 First Street, NE, Washington, DC 20426

FERC/DEIS-D0272 **September 2016**

Mountain Valley Project and Equitrans Expansion Project

Draft Environmental Impact Statement

Mountain Valley Pipeline, LLC and Equitrans, LP
FERC Docket Nos.: CP16-10-000 and CP16-13-000

Cooperating Agencies:

U.S. Forest Service; U.S. Army Corps of Engineers; U.S. Bureau of Land Management; U.S. Environmental Protection Agency; Pipeline Hazardous Materials Safety Administration; West Virginia Department of Environmental Protection; West Virginia Division of Natural Resources

Here is the cover of the BLM's Draft EIS (DEIS). The EIS is 781 pages long, and it was drafted in collaborations with multiple agencies (as you can see at the bottom of the cover).

27. Friends of the Boundary Waters Wilderness, *EPA Letter Demonstrates Significant Flaws Remain in PolyMet Environmental Impact Statement* (Aug. 15, 2013), http://www.friends-bwca.org/2013/08/epa letter demonstrates significant-flaws-remain-in-polymet-environmental-impact-statement/ (last visited Nov. 28, 2016).

28. *Id.*

b. Limitations of NEPA EIS Research

Although EIS are replete with valuable information, the sheer volume of information found within many of the statements makes them difficult to use. EIS documents can be thick, multi-thousand-page tomes full of jargon and complex engineering and scientific data.

Conversely, some government projects do not undergo EIS analysis. Not all government projects that impact the environment require EIS drafting. If the government determines that a project will not have a significant environmental impact, it may be labeled with a finding of no significant impact (FONSI).[29] If a project has a FONSI, then lead agencies do not have to draft a full EIS, instead substituting it for a much briefer document called an environmental assessment (EA).[30] Although EAs and FONSI documents are often posted online, EAs offer far less information about projects' potential environmental impacts, leaving information seekers largely uninformed. In contrast to EIS, there is no special database housing EAs.

2. EPCRA

As part of EPCRA's right-to-know mandate,[31] the law requires information about potential chemical disasters to the public. EPCRA makes two major types of information available to the public: (1) emergency plans; and (2) information about toxic releases. The law's focus is narrowly set on providing information to help reduce acute health effects from short-term exposure to chemical releases, so EPCRA data only concerns information about potential disasters involving hazardous substances.

a. How to Access Local Emergency Plans

EPCRA requires a local emergency planning committee (LEPC) to develop emergency response plans, to review those plans annually, and to provide information about chemicals in the community to citizens.[32] LEPCs are made up of local participants and community leaders, including elected officials; police, fire, civil defense, and public health professionals; environmental, transportation, and hospital officials; facility representatives, members

---

29. 40 C.F.R. §1508.13.
30. *Id.* §1508.9.
31. Emergency Planning and Community Right-To-Know Act of 1986, Pub. L. No. 99-499, 100 Stat. 1613 (codified at 42 U.S.C. §§11001-11050, ELR Stat. EPCRA §§301-330).
32. *Id.* §11001(a).

of the media and community groups. On the state level, state emergency response commissions (SERCs) are created to implement EPCRA. You can obtain more information about LEPCs and their emergency response plans through the SERC in your state.[33] Each SERC has varying amounts of information available on its website, suggesting there are no standard requirements for information content on these websites.

An LEPC community emergency response plan generally contains:

- Identification of facilities and transportation routes for extremely hazardous substances;
- Description of on-site and off-site emergency response procedures;
- Emergency notification procedures;
- Predictions about areas and populations that may be affected by a hazardous substances event and evacuation plans;
- Description of local emergency equipment and facilities and the persons responsible for them;
- Training programs for emergency responders; and
- Methods and schedules for exercising emergency response plans.[34]

There is also an organization called the National Association of SARA Title III Program Officials (NASTTPO) that collects some information about local emergency preparedness plans and provides guidance documents and handbooks for LEPCs and SERCs.[35]

Information seekers should note that facilities subject to EPCRA's emergency planning requirements are often also required to follow other federal emergency planning and response laws that require the creation and dissemination of emergency plans.[36] These programs include EPA's oil pollution prevention regulations[37]; the U.S. Department of Transportation's Pipeline Response Plan regulation[38]; EPA's risk management program (RMP)[39]; the

33. EPA maintains an online list of SERC contacts for each state. *See* U.S. EPA, *State Emergency Response Commissions Contacts*, http://www.epa.gov/epcra/state-emergency-response-commissions-contacts (last visited Nov. 29, 2016).
34. U.S. EPA, *Local Emergency Planning Committees*, https://www.epa.gov/epcra/local-emergency-planning-committees (last visited Nov. 29, 2016).
35. NASTTPO, *Homepage*, http://www.nasttpo.com/ (last visited Nov. 29, 2016).
36. Community Right-To-Know Manual ¶ 170 Relationship to Other Laws (Thompson 2015), *available at* 2003 WL 25322935 (2015).
37. 40 C.F.R. §§112.7(d) & 112.20-21 (2016).
38. 49 C.F.R. pt. 194.
39. 40 C.F.R. pt. 68.

Occupational Safety and Health Administration's (OSHA's) Emergency Action Plan Standard[40]; OSHA's process safety management standard[41]; and the U.S. Department of Homeland Security's chemical facility anti-terrorism standards (CFATS).[42] Thus, more information about the facilities and industries regulated by EPCRA may be available through reports created under these regulatory programs.

b. How to Access Toxics Release Inventory

EPCRA's most well-known information access provision is the requirement that the EPA create and maintain an inventory of toxic chemicals, commonly called the toxics release inventory (TRI).[43] The TRI compiles data about toxic releases: times when hazardous chemicals are discharged into the air or water or placed in some type of land disposal. This data is submitted annually to EPA by industrial facilities and compiled into a publicly accessible online EPA database.[44] The database is one of EPA's most streamlined and user-friendly information sources, supporting "informed decisionmaking by industry, government, non-governmental organizations and the public."[45]

The TRI data is well organized and easy to navigate.[46] The database provides several search options including searches by state, ZIP Code, chemical type, or even industrial facility, giving users several ways to look for information about toxic releases.[47] Environmental information seekers can find TRI fact sheets for their community by selecting a state, county, or ZIP Code.[48] The TRI also has a national analysis website, which contains videos and graphic distribution maps of toxic releases.[49] Further, the TRI "pollution prevention" search function can be used to learn about how facilities manage their toxic chemical releases and waste.

---

40. 29 C.F.R. §1910.38(a) (2016).
41. *Id.* §1910.119.
42. 6 C.F.R. pt. 27.
43. 42 U.S.C. §11023.
44. U.S. EPA, *TRI Data and Tools*, http://www2.epa.gov/toxics-release-inventory-tri-program/tri-data-and-tools (last visited Nov. 29, 2016) [hereinafter *TRI Data and Tools*].
45. U.S. EPA, *TRI Overview*, http://www.epa.gov/enviro/tri-overview (last visited Nov. 29, 2016).
46. The TRI component of EPCRA is the most lauded and written about form of publicly available environmental information. It was first conceptualized in a *New York Times* op-ed piece that called for more streamlined toxic waste reporting—the TRI forces industry to supply current information about toxic hazards on worksites. David Sarokin & Warren Muir, *Too Little Toxic Waste Data*, N.Y. Times, Oct. 7, 1985, at A31.
47. *See TRI Data and Tools*, *supra* note 44.
48. U.S. EPA, *Toxics Release Inventory (TRI) Program*, http://www2.epa.gov/toxics-release-inventory-tri-program (last visited Nov. 29, 2016).
49. *Id.*

Although the TRI is highly regarded, and even used as a model of environmental information access by other nations,[50] its information access is not seamless. Many facilities are not covered by TRI requirements because they fall below the threshold reporting requirements.[51] Also, TRI data can be difficult for laypeople to interpret, because it is raw data—not explanations—about the actual hazard posed by the chemicals, like the toxicity levels or potential effects of contact with the substances. Some claim that TRI data provide "a distorted picture of the health and environmental effects of chemical use,"[52] because it does not distinguish between safer and riskier chemicals or types of chemical application and use. Without more information about how various chemicals and release amounts affect public health and safety, the public cannot determine the true risks (or lack thereof) of the reported chemical releases.

V. Access to Hazardous Materials Tracking Information

The three major laws regarding toxins—the Resource Conservation and Recovery Act (RCRA),[53] the Federal Insecticide, Fungicide, and Rodenticide Act (FIFRA),[54] and the Toxic Substances Control Act (TSCA)[55]—mandate the collection of information about chemical products sold in the United States. Throughout the chemical testing, screening, and disposal processes, EPA collects information about chemicals like pesticides, fertilizers, fuels, and industrial fluids. Historically, information about toxic chemicals—like the revelation that dichlorodiphenyltrichloroethane (DDT) a widely used pesticide, posed a danger to human health and safety—served as an catalyst for Rachel Carson's landmark book *Silent Spring*.[56] How can the public access this type of information today? The types and amounts of information available depend on which statute governs the marketing and use of the chemical in question.

---

50. Kathryn Harrison & Warren Antweiler, Information Regulation vs. Environmental Information: A View From Canada's National Pollutant Release Inventory (Univ. of British Columbia 2001), http://wernerantweiler.ca/public/npri+regulation-1.pdf (noting that the United States was the first to develop a pollutant release and transfer registry, now the most prominent model of environmental information provision in its TRI system).
51. Alexander Volokh, *The Pitfalls of the Environmental Right-To-Know*, 2002 Utah L. Rev. 805, 836.
52. *Id.* at 807.
53. Resource Conservation and Recovery Act of 1976, Pub. L. No. 94-580, 90 Stat. 2395 (codified at 42 U.S.C. §§6901-6992k, ELR Stat. RCRA §§1001-11011.
54. Federal Insecticide, Fungicide, and Rodenticide Act Amendments of 1964, Pub. L. No. 88-305, 78 Stat. 190 (codified at 7 U.S.C. §§136-136y, ELR Stat. FIFRA §§2-34.
55. Toxic Substances Control Act of 1976, Pub. L. No. 94-469, 90 Stat. 2003 (codified at 15 U.S.C. §§2601-2692, ELR Stat. TSCA §§2-412.
56. Rachel Carson, Silent Spring (1962).

A. *RCRA Data*

RCRA, the law governing solid and hazardous waste disposal, requires companies that deal with certain types of toxic chemicals to track the chemicals throughout their life cycles. EPA has two online venues for RCRA information: (1) the RCRA biennial report, which provides information on trends and analysis in hazardous waste disposal; and (2) the RCRAInfo system, which compiles and reports the activities of hazardous waste generators, transporters, and facilities that treat, store, and dispose of hazardous waste.

1. RCRA Biennial Report

RCRA biennial reports provide data for hazardous waste managed under RCRA since 1991.[57] The reports contain compiled statistics about the number of hazardous waste generators and the quantities of waste they generate, as well as data regarding the transport, management, disposal, and other processes in the cradle-to-grave lifecycle of hazardous waste in the United States. The information available in these reports is not facility-specific, but it does offer a good overview of hazardous waste outlooks around the country and by state.

2. RCRAInfo

RCRAInfo[58] is now part of the larger Envirofacts research system. The database contains hazardous waste information collected pursuant to RCRA.[59] This collection includes the data sent in by generators, transporters, treaters, storers, and disposers of hazardous waste (that meet RCRA thresholds). These groups are required by the statute to provide information about their activities to EPA.[60] As a research note, this database is difficult to use, and results vary. The information available for many facilities within the RCRAInfo database is limited.

57. U.S. EPA, *Biennial Hazardous Waste Report*, https://www.epa.gov/hwgenerators/biennial-hazardous-waste-report (last updated Oct. 31, 2016) (listing biennial reports, amendments, and other related information from 1991 through 2015).

58. U.S. EPA, *RCRAInfo*, http://www.epa.gov/enviro/facts/rcrainfo/search.html (last visited Nov. 28, 2016).

59. 40 C.F.R. pt. 262 requires waste generators to comply with recordkeeping and reporting requirements. 40 C.F.R. pt. 263 requires waste transporters to comply with recordkeeping requirements. *Id.* pts. 264 and 265 applies to treatment, storage, and disposal facilities.

60. "Pursuant to RCRA Section 3002(a)(6), Congress required EPA to develop a program for hazardous waste generators to report the nature, quantities, and disposition of hazardous waste generated." U.S. EPA, *RCRAInfo Frequently Asked Questions*, https://rcrainfo.epa.gov/rcrainfoweb/action/modules/faq/view (last visited Nov. 29, 2016).

Prior to the development of RCRAInfo, RCRA materials were made available through the Resource Conservation and Recovery Information System and the Biennial Reporting System. Both of those services are no longer operational.

B. *FIFRA Data*

In contrast to other environmental statutes discussed in this chapter, FIFRA has several distinct data access points. Some are government created, while others are maintained by nonprofit organizations or educational institutions. EPA's Office of Pesticide Programs (OPP) oversees FIFRA's program,[61] creating general consumer publications about pesticides, pest control safety, bed bugs, and other FIFRA-related issues relevant to consumers.[62]

Pursuant to FIFRA, the EPA collects a huge volume of information; however, its organization is vague and unclear. The databases available on the *EPA Pesticide: Science and Policy* website include a variety of government databases, private databases requiring subscriptions, and hybrid public-private database collaborations.[63] There are also four major database categories each containing their own collection of websites and databases:

- Environmental Effects Databases;
- Water Databases;
- Health Effects Databases; and
- Regulatory Information Databases[64]

The many databases available for FIFRA-related data are organized into tables:

61. U.S. EPA, *Office of Pesticide Programs (OPP)*, http://www.epa.gov/aboutepa/about-office-chemical-safety-and-pollution-prevention-ocspp#opp (last visited Nov. 29, 2016).
62. U.S. EPA, *National Service Center for Environmental Publications*, https://www.epa.gov/nscep (last visited Nov. 29, 2016).
63. U.S. EPA, *Database Related to Pesticide Risk Assessment*, https://www.epa.gov/pesticide-science-and-assessing-pesticide-risks/databases-related-pesticide-risk-assessment (last visited Nov. 29, 2016).
64. *Id.*

Environmental Effects Databases

ECOTOX Knowledgebase[a]	EPA-maintained database provides single chemical toxicity information for aquatic and terrestrial life.
EXTOXNET, the Extension Toxicology Network[b]	Created in 1997 by a collection of universities and continues to be maintained by Cornell University. It contains toxicology information fact sheets about some pesticides. This collection does not appear to be regularly updated, so be aware that it may contain outdated information or lack new information.

a. U.S. EPA, *ECOTOX Knowledgebase*, http://cfpub.epa.gov/ecotox/ (last visited Dec. 1, 2016).
b. EXTOXNET, *Extension Toxicology Network*, http://pmep.cce.cornell.edu/profiles/extoxnet/ (last visited Nov. 29, 2016).

Water Databases

National Water Quality Assessment Data Warehouse (NAWQA Data Warehouse)[a]	The U.S. Geological Survey (USGS) and EPA started the NAWQA program in 1991. Source data on 3,000 types of chemical concentrations in water, bed sediments, and aquatic organisms is extracted daily from 51 "study units" or basins across the country. More than 4,700 surface water sites and 9,500 wells are included in this survey.[b] The NAWQA program recently moved and is now available via the National Water Quality Monitoring Council's portal.

a. National Water Quality Council, *Working Together for Clean Water*, https://acwi.gov/monitoring/ (last visited Nov. 29, 2016). For information on how to search, see *USGS, National Water Quality Assessment (NAWQA) Program*, http://cida.usgs.gov/nawqa_public/apex/f?p=136:HOME:0 (last visited Nov. 29, 2016).
b. USGS, *About the Program*, http://water.usgs.gov/nawqa/about.html (last visited Nov. 29, 2016).

Health Effects Databases

Integrated Risk Information System (IRIS)[a]	EPA's IRIS Program identifies and describes health hazards related to chemicals found in the environment. Each IRIS assessment can cover a chemical, a group of related chemicals, or a complex mixture.[b]
Human Health Risk Assessment Program (HHRA)[c]	This program is run by EPA's National Center for Environmental Assessment within the Office of Research and Development (ORD). It contains information on more than 550 chemical substances and the results of human exposure to those substances, divided into "non-cancer" and "cancer" effects.[d]

National Pesticide Information Retrieval System (NPIRS)[e]	Purdue's NPIRS is a subscription service composed of three databases containing information about all pesticides sold and marketed in the United States. The Pesticide Document Management System is a federal products database and a state products database. These databases can be searched in a variety of ways, and all information about application sites and pests to be controlled are derived from the federal master labels.
Pesticide Products Database[f]	NPIRS also houses a State Pesticide Products Database (NSPIRS) that they created in partnership with the OPP and the California Department of Pesticide Regulation. Pesticide Products Database provides brief registration information on approximately 90,000 products.[g] This database is only available through subscription.
The Occupational Pesticide Handler Exposure Database[h]	EPA uses this database to provide use and exposure risks to occupational pesticide handlers.
ATSDR's Toxicological Profiles[i]	The Center for Disease Control's Agency for Toxic Substances and Disease Registry (ATSDR) created a database containing toxicological profiles for about 275 substances pursuant to a congressional mandate that requires the agency to produce these profiles for hazardous substances found at hazardous waste sites.[j]
TOXNET,[k] the Toxicology Data Network	A group of databases for consumers, managed by the Toxicology and Environmental Health Information Program in the Division of Specialized Information Services of the National Library of Medicine containing information about specific chemicals and products, and the effects of chemicals on humans and animals.[l]

a. U.S. EPA, *Integrated Risk Information System (IRIS)*, http://www.epa.gov/iris (last updated Feb. 16, 2016).
b. U.S. EPA, *Basic Information About the Integrated Risk Information System*, https://www.epa.gov/iris/basic-information-about-integrated-risk-information-system (last updated Dec. 16, 2015).
c. U.S. EPA, *About the Human Health Risk Assessment Program*, http://www.epa.gov/aboutepa/about-human-health-risk-assessment-program (last updated Nov. 12, 2015).
d. ORD, U.S. EPA, Human Health Risk Assessment Strategic Research Action Plan 2012-2016 (2015) (EPA 601-K-15-002), https://www.epa.gov/sites/production/files/2014-06/documents/hhra-strap.pdf.
e. NPIRS, http://ppis.ceris.purdue.edu/ (last visited Nov. 29, 2016).
f. NPIRS, *NSPIRS Subscription Information*, http://npirspublic.ceris.purdue.edu/nspirs.html (last visited Nov. 29, 2016).
g. *Id.*
h. U.S. EPA, *Occupational Pesticide Handler Exposure Data*, http://www.epa.gov/pesticide-science-and-assessing-pesticide-risks/occupational-pesticide-handler-exposure-data (last visited Nov. 29, 2016).
i. ATSDR, *Toxic Substances Portal*, http://www.atsdr.cdc.gov/toxprofiles/index.asp (last updated Mar. 2, 2016).
j. ATSDR, *About Us*, http://www.atsdr.cdc.gov/about/ (last visited Nov. 29, 2016).
k. NIH, *Welcome to TOXNET*, http://toxnet.nlm.nih.gov/ (last visited Nov. 29, 2016).
l. NIH, *Fact Sheet, TOXNET: Toxicology Data Network*, https://www.nlm.nih.gov/pubs/factsheets/toxnetfs.html (last visited Nov. 29, 2016).

Regulatory Information Databases

Compendium of Pesticide Common Names[a]	Alan Wood maintains a Compendium of Pesticide Common Names. There is no official mechanism for announcing new International Organization for Standardization (ISO)[b] names when they are provisionally approved, and the only resource that publishes them, The Pesticide Manual, is only published every three to four years. Wood's electronic compendium provides information about pesticide common names, systematic chemical names, molecular formulae, structural formulae, and CAS Registry Numbers. The compendium contains data for more than 1,700 active ingredients of pesticides. Alan Wood is not affiliated with any government or scientific organization,[c] and it is unclear how frequently the site is updated, or if it is updated at all.
Label Review Manual[d]	Published by EPA and available in PDF on their website as a tool for understanding the pesticide labeling process and how labels should be drafted pursuant to FIFRA.[e]
Pesticide Data Program[f]	Since 1991, the Pesticide Data Program has collected data on pesticide residues in food, after the U.S. Department of Agriculture (USDA) was charged with doing so. EPA uses Pesticide Data Program information to support both its dietary risk and pesticide registration processes.[g]
The Pesticide Data Submitters List (PDSL)[h]	This is EPA's list of names and addresses of FIFRA registrants who choose to be listed because they have used data in their sale or distribution registration processes, and they want to be compensated by future applicants that may cite the previously submitted data.[i] Being on the list creates a venue for them to receive offers of compensation as owners of the data.[j]
Pesticide Product Information System (PPIS)[k]	Contains information about all registered pesticide products in the United States including registrant names, addresses, chemical ingredients, toxicity category, product names, distributor brand names, site/pest uses, pesticide type, formulation code, and registration status, in ASCII file format. Getting to the information requires using database and spreadsheet software compatible with the American Standard Code for Information Interchange or what's better known as ASCII.[l]
Pesticide Product Label System[m]	Contains PDF images of pesticide labels that have been approved by the OPP. The images are indexed by EPA registration number and the date on which the label was registered or amended.[n]

Scientific Advisory Panel (SAP)[o]	FIFRA's advisory board, which is composed of scientists, statisticians, toxicologists, and other experts, provides advice to EPA on various health and safety issues related to pesticides. SAP is governed by FACA, and SAP meeting materials are available through EPA's website.[p]

a. Alan, Wood, *Compendium of Pesticide Common Names*, http://www.alanwood.net/pesticides/ (last visited Nov. 29, 2016).
b. ISO, *Homepage*, http://www.iso.org/iso/home.html (last visited Nov. 29, 2016).
c. *See* AlanWood.net, *Biography*, http://www.alanwood.net/alan-wood.html (last visited Nov. 29, 2016).
d. U.S. EPA, Label Review Manual, Pesticide Registration, http://www2.epa.gov/pesticide-registration/label-review-manual (last updated Nov. 7, 2016) [hereinafter Label Review Manual].
e. *Id.*
f. USDA, *Pesticide Data Program, https://www.ams.*usda.gov/datasets/pdp (last visited Nov. 29, 2016).
g. USDA, *Pesticide Data Program: Helping Monitor the Safety of America's Food Supply*, https://www.ams.usda.gov/sites/default/files/media/PDP%20factsheet.pdf (last visited Nov. 29, 2016).
h. U.S. EPA, *Pesticide Data Submitters List (PDSL)*, http://www2.epa.gov/pesticide-registration/pesticide-data-submitters-list-pdsl (last updated Sept. 29, 2016) [hereinafter *PDSL*].
i. *See* 7 U.S.C. §§136a(c)(1)(F), (c)(2)(D) (2014); 40 C.F.R. pt. 152 subpt. E (2016) (discussing ownership of data used to support registration).
j. *See PDSL, supra* note h.
k. U.S. EPA, *Pesticide Product Information System (PPIS)*, http://www2.epa.gov/ingredients-used-pesticide-products/pesticide-product-information-system-ppis (last updated Dec. 14, 2015).
l. *Id.*
m. *Pesticide Product Label System (PPLS)*, U.S. EPA, https://iaspub.epa.gov/apex/pesticides/f?p=PPLS:1 (last updated Mar. 4, 2016).
n. *Id.*
o. U.S. EPA, *FIFRA Scientific Advisory Panel*, http://www.epa.gov/sap (last visited Nov. 29, 2016).
p. U.S. EPA, *FIFRA Scientific Advisory Panel (SAP) Basic Information*, https://www.epa.gov/sap/fifra-scientific-advisory-panel-sap-basic-information (last visited Nov. 29, 2016).

With all of this chemical information separated into so many different databases, finding what you need can prove tedious and time consuming. Hopefully, in the future, EPA, or another government agency or program, will create a more streamlined center for pesticide and toxic chemical research and receive adequate funding to do so.

C. *TSCA Data*

In contrast to the multiple databases containing FIFRA information, TSCA, which governs similarly hazardous toxins like asbestos, chlorofluorocarbons, and polychlorinated biphenyls has only one major database, popularly called the TSCA Inventory. EPA is required to maintain this inventory under §8 of the law, and the TSCA Inventory is publicly accessible as an EPA database. As new chemicals are commercially manufactured or imported, EPA adds them to the TSCA Inventory list. Pre-manufacturing notices (PMNs) are also required under TSCA. Manufacturers must submit these notices to EPA before manufacturing new chemicals for commercial purposes; however, PMNs are not readily available on any EPA.gov page.

The non-confidential portion of the TSCA Inventory is available on EPA's website.[65] Some of the information is trade secret and not available in list. (For example, some PMNs contain trade secret design and compound information about chemicals that have not yet entered the U.S. marketplace). The format is a large,[66] downloadable, regularly updated zip file. Data.gov also provides the non-confidential portion of the TSCA Inventory as both a data set and an extraction tool which allows users to "select a basket full of variables and then recode them in a form you desire" to develop and customize tables.[67]

EPA's website also directs TSCA information seekers to the National Technical Information Service (NTIS) website, "the largest central resource for federal government information publications."[68] It is important to note, however, NTIS products cost money, including TSCA Inventory data stored on CD-ROM. The U.S. Government Printing Office no longer provides paper copies of the original 1985 TSCA Chemicals Substances Inventory or its 1990 supplement, but federal depository libraries and EPA libraries may have copies of these historic versions in their stacks.

EPA's Substance Registry Services provides basic information about every substance that is regulated or tracked by the Agency (including Chemical Abstract and Taxonomical Serial Numbers for the chemicals), so while it is not solely dedicated to TSCA toxics, it does contain information about those substances.[69] EPA updates this database every six months. On its website, EPA also mentions that some commercial services will search the non-confidential inventory for a fee, using the NTIS, although the federal agency does not endorse any of these services.[70]

---

65. U.S. EPA, *How to Access the TSCA Inventory*, https://www.epa.gov/tsca-inventory/how-access-tsca-inventory (last visited Nov. 29, 2016) [hereinafter *How to Access the TSCA Inventory*].
66. For example, the April, 2015 file is 3.40 MB.
67. Data.gov, *TSCA Inventory Data Extraction Tool*, https://catalog.data.gov/dataset/tsca-inventory-data-extraction-tool (last updated Aug. 1, 2016).
68. U.S. EPA, *How to Access the TSCA Inventory*, https://www.epa.gov/tsca-inventory/how-access-tsca-inventory (last updated Mar. 10, 2016).
69. U.S. EPA, *Search and Retrieve, Substance Registry Services (SRS)*, http://ofmpub.epa.gov/sor_internet/registry/substreg/searchandretrieve/substancesearch/search.do (last updated Sept. 1, 2015).
70. *See How to Access the TSCA Inventory*, *supra* note 65.

VI. Access to Statutory Reporting Information

The CWA[71] and Clean Air Act (CAA)[72] require polluters to apply for licenses to engage in polluting activities and to participate in reporting programs to maintain their licenses. Requiring polluters to report regularly ensures that the polluters do not exceed their pollution limits. Thus, these laws do not track polluting materials from cradle to grave or provide pollution prevention information, but instead monitor pollutant dischargers and retrieve pollution reports from them on a regular basis, and during some extraordinary pollution events.

For research purposes, both the CAA and the CWA require comparable types of data reporting in their licensing and reporting provisions, so the types of records disseminated to the public are similar. Each law requires stationary source polluters to regularly report pollution amounts to EPA (air pollution and water pollution, respectively).

Specifically, the CAA requires a stationary source air polluter to alert the authorized permitting authority when its emissions center experiences an unexpected startup, shutdown, or malfunction, or an emergency that causes it to exceed permitted emission limitations.[73] The CWA requires stationary source water polluters to notify authorized permitting authorities within 24 hours when they violate their permits in a way that may endanger health or the environment.[74] To that effect, both laws also require polluters to develop plans in case of these emission events. The CAA requires the creation of startup, shutdown, and malfunction plans,[75] and the CWA requires oil storage facilities to prepare spill prevention, control, and countermeasure plans.[76]

Laws like the CAA and the CWA obligate EPA to delegate authority to operate the underlying environmental programs to the states. Once a program is delegated to a state, activities like permitting, inspections, monitoring, and enforcement, and sometimes even setting the pollution standards, are state responsibilities. Thus, always check state agency materials for information about environmental law programs like those enacted under the CAA, the CWA, and several other environmental laws. States also have some reporting obligations under the CAA and the CWA. The CAA requires states to write

71. 33 U.S.C. §§1251-1387, ELR Stat. FWPCA §§101-607.
72. 42 U.S.C. §§7401-7671q, ELR Stat. CAA §§101-618.
73. 40 C.F.R. §60.7 (2016).
74. *Id.* §122.41(l)(6)(ii).
75. *Id.* §60.11(d).
76. *Id.* §12.1(c).

state implementation plans (SIPs)[77] and submit them to EPA for approval. EPA posts information about the status of states' CAA SIPs on its website.[78] Similarly, the CWA requires states to compile lists of waterways and their pollution status.[79] EPA also maintains a website about these lists and their permissible TMDLs.[80]

Additionally, CAA and CWA permitting processes provide some background into the polluters' plans and pollution limits. All facilities that are considered point sources discharging pollutants into U.S. waters must participate in the national pollutant discharge elimination system (NPDES) permitting process.[81] Similarly, the CAA requires stationary sources of air pollution to obtain and comply with operating permits. The permits granted under both of these programs can help environmental information seekers learn more about the various polluters who participate in them.

A. *Accessing the CWA's Data*

The CWA requires EPA to provide Congress with a *National Water Quality Inventory Report*[82] and maintains a Water Quality Reporting Database (ATTAINS).[83] According to the EPA, ATTAINS is the regularly updated state water quality information source.[84] The ATTAINS database lets you see maps showing impaired waterways, and allows you to search by state or waterway, or impaired status. It also provides charts showing the percentage of impaired lakes, ponds, and reservoirs. ATTAINS and the congressional *National Water Quality Inventory Report* are the primary vehicles for informing the public about water quality conditions.

EPA has also created a database called Enforcement and Compliance History Online (ECHO),[85] where users can search for facilities to assess their

77. 42 U.S.C. §7407.
78. U.S. EPA, *State Designated Area Reports*, http://www3.epa.gov/airquality/urbanair/sipstatus/reports/map_s.html (last visited Nov. 29, 2016).
79. These are called 303(d) lists because they fall under CWA §303(d) requirements. 33 U.S.C. §1313(d).
80. U.S. EPA, *Implementing Clean Water Act Section 303(d): Impaired Waters and Total Maximum Daily Loads (TMDLs)*, http://www.epa.gov/tmdl (last visited Nov. 29, 2016).
81. 33 U.S.C. §1342.
82. This report is often called a 305(b) report because it is required by CWA §305(b). 33 U.S.C. §1315(b).
83. U.S. EPA, *Assessment and Total Maximum Daily Load Tracking and Implementation System (ATTAINS)*, http://www.epa.gov/waterdata/assessment-and-total-maximum-daily-load-tracking-and-implementation-system-attains (last visited Nov. 29, 2016).
84. U.S. EPA, *National Water Quality Inventory Report to Congress*, http://www.epa.gov/waterdata/national-water-quality-inventory-report-congress (last updated Aug. 18, 2016) (listing reports for the following years: 1992, 1994, 1996, 1998, 2000, 2002, and 2004).
85. U.S. EPA, *ECHO: Enforcement and Compliance History Online*, https://echo.epa.gov/ (last visited Nov. 29, 2016).

compliance with environmental regulations. The database also provides a great starting point for people trying to initiate lawsuits for CWA violations, showing which nearby facilities have permits, and which of those facilities have current or recent permit violations.

Aside from information about waterway pollution and actions about polluters, you can search the NPDES[86] for hundreds of thousands of waterway polluters. NPDES-regulated entities submit information regularly to EPA, including discharge monitoring reports (DMRs),[87] notices of intent to discharge in compliance with their permits, and program reports. EPA has created a tool called the DMR Pollutant Loading Tool to access DMR data.[88]

EPA maintains an NPDES General Permit Inventory with search tools to allow users to search for permits by permit number, name, state, date, category, and permitting authority.[89] The Agency advises that you should contact your permitting authority to get any permits not included in the inventory.

Also, to learn more about anticipated or ongoing dredging and filling projects in the U.S. waterways, some CWA §404 dredge and fill permits[90] are available on the U.S. Army Corps' website.[91] The website also links to information about other waterway-related programs, like the National Wetland Plant List.[92] Finally, although the NPDES inventory and the Corps' website are good starting points for license searches and information about local programs, more in-depth searches may require seeking permit information from local divisions of the agency and their specific websites. You may have to contact an individual regional office to get the local information you need for a particular waterway project.

If you are interested in seeing the water quality standards for a particular area, EPA maintains a website with links to federally promulgated water quality standards and state-specific water quality standards.[93] You may need to look at both state and federal standards to understand the water quality requirements for a particular waterway.

86. 33 U.S.C. §1342(b); 40 C.F.R. pt. 123 (2016).
87. 33 U.S.C. §1318; 40 C.F.R. §122.41(k)(4) (2014).
88. U.S. EPA, *Discharge Monitoring Report (DMR) Pollutant Loading Tool*, http://cfpub.epa.gov/dmr/ (last visited Nov. 29, 2016).
89. U.S. EPA, *NPDES General Permit Web Inventory*, https://ofmpub.epa.gov/apex/aps/f?p=GPWI:HOME (last visited Nov. 29, 2016).
90. 33 U.S.C. §1344.
91. U.S. Army Corps of Eng'rs, *ORM Jurisdictional Determinations and Permit Decisions*, http://corpsmapu.usace.army.mil/cm_apex/f?p=340:11:0::NO (last visited Nov. 29, 2016).
92. U.S. Army Corps of Eng'rs, *NWPL—National Wetlands Plant List*, http://rsgisias.crrel.usace.army.mil/NWPL/# (last visited Nov. 29, 2016).
93. U.S. EPA, *Water Quality Standards: Regulations and Resources*, http://www.epa.gov/wqs-tech (last updated Feb. 26, 2016).

B. *Accessing the CAA's Data*

CAA information is difficult to find. Some information can be gleaned from EPA websites, but there is no easy-access, easy-to-find portal or center for CAA information. While there is no central online CAA data source, several sources provide general air quality statistics. EPA has produced retrospective and prospective studies related to clean air in the United States.[94] The Agency also maintains an AirTrends[95] webpage that links to various statistics gleaned from a nationwide network of monitoring sites that EPA has aggregated into ambient air quality trends for the six types of criteria air pollutants: (1) carbon monoxide; (2) ozone; (3) lead; (4) nitrogen dioxide; (5) particulate matter; and (6) and sulfur dioxide—the common air pollutants for which EPA has set national air quality standards.

The CAA has three different types of permits that require the collection of certain data points[96]: (1) acid rain permits (Title IV permits)[97]; (2) preconstruction permits (new source review permits)[98]; (3) and operating permits (Title V permits).[99] While EPA does not have central databases for these permits, many state environmental agencies and EPA's regional offices do contain online information about the CAA permits. For example, EPA's Region 9 provides some of its issued permit dockets online.[100]

EPA also collects air quality data annually, tracking the level of emissions from various pollution sources (CAA only tracks stationary sources) and notes increases and decreases over numbers of years. EPA's Office of Air Quality Planning and Standards sets national ambient air quality standards that determine which pollutants get tracked.[101] EPA provides information about the criteria pollutants online. The website notes whether each criteria pollutant is assigned primary (public health concerns) or secondary

---

94. U.S. EPA, *Benefits and Costs of the Clean Air Act*, http://www.epa.gov/clean-air-act-overview/benefits-and-costs-clean-air-act (last updated Oct. 19, 2015).
95. U.S. EPA, *National Air Quality: Status and Trends of Key Air Pollutants*, https://www.epa.gov/air-trends (last visited Nov. 29, 2016) [hereinafter *AirTrends*].
96. U.S. EPA, *Permitting Under the Clean Air Act*, https://www.epa.gov/caa-permitting (last updated Oct. 21, 2015).
97. U.S. EPA, *Acid Rain Permits Program*, https://www3.epa.gov/region1/eco/acidrain/permitsprog.html (Feb. 23, 2016); 40 C.F.R. §72.1-72.96 (2016).
98. 42 U.S.C. §7475; U.S. EPA, *New Source Review (NSR) Permitting*, https://www.epa.gov/nsr (Feb. 18, 2016).
99. 42 U.S.C. §7661a(a); U.S. EPA, *Operating Permits Issued Under Title V of the Clean Air Act*, https://www.epa.gov/title-v-operating-permits (last updated Nov. 24, 2015).
100. U.S. EPA, *CAA Permitting in EPA's Pacific Southwest Region*, https://www.epa.gov/caa-permitting/caa-permitting-epas-pacific-southwest-region (last updated Mar. 31, 2016).
101. 40 C.F.R. pt. 50 (2016).

(public welfare concerns) standards, or both, and how many parts per million or billion are permissible according to the current standards.[102]

The AirTrends webpage offers a map view where users can click on localities to see the air trends for the six common air pollutants in that location, and they also link to EPA's report card on the status of air quality and air pollutant emissions. Some reports about air quality and special studies are also linked on the AirTrends website.[103]

The bottom of the AirTrends report webpage provides a list of links to more information about air quality in the United States. Like the FIFRA links, however, they do not appear to be officially collected by the government, but rather they are databases maintained by private and academic institutions and organizations.

National Emission Inventory (NEI)[a]	Contains estimates of criteria pollution air emissions and other hazardous air emissions, from data provided by state, local, and tribal air agencies. The NEI is built from the Emissions Inventory System (EIS) which is the first to collect data from the local agencies, and then the information is blended into the NEI.[b]
AirNow[c]	Provides current Air Quality Index (AQI)[d] data to inform the public about outdoor air quality in real time.
AirData[e]	Provides not only the AQI data, but additional data from outdoor monitors across the United States. AirData also provides users with query and visualization tools to synthesize the data.[f]
AirCompare[g]	Allows users to compare AQIs in different counties and states.
CASTNET[h]	This program monitors 95 sites to assess trends in acidic deposition due to emission reduction programs.[i]
National Atmospheric Deposition Program (NADP)[j]	This program also monitors precipitation chemistry due to air emissions. It is housed on the University of Illinois' website.[k]
Visibility Information Exchange Web System (VIEWS)[l]	This Colorado State program was designed to help the public and municipalities to evaluate air quality in their areas by providing data from more than two dozen monitoring networks.[m]

a. U.S. EPA, *National Emissions Inventory*, http://www.epa.gov/air-emissions-inventories/national-emissions-inventory (last updated Oct. 14, 2015).
b. *Id.*
c. U.S. EPA, *AirNow*, https://www.airnow.gov/ (last visited Nov. 29, 2016).
d. AirNow, *Air Quality Index (AQI) Basics*, https://www.airnow.gov/index.cfm?action=aqibasics.aqi (last visited Nov. 29, 2016).
e. U.S. EPA, *Air Quality Data Collected at Outdoor Monitors Across the US*, https://www.epa.gov/outdoor-air-quality-data (last visited Nov. 29, 2016).

102. *Id.*
103. *See AirTrends*, *supra* note 95 (providing information on materials from 1994 through 2011).

f. *Id.*
g. AirNow, *AirCompare*, http://www3.epa.gov/aircompare/ (last visited Nov. 29, 2016).
h. U.S. EPA, *Clean Air Status and Trends Network (CASTNET)*, http://www.epa.gov/castnet (last visited Nov. 29, 2016).
i. U.S. EPA, *Clean Air Status and Trends Network (CASTNET) Overview*, https://www3.epa.gov/castnet/docs/CASTNET-Factsheet-2015.pdf (last visited Nov. 29, 2016).
j. NADP, *Monitoring Our Changing Chemical Climate*, http://nadp.sws.uiuc.edu/ (last visited Nov. 29, 2016).
k. NADP, *About NADP*, http://nadp.sws.uiuc.edu/NADP/ (last visited Nov. 29, 2016).
l. Views 2.0, *Air Quality Data, Tools, and Resources*, http://views.cira.colostate.edu/web/ (last visited Nov. 29, 2016).
m. Views 2.0, *Views Overview*, http://views.cira.colostate.edu/web/About/Overview.aspx (last visited Nov. 29, 2016).

VII. Accessing Public Notice Statutory Information

A. *Comprehensive Environmental Response, Compensation, and Liability Act*

According to the Comprehensive Environmental Response, Compensation, and Liability Act of 1980 (CERCLA),[104] EPA records a list of all of the nation's Superfund sites. This list is a useful tool to notify the public of nearby sites containing hazardous waste.[105] EPA has created a publicly accessible database for Superfund information. Until recently, CERCLIS was the major Superfund database, but CERCLIS has been replaced.[106] This database has been replaced with the Superfund Enterprise Management System (SEMS).[107]

SEMS provides extensive site information for Superfund sites on the national priorities list (NPL) and sites deleted from the final NPL. SEMS supplies information for sites with Superfund Alternative Approach agreements. Finally, SEMS also catalogs information about non-NPL and archived sites, but there is less information about those sites. You can search for sites by name, EPA identification number, address information, congressional district, federal agency, and Indian entity. Search results include an overview of the site and its current status, as well as what is being done to protect the health and human environment near the site. The information provided through SEMS is generally thorough and written in terms laypeople can understand, which is very helpful for decisionmaking purposes.

104. Pub. L. No. 96-510, 94 Stat. 2767 (codified at 42 U.S.C. §§9601-9675, ELR Stat. CERCLA §§101-405).
105. U.S. EPA, *Search for Superfund Sites Where You Live*, http://www.epa.gov/superfund/search-superfund-sites-where-you-live (last visited Nov. 29, 2016).
106. U.S. EPA, *SEMS Search*, http://www.epa.gov/enviro/cerclis-search (last visited Nov. 29, 2016).
107. U.S. EPA, *Search Superfund Site Information*, https://cumulis.epa.gov/supercpad/CurSites/srchsites.cfm (last visited Nov. 29, 2016).

There is also EPA's database called Cleanups in My Community (CIMC) that provides access to CERCLA and RCRA cleanup progress profiles.[108] With CIMC, you can map and list hazardous waste cleanup locations and grant areas and get more details about the cleanups, although it is not as in-depth as the SEMS information, so if the site is listed on the NPL, you may want to default to SEMS.

B. Labelling Laws

There is also a series of labeling laws enforced by a variety of agencies, including EPA, the U.S. Food and Drug Administration, and OSHA. These laws require the placement of labels on items from toxic substances (FIFRA, RCRA, and TSCA have product labeling requirements) to edible items (organic labeling, for instance) to materials used in the workplace. These labels contain safety warnings for hazardous chemicals, describe the most environmentally safe way to use the items, and provide the public with information to help them make informed purchasing and use decisions (energy efficiency labels on electronic products, etc.). The best place for information seekers to find the labels is usually on the item itself, but there are agency guidance documents that provide images of the labels and describe the label meanings and chemical uses in greater detail.[109]

EPA and other agencies create guides and handbooks detailing the appearance and contents of various labels required by the federal laws. For example, EPA publishes a *Label Review Manual* (both in paper and online) for labels required by FIFRA.[110] OSHA and the U.S. Department of Transportation require labeling under RCRA[111] for hazardous chemicals distributed, imported, and shipped in the United States. Both agencies produce guidance materials containing examples of label pictograms representing certain hazards.[112]

There are also labels designed not to warn of hazardous materials, but to enhance the public's environmental decisionmaking processes as they purchase goods and services. The federal government allows manufacturers and producers to label their products as "green." The green vehicle program

108. U.S. EPA, *Cleanups in My Community*, http://www.epa.gov/cleanups/cleanups-my-community (last updated Mar. 7, 2016).

109. For instance, here is an OSHA guide about its labels and pictograms, see OSHA, HAZARD COMMUNICATION STANDARD: LABELS AND PICTOGRAMS (OSHA BRIEF 2/2013), https://www.osha.gov/Publications/OSHA3636.pdf (last visited Nov. 29, 2016).

110. *See* LABEL REVIEW MANUAL, *supra* page 97, note d.

111. C.F.R. 1910.1200 C.2.3.3.

112. *See, e.g.*, OSHA, *OSHA QuickCard*, https://www.osha.gov/Publications/HazComm_QuickCard_Pictogram.html (last visited Nov. 29, 2016).

informs car purchasers about the fuel efficiency of their potential automobile purchases.[113] "Safer choices" labels can be placed on products that meet certain heightened EPA safety standards.[114] These labels do not instruct people on how to use, transport, or store chemicals or otherwise instruct people, but they do educate potential consumers about the products' environmental efficacy.

C. *Research Challenges*

As this chapter demonstrates, U.S. federal environmental law generates a variety of data types, and federal agencies, as well as private organizations and institutions, have created many access points for the information gathered pursuant to these laws. There are, however, a few issues that limit the availability and reliability of this type of information:

- Information Collection Is Limited to Regulated Entities. Environmental laws only apply to the regulated entities outlined by each law. Plenty of polluting forces do not have to report their activities, e.g., non-stationary sources under CWA, chemical manufacturers that do not meet the threshold of consumer activities required by the toxics laws, and non-stationary sources under the CAA.

  Also, certain populations and industries are exempt from information disclosure, like agriculture under the CWA, or TRI requirements that only cover companies with more than 10 employees. One environmental justice report focused on this issue, pointing out that "[i]n many communities, the most egregious offenders are the small electroplating plants, auto-body welding shops, dry cleaners, and waste transfer stations[ ]" whose effects are not monitored by environmental data collection law provisions.[115]

- Information Provided by EPA Is Often Voluminous and Complicated. One EIS can be multiple thousand-page packets. For example, the EIS for the SpaceX privacy spaceflight launch site is three volumes and 1,576 pages (89.4 MBs of data) long. More comprehensive projects are far longer. Diving into the TRI database involves sifting through tens

113. U.S. EPA, *Learn About Green Vehicles*, https://www.epa.gov/greenvehicles/learn-about-green-vehicles (last visited Nov. 29, 2016).

114. U.S. EPA, *Learn About the Safer Choices Label*, https://www.epa.gov/saferchoice/learn-about-safer-choice-label (last visited Nov. 29, 2016).

115. Julianna Mantaay, *Mapping Environmental Injustices: Pitfalls and Potential of Geographic Information Systems in Assessing Environmental Health and Equity*, 110 Envtl. Health Persp. 161, 163 (2002) (suppl. 2).

of thousands of bits of data, each labelled with complex chemical names and measurements, with no additional explanations for laypeople. As one legal scholar notes: "More information is not always better."[116] Additionally, the information retrieved pursuant to environmental laws is often full of complex terms and jargon, requiring interpretations and analysis in order for the public to use it in their environmental decision-making processes.

- EPA's Website Is Massive and Confusing. It has been well documented that the federal government has trouble adopting new technology, especially in their websites. Bureaucratic factors like long procurement cycles for technology projects, delays in approval to implement those programs, and a vast network of preexisting systems that information technology managers must untangle and update all contribute to federal agencies being stuck in a technology time warp. EPA is no exception to this general truism. EPA's overlapping and duplicative webpages are full of dead links, conflicting and repeating information, and links that take users on circular and dead-end tours across many EPA webpages without ever leading to the information they need. In short, EPA's system of Internet resources is a huge maze with no guide.

  The complexity of the websites and databases that compose EPA's gargantuan online information presence can only be matched by the complexity of the search functions on those webpages. Instead of the Google-like ease of many of today's online search mechanisms, EPA database searches tend to be old and clunky, and search results often time-out before loading due to their bulky sizes.

- Data Is Not Up to Date. Although at press time, it is 2016, the most recent TRI data accessible in EPA's TRI database is from 2013. Similarly, CERCLA's Superfund Search website is only current through Nov. 12, 2013. In addition, according to EPA's September 11th response website, EPA archives its old data, putting it out of sight from the public. The lack of current data and historical data on EPA's webpages leaves information seekers with incomplete information.

- Self-Reporting Inaccuracies. Because of the amount of time and effort it takes to gather all of the points of information required by environmental laws and their supporting regulations, the federal government leaves much of its information gathering responsibilities up to

116. Volokh, *supra* note 51, at 807.

the polluters. Statutes like the CWA, the CAA, RCRA, and EPCRA's TRI program do not have EPA scientists in the field gathering data on hazardous chemicals and pollutants. Rather, the companies creating, storing, delivering, and otherwise handling the pollutants and hazardous materials report data to EPA. This means that any information gathered under these laws is an act of good faith between EPA and the regulated entity.

Self-reporting is an imperfect data collection method, and regulated entities often have superior information that they do submit to the regulating authorities.[117] While regulated entities are often better equipped to collect and provide correct information, the government authorities (who often have the power to issue enforcement actions against the regulated entities), are not well positioned to obtain information from them. As one legal scholar surmises: "Government cannot count on self-interested holders of information to reveal it fully and without bias. Neither can government count on its power to compel the disclosure of information."[118] One study following TRI data collection found that regulated entities violate reporting requirements both intentionally, by underestimating pollution levels and even submitting false data to avoid pollution controls, and unintentionally, due to ignorance of the law and lack of expertise.[119] Thus, the end-user of the data—the public—is often receiving imperfect information.

117. This phenomenon is widely recognized, and some experts have even proposed improvements to EPA audit policies to combat this problem. For more information, see Jodi L. Short & Michael W. Toffel, *Coerced Confessions: Self-Policing in the Shadow of the Regulator*, 17 J. L. ECON. & ORG. 239-56 (2007), http://www.environment.harvard.edu/docs/faculty_pubs/toffel_coerced.pdf.

118. Cary Coglianese et al., *Seeking Truth for Power: Informational Strategy and Regulatory Policymaking*, 89 MINN. L. REV. 277, 341 (2004).

119. David J. Abell, *Emergency Planning and Community Right-To-Know: The Toxics Release Inventory*, 47 SMU L. REV. 581, 592-93 (1994).

Chapter 5: Environmental Information and the Freedom of Information Act

Law in Action

Sandra Cart, an investigative journalist for the Shawnee Gazette got a call from a Shawnee resident complaining about a bad smell coming from the Chickadee River, a waterway running through the east side of town. This was one of several complaints Cart had received involving odors coming from the river, and she suspected that Patterson's Poultry Plant, situated on the waterway, might be the source. She asked a biologist from Shawnee University to investigate and several tests were conducted. The tests revealed alarmingly high amounts of poultry processing waste, including fecal matter, in the water. Cart located the Patterson Plant permits online, and wondered whether the plant had received any other complaints or citations or been subject to investigations recently that could shed light on the residents' complaints or reveal a deeper public health issue. Cart decided to contact the U.S. Environmental Protection Agency and the state environmental protection agency to see whether she could obtain more information about Patterson's plant and its practices.[1]

I. What Is the Freedom of Information Act?

The Freedom of Information Act (FOIA)[2] is the crown jewel of federal transparency laws, providing the public with access to a myriad of federal agency records. FOIA is an environmental researcher's best tool for accessing otherwise nonpublic government documents. Environmental advocacy groups and nonprofit organizations, journalists, scientists, lawyers, and private concern groups often use it to obtain government information related to environmental issues. The issues involved can be large and broad such as those related to climate change or multistate pipeline projects,

1. Annie Ropeik, *Locals Dismayed as Allen Harim Prepares to Double Chicken Capacity in Harbeson*, Delaware Public Media, Feb. 22, 2016.
2. Freedom of Information Act of 1966, Pub. L. No. 89-487, 80 Stat. 250 (codified at 5 U.S.C. §552 (2014)).

or issues far more local and isolated like water contamination in a single Marine Corps base.[3]

But for researchers, is FOIA really a magical law that can unlock every kind of environmental information? On paper, it seems like the ideal tool for government information retrieval, with flexible terms and broad permission to access government records. In practice, however, the short answer is that FOIA is not without its access issues and obstacles.

Notably, FOIA can be utilized by any "person," including U.S. citizens, foreign nationals, organizations, associations, and universities.[4] Access under FOIA is provided in two ways:

1. Proactively disclosing agency records including final agency opinions, policy statements, and certain staff manuals that affect the public.[5]
2. Enabling the public to request agency records that the agency does not voluntarily provide to the public through proactive disclosure.[6]

Administrative agencies under the executive branch fulfill the proactive disclosure requirements of the law by creating publicly accessible reading rooms, available both in print and online through agency website portals and by providing online methods for the public to file FOIA requests.[7] Although agencies do not implement online FOIA policies uniformly, as technical capabilities expand, so do FOIA disclosure venues. Federal agencies continue

---

3. James LaPorta, *Veterans Group to Seek Information on Contaminated Water at Camp Lejeune*, JDNews.com, Dec. 9, 2015, http://www.jdnews.com/article/20151209/NEWS/151209146.
4. The definition of "person" comes from the Administrative Procedure Act, 5 U.S.C. §551(2) (2014), Administrative Procedure Act of 1946, Pub. L. No. 79-404, 60 Stat. 237 (codified at 5 U.S.C. §§551-559 (2014)). According to the U.S. Department of Justice (DOJ):

   > Under existing law, foreign nationals have the same rights of access under the Freedom of Information Act as do United States citizens. The only point of difference in this regard is that when an agency advises a foreign national requester of his right to review in federal district court, he need not be advised that he can bring suit in a judicial district of his residency (unless he has been lawfully admitted for permanent residence). By contrast, the Privacy Act of 1974 limits its access rights to United States citizens and aliens lawfully admitted for permanent residence.

   U.S. DOJ, *FOIA Update: FOIA Counselor: Questions & Answers*, https://www.justice.gov/oip/blog/foia-update-foia-counselor-questions-answers-25 (last visited Dec. 2, 2016).
5. 5 U.S.C. §552(a)(2).
6. *Id.* §552(a)(3).
7. *See infra* note 15 (discussing how the E-FOIA Amendments of 1996 requires agencies to make proactively disclosed records created after Nov. 1, 1996 available by "electronic means." 5 U.S.C. §552(a)(2). Disclosing information to satisfy §552(a)(2) of FOIA is commonly done in spaces (either in person or online) called "reading rooms" (online reading rooms are often called "electronic reading rooms"). U.S. DOJ, FOIA Reading Rooms, FOIA Guide (2004), https://www.justice.gov/oip/foia-guide-2004-edition-foia-reading-rooms (last visited Dec. 2, 2016). The term of art "reading room" will be used throughout this chapter to differentiate information disclosed pursuant to FOIA from general online information that agencies make available on their websites.

to create improved access to their records online, although their offerings vary by agency.

II. History of FOIA

FOIA was passed in response to government secrecy scandals and a growing sense of distrust in the government in the 1960s.[8] Prior to FOIA's passage, the only statutorily created means of accessing agency records was through the Administrative Procedure Act (APA).[9] The APA required federal agencies to make "matters of the official record" like public procedures, rules, and policies available to people "properly and directly concerned" with them.[10] This narrow provision was a step forward for information access, but still limited information disclosed to the general public to only "concerned" parties. Moreover, its broad exemptions and loopholes allowed government agencies to withhold information for so many reasons the provisions were once sarcastically referred to as a "withholding statute" used as an administrative excuse for secrecy.[11]

After the Watergate scandal renewed concerns about government secrecy, the law was amended in 1974.[12] The amendments made FOIA an even more effective transparency tool by including time requirements for the release of agency records, fee waivers and fee limitations for certain types of FOIA requesters, and sanctions for agencies improperly withholding information.[13] Since 1974, each presidential administration has administered FOIA's transparency provisions via executive order.[14] Various administrations have cre-

8. Lyndon B. Johnson, Statement by the President Upon Signing the Freedom of Information Act (July 4, 1966), 11 Pub. Papers 699 (July 4, 1966) (noting that "[a] democracy works best when the people have all the information that the security of the Nation permits").
9. *See* Administrative Procedure Act of 1946, Pub. L. No. 79-404, §3, 60 Stat. 237, 238-39.
10. *See* H. Rep. No. 89-1497, at 1, 6, 8, 9 (1966); S. Rep. No. 89-813, at 5, 9 (1965), *reprinted in* 1966 U.S.C.C.A.N. 2418, 2426.
11. S. Rep. No. 89-813, at 38 (1965). *See also* 5 U.S.C. §552.
12. Freedom of Information Act and Amendments of 1974, Pub. L. No. 93-502, 88 Stat. 1561; H. Subcomm. on Gov't Info. & Individual Rights & S. Subcomm. on Admin. Practice & Procedure, 94th Cong., Freedom of Information Act and Amendments of 1974 (P.L. 93-502) (J. Comm. Print 1975) (discussing the amendments in further detail), http://www.loc.gov/rr/frd/Military_Law/pdf/FOIA-1974.pdf (last visited Dec. 2, 2016).
13. *U.S. Government Information Policies and Practices—Administration and Operation of the Freedom of Information Act (Part 4): Hearing Before the H. Comm. on Gov't Operations*, 92nd Cong. 1223-24 (1972). Testimony discussed agency copying charges of $1.00 per page, with searching costs of $7.00 per hour, and the manipulation of tasks to increase fees. For instance, a requestor seeking the names of 26 people who directly reported to a Federal Aviation Administration official accrued a separate search fee for each of the 26 names.
14. For example, see the Obama Administration's FOIA memorandum: Presidential Memorandum for Heads of Executive Departments and Agencies Concerning the Freedom of Information Act, 74 Fed. Reg. 4683 (Jan. 21, 2009). For Attorney General Eric Holder's response, see President Obama's FOIA Memorandum and Attorney General Holder's FOIA Guidelines (July 23, 2014), http://www.justice.gov/sites/default/files/oip/legacy/2014/07/23/foia-memorandum.pdf.

ated a cycle of expansion and contraction as the law's scope, parameters, and procedure are effectively changed. In 1996, the Electronic Freedom of Information Act (popularly referred to as the E-FOIA Act)[15] explicitly added agencies' computer databases and other electronic records to FOIA's purview, and also required agencies to provide records electronically if so requested.[16] The E-FOIA Act also required agencies to create online access to their records, leading to the widespread adoption of electronic reading rooms available through federal agency websites.

In addition to the federal FOIA statute, each state has passed their own version of a freedom of information law (FOI) law for state agency records.[17] These state laws are especially useful in an environmental information context, as many environmental projects are carried out at the state level and many state agencies are charged with implementing and enforcing federal environmental statutes. In the world of environmental law, state and federal agency intermingling is the norm, so there may be situations where you must obtain information from both state and federal agencies using both federal FOIA provisions and also relying on state FOI laws.

III. How FOIA Works

FOIA provides information to the public in two specific ways. First, the law provides for mandatory proactive disclosure of certain types of agency information.[18] Second, for information not proactively disclosed, the law allows the public to request specific records from federal agencies.[19] These two FOIA-mandated federal agency actions: (1) proactive disclosure of certain records; and (2) agency responses to public requests for information, work in tandem.

A. *Proactive Disclosure*

FOIA's proactive disclosure provisions require agencies to make material that falls into certain pre-determined categories available to the public. Types of information that are automatically made available to the public include: gen-

15. Electronic Freedom of Information Act Amendments of 1996, Pub. L. No. 104-231, 110 Stat. 3048 (codified at 5 U.S.C. §552(a)(2)).
16. *Id.* §552(a)(2)(E).
17. See National Freedom of Information Coalition, *State Freedom of Information Laws*, http://www.nfoic.org/state-freedom-of-information-laws (last visited Dec. 2, 2016) for a list of links to state laws.
18. 5 U.S.C. §§552(a)(1)-(2).
19. *Id.* §552(a)(3).

eral information about agency organization and process[20]; final administrative adjudication orders and opinions[21]; agency guidance materials (policy statements and interpretations adopted by the agency but not published in the *Federal Register*)[22]; administrative staff manuals and instructions to staff that affect the public[23]; and copies of all records that have been released via previous FOIA requests.[24] All of these records must be available for the public to inspect and copy.

There is no central repository where agencies disclose FOIA records. Instead, they are scattered throughout the federal agencies' various online and paper repositories.[25] Information seekers can access FOIA records through online agency reading rooms, which are increasingly providing more records. A relatively new website, FOIAonline, tracks and processes FOIA requests for some participating federal agencies, providing FOIA records to the public in one online location. The U.S. Environmental Protection Agency (EPA) does participate in FOIAonline, however, at this point the majority of agencies do not, so the website is still not an ideal central FOIA repository.[26] (See Figure 5.1). When information seekers are unable to find what they need in reading rooms, they often file a FOIA request to obtain the missing information. It is a best practice to search online, including in FOIA reading rooms, to see whether information has been proactively released before filing a FOIA request.

B. *Disclosure by Request*

FOIA requests are the most well-known function of the FOI law. Usually, when people refer to FOIA, they are referring to the FOIA request process.[27] The "information upon request" section of FOIA is the primary way to

---

20. *Id.* §552(a)(1).
21. *Id.* §552(a)(2)(A).
22. *Id.* §552(a)(2)(B).
23. *Id.* §552(a)(2)(C).
24. *Id.* §552(a)(2)(D) (stating that records must be released regardless of form or format and should include information that is likely to become the subject of subsequent requests for substantially the same records; requests should also include and a general index of those FOIA request records).
25. The issues surrounding finding these records increases as proactively disclosed materials are migrated to new versions of government websites, leading to missing links and link rot.
26. FOIAonline, *About FOIAonline*, https://foiaonline.regulations.gov/foia/action/public/home/about (last visited Dec. 1, 2016) [hereinafter *About FOIAonline*]. (Thus far, only 12 agencies participate in this online FOIA database.)
27. Today, 5 U.S.C. §552(a)(3) is "by far the most commonly utilized part of the Act." U.S. DOJ, Department of Justice Guide to the Freedom of Information Act 2 (2016), http://www.justice.gov/oip/doj guide-freedom-information-act-0 (last visited Dec. 2, 2016) [hereinafter DOJ FOIA Guide].

Figure 5.1 BLM Online Reading Room

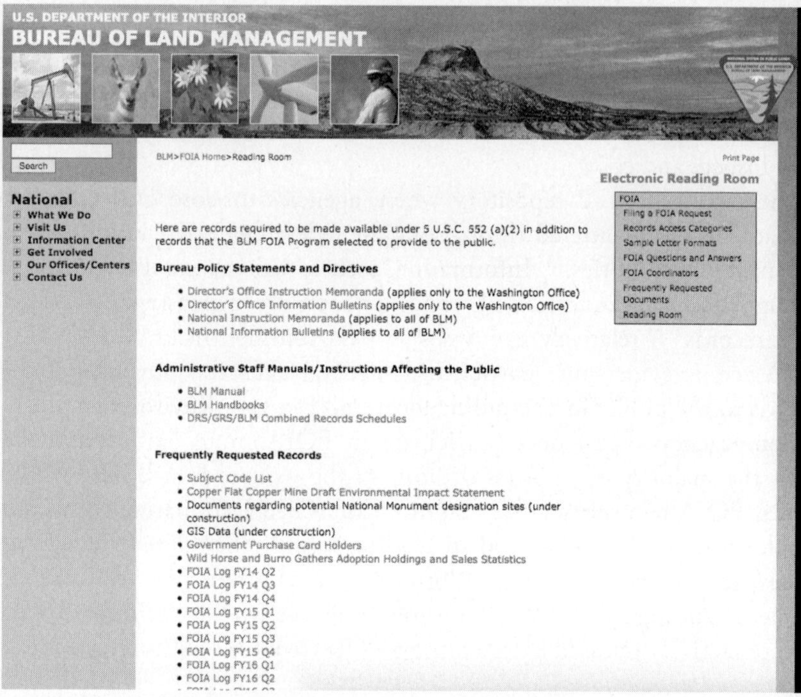

Although EPA now links to FOIAonline to fulfill their online reading room provisions, some agencies have separate online reading rooms. For instance, the Bureau of Land Management (BLM) electronic reading room contains agency statements and directives, manuals and instructions, and links to records that are frequently requested via FOIA. *See* https://www.blm.gov/wo/st/en/res/FOIA/Reading_Room.html.

obtain government information that is not otherwise disclosed.[28] The statute itself does not provide many requirements for making FOIA requests beyond requiring that the request must reasonably describe the records and follow any published rules about time, place, fees, and procedures.[29] The statute is intentionally broad, allowing any person to make a request without having to explain why they want the information. The statute says that once a request is made, the agency shall make the records promptly available.[30]

The FOIA request process has three major components: (1) determining what agency has the records sought; (2) creating a request; and (3) submitting the request. Although the statutory language provides few guidelines

28. 5 U.S.C. §552(a)(3)(A).
29. *Id.* §§552(a)(3)(A)(i) & (ii).
30. *Id.* §552(a)(3)(A)(ii).

and offers no format or content requirements for FOIA requests, there are many resources to help researchers draft FOIA requests that improve their chances of getting quick and complete results. These tips and resources are described in detail in section VI of this chapter.

Even though FOIA requests are the most prevalent mechanism for unearthing government information, the statute does contain some limitations and bars to information access. FOIA requesters should understand the law's restrictions and constraints before making a FOIA request to save time (if the information they seek is not obtainable), and also to craft FOIA requests using language that ensures the request falls within the parameters of permissible FOIA disclosures (for example, requesting "records" instead of "materials").

IV. FOIA Request Limitations: Statutory Definitions and Exemptions

FOIA's parameters are somewhat narrow when one considers all of the information created and utilized by the government in its myriad daily operations. The statute provides access only to "federal agency records," and also creates nine categories of records exemptions to restrict agencies from releasing certain types of information, like data that may endanger national security, peoples' private information, or companies' trade secrets.

A. *Must Be an Agency*

FOIA defines an "agency" as "any executive department, military department, Government corporation, Government controlled corporation, or other establishment in the executive branch of the Government (including the Executive Office of the President), or any independent regulatory agency."[31] This means that FOIA applies to all 15 federal departments, and over 70 federal agencies within the executive branch of the U.S. government.[32]

While much of the government's work does occur at the agency level, many important government activities occur in other branches or parts of the government that fall outside FOIA's scope. In particular, FOIA does not apply to the president, to Congress, or to the courts. Also, the federal FOIA

---

31. *Id.* 5 U.S.C. §552(f)(1).
32. DOJ FOIA Guide, *supra* note 27.

does not apply to state government agencies. (Although it should be noted that each state has state FOI laws, which we discuss in Chapter 9).

As an illustration of the interaction between federal and state FOI laws, in 2015-2016 local scientists discovered unsafe levels of lead contamination in Flint, Michigan's, drinking water supply. The state's governor initially refused to hand over documents and correspondence related to the toxic water supply. Because the documents were held by a state governor, and not a federal agency, the public was unable to access this information through FOIA (nor through Michigan's state FOI law) and local public interest groups were left begging state leaders to hand over the documents, to no avail.[33] Michigan Gov. Rick Snyder eventually did turn over some of his e-mails after succumbing to public pressure long after the crisis was initially discovered.[34]

Information created and kept by government contractors frequently falls outside of FOIA's coverage.[35] This lack of access is especially important for environmental information seekers, as much of the environmental work in the United States is managed, supervised, and conducted by private entities contracted by the government. FOIA's limited scope may remove swaths of environmentally relevant records from public access.

B. *Must Be a Record*

Under the definition section of the Act, the term "record" is fairly inclusive. Records, according to FOIA, include "any information [...] maintained by an agency in any format, including an electronic format."[36] Under FOIA's "record" definition, many environmental information resources qualify as records: paper files, online repositories or databases, videos, datasets, etc. fall under FOIA's transparency purview, so long as one of the nine FOIA exemptions do not apply.

While FOIA defines "record" broadly, it does not require agencies to create new records or maintain existing ones.[37] FOIA does not force agencies to aggregate data, put together excerpts, or otherwise create new records in order to provide information for requesters. It also does not require agencies

33. Ron Fonger, *Watchdog Group Calls on Gov. Snyder to Release All Flint Water Documents*, Mich. Live, Jan. 5, 2016, http://www.mlive.com/news/flint/index.ssf/2016/01/common_cause_calls_on_gov_snyd.html (last visited Dec. 2, 2016).

34. Press Release, Gov. Rick Snyder Releases Departmental E-Mails Produced Regarding Flint Water Crisis (Feb. 12, 2016).

35. For more information, see Sarah Shik Lamdan, *Sunshine for Sale: Environmental Contractors and the Freedom of Information Act*, 15 Vt. J. Envtl. L. 227-61 (2014).

36. 5 U.S.C. §552(f)(2)(A).

37. FOIA.gov, *Frequently Asked Questions*, http://www.foia.gov/faq.html (last visited Dec. 2, 2016) ("What can I ask for under the FOIA?").

to conduct research, analyze data, or answer questions when responding to FOIA requests. Simply, agencies are only required to hand over pre-made documents from their files. They are under no obligation from FOIA to clarify, expand upon, create, or explain content.[38]

Additionally, FOIA's definition of "record" does not include any tangible, evidentiary objects like computer hardware, or archival exhibits like guns, bullets, or clothing.[39] So, while the definition of records includes photographs and electronic document delivery mechanisms like compact discs and USB drives, there are some things that are outside of FOIA's bounds. For example, you are unlikely to be able to obtain water samples used in EPA testing through FOIA, even if you can likely get the data collected through testing of those samples.

Because FOIA defines "agency" and "record," but not "agency record" as a singular term, the U.S. Supreme Court created a two-part test to determine whether something is an agency record. Under this test, an agency must: (1) create or obtain the requested materials; and (2) be in control of the requested materials at the time the FOIA request is made.[40] Thus, to fall within FOIA's scope, environmental information must be either created or obtained by an agency or under some agency's control at the time of the FOIA request.[41]

C. *FOIA Exemptions*

When Pres. Lyndon B. Johnson signed FOIA into law in 1966, he worried that government transparency would threaten national security, privacy, and other sensitive government information[42]—so much so that he almost vetoed the law, and wrote a signing statement undercutting its transparency provisions.[43] Despite the fact that Congress created a number of exemptions to bar certain types of information from public disclosure,[44] some officials, includ-

38. *Id.* (noting "[t]he FOIA does not require agencies to create new records or to conduct research, analyze data, or answer questions when responding to requests").
39. DOJ FOIA Guide, *supra* note 27, at 33 (referring to a case where a requester wanted evidentiary items pertaining to the assassination of President John F. Kennedy).
40. Department of Justice v. Tax Analysts, 492 U.S. 136, 144-45 (1989). *See also* Reporters Comm. for Freedom of the Press, *Not an "Agency Record,"* http://www.rcfp.org/federal-foia-appeals-guide/not-agency-record#sthash.6S2Wm3Th.dpuf (last visited Dec. 2, 2016).
41. DOJ FOIA Guide, *supra* note 27, at 34; *Tax Analysts*, 492 U.S. at 144-45.
42. *See* Statement of Lyndon B. Johnson, *supra* note 8.
43. National Security Archive, *Freedom of Information at 40*, July 4, 2006, http://nsarchive.gwu.edu/NSAEBB/NSAEBB194/ (last visited Dec. 2, 2016) (explaining that the "normally gregarious" president refused to even have a ceremony for the statute, and that he removed the openness language from his signing statement).
44. *See* 89 S. 1160—Introduced in Senate Feb. 17, 1965 (containing enumerated exemptions).

ing Johnson, remained uneasy with the thought of government information being "available to all and sundry including the idly curious."[45]

FOIA does limit certain types of material from its public access provisions. The nine basic exemptions are set out in subsection (b) of the statute.[46] These exemptions are:

- **Exemption 1:** Information that is classified to protect national security.
- **Exemption 2:** Information related solely to the internal personnel rules and practices of an agency.
- **Exemption 3:** Information that is prohibited from disclosure by another federal law.
- **Exemption 4:** Trade secrets or commercial or financial information that is confidential or privileged.
- **Exemption 5:** Privileged communications within or between agencies, including:
    - Deliberative Process Privilege.
    - Attorney-Work Product Privilege.
    - Attorney-Client Privilege.
- **Exemption 6:** Information that if disclosed, would invade another individual's personal privacy.[47]
- **Exemption 7:** Information compiled for law enforcement purposes which:
    - 7(A). Could reasonably be expected to interfere with enforcement proceedings.
    - 7(B). Would deprive a person of a right to a fair trial or an impartial adjudication.
    - 7(C). Could reasonably be expected to constitute an unwarranted invasion of personal privacy.
    - 7(D). Could reasonably be expected to disclose the identity of a confidential source.

45. Letter to Bill Moyers, from Lee White, re: proposed FOIA legislation, http://nsarchive.gwu.edu/NSAEBB/NSAEBB194/Document%201.pdf.
46. 5 U.S.C. §§52(b)(1)-(9) (often referred to by number as "Exemption 1" or "Exemption 6" etc.).
47. Researchers should note that individuals can request their personal records under the Privacy Act of 1974, which is described in Chapter 6.

- 7(E). Would disclose techniques and procedures for law enforcement investigations or prosecutions.
- 7(F). Could reasonably be expected to endanger the life or physical safety of any individual.

- **Exemption 8:** Information that concerns the supervision of financial institutions.
- **Exemption 9:** Geological information on wells.

In the decades since FOIA's passage, a body of case law has developed around each exemption, with many court-created interpretations and tests further delineating and explaining what types of records are allowed or prohibited under each unique exemption.[48]

When looking at the exemptions through the lens of environmental information access, it is clear that while all may impact the disclosure of environmental information records, some are more commonly utilized to prevent the disclosure of environmental information. Beyond Exemption 9, geological information on wells, which obviously withholds a very specific type of information about the environment, several of the other exemptions often withhold environmental information from the public. According to FOIAonline records, Exemptions 5 and 6, which restrict trade secrets and personal information from disclosure, are EPA's most commonly used FOIA exemptions.[49]

Another commonly invoked exemption that environmental information seekers encounter is Exemption 4, which bars the release of trade secrets or commercial and financial information from release. Frequently, environmental information involves chemical compounds and processes created and used by industrial companies. Under FOIA, the definition of trade secret includes "a secret, commercially valuable plan, formula, process, or device that is used for the making of preparing, compounding, or processing of trade commodities" where there is a "direct relationship" between the trade secrets and the productive process.[50] This definition includes everything from lists of chemicals used in the hydraulic fracturing or fracking process to descriptions of tools and processes used in harbor dredging projects. Because many environmental laws require or encourage polluters to self-report, the agency often allows polluters to mark their documents as

48. For a list of major FOIA cases, see U.S. DOJ, *Court Decisions Overview*, https://www.justice.gov/oip/court-decisions-overview (last visited Dec. 2, 2016) (listing almost 900 cases by topic with keyword search capabilities).
49. Reports are available at https://foiaonline.regulations.gov/foia/action/public/report/createReport.
50. Public Citizen Health Research Group v. FDA, 704 F.2d 1280, 1288 (D.C. Cir. 1983).

"confidential" on an ad hoc basis to ensure that the information will not be publicly released, despite the fact that the self-reported records become agency records under FOIA.[51]

In short: if it is not an agency record, you are not going to get it with a FOIA request, and even if it is an agency record, it may be exempt from disclosure. The scope of FOIA, as well as its explicit exemptions, creates transparency, albeit transparency with limitations, on certain types of federal environmental information records.

V. FOIA Best Practices

FOIA prescribes no particular way to make a request. It is generally understood that "[t]here is no magic formula for making a successful [FOIA] request,"[52] however, there *are* some best practices and tips that may optimize your FOIA results.

A. Step 1: Research Your Request

Before you even consider filing a FOIA request, check to see whether the information you want is already available. As the National Security Archive explains: "A FOIA request should be a research tool of last resort; it can take a long time to obtain records and can be costly."[53] Look in agencies' online FOIA reading rooms and elsewhere on their websites to see if the agency has already proactively shared the records you want. Check with government and library sources to see whether the records are available through an outside source. Do a thorough search to make sure that what you are going to request through FOIA is not already available immediately and free of charge.

Your FOIA-related research tasks do not end there. The first things you must do when drafting a FOIA request is to develop a clear idea of what agency records you will need and what agencies you should request them from. You should be able to articulate what data or documents you are seek-

51. For example, this American Bar Association (ABA) briefing paper explains how to claim confidentiality for the disclosure of information under the Toxic Substances Control Act (TSCA): Section on Environment, Energy & Resources (SEER), American Bar Association (ABA), ABA SEER TSCA Trade Secret and Confidential Business Information Briefing Paper (2014), http://www.americanbar.org/content/dam/aba/administrative/environment_energy_resources/whitepapers/tsca/TSCA_paper_CBI_briefing.authcheckdam.pdf.

52. Gidiere, *infra* note 86, at 177.

53. National Security Archive, *Making the FOIA Work for You*, http://nsarchive.gwu.edu/nsa/foia/tips.html (last visited Dec. 2, 2016).

ing with some specificity and where they are located in the federal agency system—which can sometimes seem like a labyrinth of federal offices.

1. Which Records?

Figuring out what records to request often takes some significant preliminary research. For instance, in one FOIA request, the pro-environmental Center for Biological Diversity (CBD), made a FOIA request for records containing air pollution emission data for Volkswagen automobiles.[54] The request contained a date parameter (records from 2009-present), and CBD asked for specific data types (total amount of the greenhouse gas nitrous oxide from Volkswagen vehicles with defeat devices) and types of documents (documents and communications containing information on other vehicle brands that emit nitrogen oxides in excess of Clean Air Act limits, etc.).[55] This request also specified particular events, company names, and data sets that the federal agency could identify and locate records for.

In order for FOIA requests to be processed efficiently (and for the requester to get the information they need) gathering enough information to create a clear request is vital. Even if what you want to get is a broad swath of information, e.g., all releasable records concerning a pollution event or all documents about a certain natural resource between a span of dates, it still helps to give any useful, specific information you have that will identify the records you seek.

Knowing what date a particular record was created or what person or entity may have authored the document also helps. Any information that will help a FOIA officer locate the records you need (and make it harder for a FOIA officer to deny a record's existence) will enhance your FOIA request. If you have any helpful documents (similar record types or other documents discussing the record you seek), attach those to your request.

Also, remember that although FOIA requires agencies to provide agency records, it does not require the agencies to *create* records in response to your request. Do not fear, however, as agencies have many types of information in their records. Agencies carry out most of the day-to-day functions of the government, and their records reflect the following activities: data collection, oversight of government programs and projects, and private projects with

54. Press Release, Center for Biological Diversity, After Volkswagen Scandal, Legal Petition Seeks Better Smog Testing, Bigger Fines (Oct. 1, 2015), https://www.biologicaldiversity.org/news/press_releases/2015/fuel-economy-standards-10-01-2015.html.

55. Center for Biological Diversity, *FOIA Request*, Oct. 6, 2015, http://www.biologicaldiversity.org/programs/climate_law_institute/transportation_and_global_warming/fuel_economy_standards/pdfs/FOIA_EPA_Vehicle_Emissions_Exceedance_Data.pdf.

statutorily prescribed regulatory oversight, providing advice, conducting studies, coordinating projects, investigations, and the myriad activities that agencies are tasked with doing under legislation or by executive order. That said, users should request items that actually exist in agency records, and not some amalgamation, analysis, or portion of a record.

Do not hesitate to contact the agency to help you narrow down your request. If you have questions about what certain records (inspections and complaints, etc.) are called or named or how the agency documents certain events, check with an administrator so that you use the correct terminology in your FOIA request.

2. Which Agency?

Although knowing which agency to send a FOIA request to sounds like a simple task, the agency system is larger than any simple federal organization chart indicates. For instance, when reporters filed FOIA requests for data on coal ash sludge levels after a major coal ash pond breach that led to the release of a million gallons of coal ash slurry, they did not merely send a FOIA request to EPA, instead, they located the proper regional office, and named a particular FOIA administrator in their request.[56] It is important to do the required background research to know where to direct your FOIA request.

Also, make sure your FOIA request is indeed for federal records, and not state records. Many environmental programs and projects are carried out through federal-state partnerships, and many important environmental documents are held in state agency files. For example, the recent case of toxic levels of lead in Flint, Michigan, waterways was uncovered when an academic scientist sent a state FOIA law request to the Michigan environmental agency and also to federal EPA offices, even though waterways are monitored according to the federal Clean Water Act.[57] Thus, if you are unsure whether the records will be held at a state environmental agency or a federal one send a freedom of information request to the state agency under the relevant state law as well.

In addition, be careful not to send your request to the incorrect agency. Although EPA is the main "environmental" agency, many other agencies are engaged in environmental activities, from the U.S. Army Corps of Engi-

56. FOIA Request from The Society of Environmental Journalists to EPA (n.d.), https://perma.cc/3QPN-KTMW (captured Jan. 15, 2016).

57. Andrew Emett, *FOIA Request Shows Gov't Lied About Lead in Water, Knowingly Poisoning Countless Children*, The Free Thought Project.com, Dec. 24, 2015, http://thefreethoughtproject.com/government-officials-knew-drinking-water-gave-kids-lead-poisoning-lied-safety/.

neers and the U.S. Department of the Interior to the Federal Energy Regulatory Commission. Environmental issues are dealt with by most federal agencies on some level, so research which agency holds the relevant information for your issue. If you are unable, however, to determine which precise agency may hold the information, then think about sending requests to all potential agencies.

When you are able to name specifics, including dates, record authors, statutes, and project names, your request will be easier to direct and process, more likely to get responsive records, and to get those records more quickly.

B. *Step 2: Drafting Your FOIA Request*

There are many online resources available to help you draft a FOIA request. Generally, when you write a FOIA request, include a subject, a statement defining your request, information relevant to your status for fee purposes or request a fee waiver, and contact information.[58]

Because there is no set structure for FOIA requests, they vary widely in form and depth. Some are extensive (note CBD's FOIA request is six pages long, proactively rebutting fee waiver arguments and containing public interest arguments and citations to case law)[59] and some are quite succinct, simply describing the requested information and briefly addressing fee waivers and document format options.[60] Some online submission forms have word limits for FOIA request descriptions.[61]

The main marks of a good FOIA request are that it is written clearly and specifically. FOIA administrators have to sift through thousands of FOIA requests every year. The simplest requests are sometimes answered first, so the easier you can make your request for the agency FOIA staff to locate and retrieve your records, the better your FOIA experience will be. Thus, researchers should specifically describe the records they seek in the letter and provide some background where helpful to identify the records. It is best to assume that a FOIA officer is not familiar with what you seek. Calling a record "Mary Smith's letter from December 2014, to the South Coast Air

58. National Security Archive, Effective FOIA Requesting for Everyone: A National Security Archive Guide (2008), http://nsarchive.gwu.edu/nsa/foia/foia_guide/foia_guide_full.pdf [hereinafter Effective FOIA Requesting for Everyone].

59. *See, e.g.*, FOIA Request from Center for Biological Diversity to EPA, Re: FOIA Request—Full Clean Air Act Vehicle Emissions Exceedance Data for All Air Pollutants (Oct. 6, 2015), https://perma.cc/ALC5-MX9L.

60. Muckrock, *Create a FOIA Request*, https://www.muckrock.com/foi/united-states-of-america-10/101st-airborne-division-ebola-mission-malaria-prophylaxis-procedures-13669/ (last visited Dec. 2, 2016).

61. For example, FOIAonline has a 2,000-word limit for the FOIA description (although additional documents are allowed to be uploaded to expand the available space).

Quality Management District regarding the Edgington Oil Company permit," for example, makes it much easier to find than calling it "a letter written about air pollution from oil companies."

Another good practice is to incorporate keywords that administrators can enter into database searches, as much of the material is stored electronically. Also provide accurate titles, dates, names, and pertinent news stories discussing the subject of your request. Help FOIA staff by being as assertive, clear, and organized as possible.[62] That being said, you should try to keep your FOIA request concise. Just as you do not appreciate reading long-winded, confusing e-mails, FOIA administrators do not have the time or desire to parse through confusing and meandering FOIA requests. Confusing and poorly researched requests likely linger in the queue for longer periods of time than clear and succinct requests.

One important aspect to remember is that you do not have to include your rationale for requesting the records, as FOIA requires agencies to disclose agency records regardless of the recipient's intent or rationale for wanting them. The only exception for offering a rationale for your search is if you are trying to obtain a fee waiver by demonstrating that the documents will benefit the public interest by contributing significantly to the public's understanding of operations or activities of the government.[63]

C. *Step 3: Send Your Request*

How and where you send your request may affect your FOIA results. Information seekers should know the agency's preferred FOIA request format. Often, this information will be available on the agency's website. If you want a request to be received quickly by the agency, send it electronically, so long as that is a preferred method for transmission. FOIA.gov provides links to contact information for all federal FOIA offices.[64]

It obviously helps to know which agency office to send your FOIA request to. If the documents deal with something in a particular geographic location, make sure to send your request to the correct regional office. EPA offers a map with addresses for each region on their website.[65] If you are trying to get information quickly, note that sometimes sending the request to the attention of a local or regional office may lead to a quicker response, however,

---

62. Effective FOIA Requesting for Everyone, *supra* note 58, at 11.
63. 5 U.S.C. §552(a)(4)(A)(iii); 40 C.F.R. §2.107(1) (2016).
64. FOIA.gov, *Make a Request*, http://www.foia.gov/report-makerequest.html (last visited Dec. 2, 2016).
65. U.S. EPA, *Contact Us About the Freedom of Information Act and FOIA Requests*, http://www.epa.gov/foia/forms/contact-us-about-freedom-information-act-and-foia-requests (last visited Dec. 2, 2016).

make sure that you are sending the request to an administrator who is accustomed to processing FOIA requests.

D. *Step 4: Track Your Request*

Upon receiving your request, the agency will usually notify you promptly that they have received, and are processing, your request. They will often assign an identification number to the particular FOIA request, which you should use in future correspondence with the agency regarding the request.

Persistence and organization are key to getting a timely FOIA response (or any response at all, in some cases). In the case of a paper FOIA submission, keep in contact with the agency and check up on the status of your request often. In the case of electronic submissions, many systems now give users the ability to check the status of their request online as it changes over time. While you should not harass FOIA administrators, you should aim to contact them regularly, and could even set a regular weekly or bi-weekly reminder for yourself to check on your request. If you can, get the name and contact information of a specific FOIA administrator, and check in with them regularly. Sometimes, the agency will contact you for clarification on your request. Keep track of the correspondence you have with the agency regarding the request, as you may need this timeline of communication should you later decide to pursue FOIA litigation in regards to the request.

If you are looking for a quick response, know that agencies sometimes set up a faster track for easy requests and a slower track for more complex requests, so breaking your request up into smaller, manageable pieces may also expedite the FOIA process.[66] If an agency cannot provide the records you request within 20 days, it may notify you and provide you an opportunity to limit the scope or reform your request in order to get a quicker response.

66. 43 C.F.R. §2.26. FOIA officers engage in multi-track processing which is defined as:

> A system that divides in-coming FOIA requests according to their complexity so that simple requests requiring relatively minimal review are placed in one processing track and more complex requests are placed in one or more other tracks. Requests granted expedited processing are placed in yet another track. Requests in each track are processed on a first in/first out basis.

FOIA.gov, *Glossary*, http://www.foia.gov/glossary.html (last visited Dec. 2, 2016).

E. *Step 5: Assess Your Request*

Ideally, you will eventually get an electronic or paper copy of records in response to your FOIA request. Hopefully, the request will contain everything you ask for. You may, however, receive a partial document or a document containing redactions (that is, blacked-out or otherwise omitted words, sentences, or paragraphs). Agencies will redact portions of documents that are exempt from disclosure, and release the remainder of the agency record. Some agencies will identify the exemption on which they are relying in redacting certain information.

Many times, environmental FOIA requests will contain complicated chemical information and scientific jargon, so they may require some decoding. If you need assistance with understanding the materials received, consider contacting environmental nonprofit organizations or scientific research and advocacy organizations dealing with the type of information found in your request (water pollution, industrial farming, etc.). They may be able to connect you with someone who can help translate your request documents. If your records are newsworthy, you may want to contact an environmental journalist or a local paper for help with sorting and reporting the information you have obtained.

F. *Step 6: The FOIA Appeals Process*

If you do not get the documents you want from a FOIA request to a federal agency or you receive an adverse determination to a FOIA request, you may file an appeal with the agency. You must undertake and exhaust the agency appeals process before you attempt to sue the agency for failing to release the document. Often, appeals are filed when:

- Agencies withhold records or portions of records.
- Agencies fail to conduct an adequate search for the records you request or determine that the records are not subject to FOIA because they are not agency records.
- Agencies charge excessive fees or deny fee waiver requests.[67]

Be mindful of your deadlines for filing a FOIA appeal. These deadlines vary from agency to agency, but for most, appeals are due 30 to 90 days after the agency denies your FOIA request. Check with your agency to see how

67. Effective FOIA Requesting for Everyone, *supra* note 58, at 29.

Figure 5.2 FOIA Process

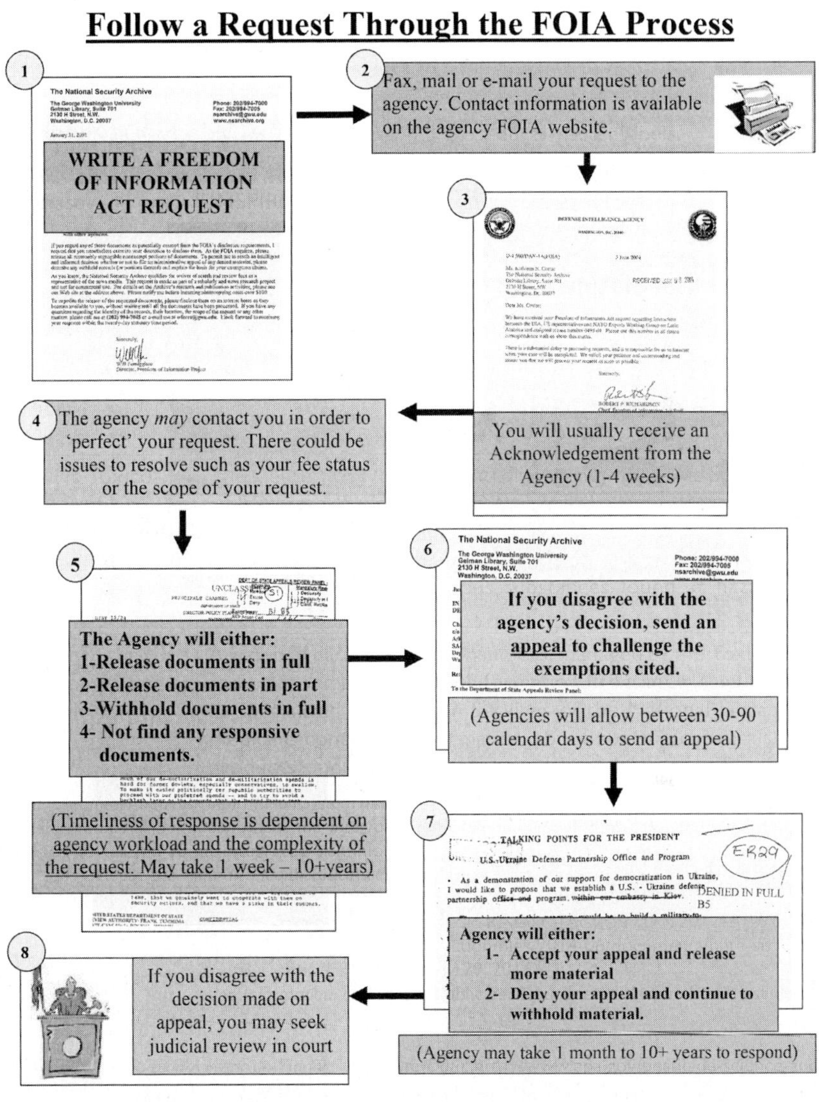

This chart, created by the National Security Archives, depicts the FOIA process from initial request to either satisfaction of the request or judicial review.

long you have to draft an appeal. Often, the response letter they send should advise you of your right to appeal and how to appeal, so that document is a good place to start.

Sample appeals letters are available online. FOIA appeals must be made in writing (a phone call is not sufficient), and they must follow the procedural requirements created by the agency, which are available in FOIA handbooks or regulations for each agency. Any FOIA response you receive may contain additional requirements as well. The major components of a FOIA agency appeal are:

- Statement of Appeal. Clearly label your letter as an appeal of the agency's adverse determination.
- Background on the Initial Request and Its Denial. Include subject details of your request and its denial, and make sure to include the identification number for the request, if one is provided.
- Arguments in Support of the Appeal. Arguments will vary depending upon why you are appealing and whether any exemptions have been invoked in the denial. Your arguments may center on public policy or FOIA-related case law.[68]

Also include any supporting materials and your contact information.

If your record is redacted or fully exempted from FOIA release, remember that each FOIA exemption has a constellation of associated case law that interprets and creates rules and threshold tests for overcoming the exemption. For example, the privacy exemption can be overcome if the public interest in shedding light on the government's activities outweighs the privacy interest.[69] Additionally, the business information and trade secrets exemption only applies to information that businesses provide the government that would cause "substantial harm" to them if released, and not to records prepared by the government using outside sources or records that are already in the hands of members of the public.[70] Similarly, the law enforcement exemption no longer applies if the case has been closed. The U.S. Department of Justice's (DOJ's) *FOIA Guide* contains a discussion of the case law that created all of these tests, and is a good source to consult to see whether your request can overcome the exemptions assessed by the federal agency.

---

68. *Id.* at 31.
69. Public Citizen Health Research Group v. Department of Labor, 591 F.2d 808, 809 (D.C. Cir. 1978).
70. National Parks & Conservation Ass'n v. Morton, 498 F.2d 765, 4, ELR 20385 (D.C. Cir. 1974).

FOIA Appeals Tips

Allow Enough Time to Pass. *Remember, you cannot appeal an agency's lack of initial response without first exhausting the available administrative remedies.*[71] *After filing an appeal, the agency has an additional 20 business days to consider your appeal before you may file a complaint against the agency in federal court.*[72]

Be Reasonable. *In addition to exercising patience and allowing the statutory permissible number of business days to pass before appealing or filing a lawsuit, also refrain from filing frivolous appeals or sending frivolous letters to the agency. Not only does bothering FOIA administrators sour them on helping you with your current request, it harms the chances of friendly relations in future FOIA requests.*

Make Sure the Agency Is Following Its FOIA Requirements. *At the same time you should be reasonable when making your request, you must also be vigilant and make sure the agency is keeping up its end of the bargain, complying with FOIA's requirements. Agencies must also make reasonable efforts to provide records in "any form or format requested by the person if the record is readily reproducible by the agency in that form or format."*[73] *Agencies are also required to search their electronic databases for responsive electronic records in all instances except where "such efforts would significantly interfere with the operation of the agency's automated information system."*[74] *Remember, you are not begging the agency for a favor; you are requesting information as part of your statutorily granted rights.*

If you exhaust your appeals process, then you can proceed to FOIA litigation by filing a complaint against the agency in federal court. FOIA fees can also be actionable, like in the case when New Mexico was charged $20,000 in fees to process a FOIA request regarding the Gold King mine spill that turned a river into a toxic yellow waterway.[75]

To pursue a cause of action under FOIA, you must be able to show that the agency has (1) improperly, (2) withheld, (3) agency records.[76] If you can

71. Oglesby v. Department of the Army, 920 F.2d 57, 61-62 (D.C. Cir. 1990).
72. 5 U.S.C. §552(a)(3)(B).
73. *Id.*
74. *Id.* §552(a)(3)(C).
75. Bruce Finley, *New Mexico Irked, Threatens to Sue EPA, Colorado Over Gold King*, DENVER POST, Jan. 14, 2016, http://www.denverpost.com/news/ci_29385299/new-mexico-may-sue-epa-over-gold-king.
76. 5 U.S.C. §552(a)(4)(B).

prove these three elements, you may consider filing a FOIA case. Beware, however, that FOIA litigation can be costly and is subject to court filing and attorneys fees. Because of the time and costs of litigation, it would benefit anyone considering FOIA litigation to consult one of the FOIA litigation discussions like those in DOJ's *FOIA Guide* or the *EPIC FOIA Litigation Manual*, which describe the many varieties of FOIA litigation and court rulings in major FOIA cases.

If your FOIA request response is incomplete, you can follow up with the agency before filing a formal appeal and proceeding to potentially time consuming and costly litigation. Contact the FOIA officer who managed your request (their name will appear on the FOIA correspondence you received for the request). You may also be able to negotiate with the agency's FOIA liaison. The National Archives Office of Government Information Service (OGIS) also provides mediation services for disputes between FOIA requestors and federal agencies.[77] You can work together to see what kinds of information can be obtained pursuant to your request; it could be possible that if you re-frame your request or refocus the types of information or documents you want, you may be able to get helpful records that do not contain exempted materials. You can also try to garner the support of a public official or news service; agencies can sometimes be swayed in favor of granting a FOIA request when outside media and political forces urge them to do so.

Well-written FOIA requests preemptively discuss possible fees an agency may charge for researching and printing out the requested records. To try to avoid exorbitant processing fees, discuss fee limits and categories in your initial FOIA request. Agencies are required to charge *reasonable* FOIA fees for searching, reviewing, and reproducing records in response to requests. Unfortunately, despite these requirements, agencies sometimes try to charge hefty fees to FOIA requesters. Search fees can be as high as $50 an hour, and printing fees can cost 35 cents a page. Most agencies' FOIA regulations clarify that by filing a request, FOIA users agree to pay threshold fees at a minimum of $25.[78]

There are certain situations where FOIA requests are free of charge or are considered low-fee requests, including requests made for records serv-

---

77. OGIS was established under the OPEN Government Act of 2007, Pub. L. No. 110-175, 121 Stat. 735. (This law also provides that "The Office of Government Information Services shall offer mediation services to resolve disputes between persons making requests under this section . . . .", 5 U.S.C. §552(h)(1)(3) (amending FOIA).

78. For example, EPA regulations set the minimum at $14 instead of $25, but require notice to the requestor if costs will exceed $25. 40 C.F.R. §2.107.

ing the public interest. FOIA's fee schedule provides for several different categories of requesters, and a sliding scale for fees across the categories.[79] Commercial requesters, those seeking information for profit, including use in litigation, pay all fees. Educational and non-commercial scientific institutions and news media representatives are entitled to no charges aside from duplication costs for responses over 100 pages, as are any other requesters (but they may have to pay additional search costs if they arise). In short, if you clarify that your interest in records is non-commercial, and that it will improve the public's understanding of government activities and operations in promotion of the public interest, you may be able to reduce your FOIA fees or waive them entirely.

Also, with the 2007 FOIA amendments, agencies cannot assess fees for FOIA requests that exceed time limits, unless there are "exceptional circumstances."[80] FOIA defines these circumstances as ones where the agency must: 1) search for and collect records from facilities or establishments outside the agency office; 2) gather and examine a large volume of records to respond to a single request; or 3) consult with another agency or two or more agency components.[81] In these situations, agencies may require fees even when the request response time passes the twenty working day time limit.[82]

Usually, agencies are not allowed to have requesters pre-pay for their requests, unless the requester has previously unpaid fees or if the request exceeds $250. Note that even though you do not have to pre-pay, you will usually have to agree to pay the agency's cost estimate or the amount you specify you are willing to pay in your initial FOIA request, once the request has been fulfilled. Also, some agencies will deny your fee category or waiver request until the processing is complete, leaving your fee assessment in limbo until your request is fulfilled. If you get a large fee estimate or charge for your request, you may try to negotiate with the agency to see if you can focus your records request to reduce costs, or you may go ahead with an administrative appeal.

VI. FOIA Resources

A number of online and print resources are dedicated to explaining FOIA and its mechanisms, as well as articulating best practices for obtaining gov-

---

79. 5 U.S.C. §552 (a)(4)(A)(ii)(I)-(III).
80. *Id.* §552(a)(4)(A)(viii).
81. *Id.* §552(a)(6)(B)(iii)
82. DOJ FOIA Guide, *supra* note 27, at 32.

ernment information using the statute. Here are brief descriptions of some of the most complete and helpful resources:

A. *Books*

Although these treatises are the most well-known of FOIA resources, researchers should note that virtually any book that discusses government transparency will discuss FOIA in at least some small way because it is the main statutory mechanism for obtaining government information in the United States.

1. Department of Justice Guide to FOIA[83]

The DOJ describes its FOIA guide as a "comprehensive legal treatise on FOIA."[84] This work provides an in-depth look at FOIA's process and its nine exemptions, however, the distinguishing feature about this resource is its intensive footnotes citing FOIA court opinions. The book has the most complete and extensive case law annotations of any FOIA resource, providing all of the benchmark FOIA cases and histories. The guide also provides an excellent history of FOIA litigation, as well as thorough discussions interpreting FOIA's statutory language.

Researchers should note that until 2009, this book was regularly published and updated in its paper format by DOJ's Office of Information Policy. Now, however, the book is updated on a rolling basis on their website, where users can view it online or download chapters or the entire text in a portable document format (PDF) version.[85] Since the book is drafted by DOJ administrators, users should be cautious that the wording may be more pro-agency than neutral and preferential to agencies' FOIA goals, which may oppose those of transparency advocates in some instances.

2. *The Federal Information Manual*: How the Government Collects, Manages, and Discloses Information Under FOIA and Other Statutes[86]

This treatise discusses FOIA from a legal practitioner's perspective. Written by an environmental law attorney who is a member of the Association's

83. *See* DOJ FOIA GUIDE, *supra* note 27.
84. *Id.*
85. *Id.*
86. P. STEPHEN GIDIERE III, THE FEDERAL INFORMATION MANUAL: HOW THE GOVERNMENT COLLECTS, MANAGES, AND DISCLOSES INFORMATION UNDER FOIA AND OTHER STATUTES (2d ed. 2013).

Section of Environment, Energy, and Resources, the book offers FOIA overviews and tips from an environmental information perspective, albeit one of an environmental defense attorney. The book's Chapter 7, "The Elements of a Successful FOIA Request," is thorough and accessible, providing a checklist for making a successful FOIA request. A bonus of the book is that in addition to discussing FOIA, it also reviews many of the other government transparency and secrecy law processes described in Chapter 6 of this book.

3. Federal Information Disclosure[87]

Researchers have called this work "the leading treatise on federal information disclosure."[88] James T. O'Reilly discusses the political aspects of information advocacy and describes how to "win requests." Part I is dedicated to the process and practice of FOIA. Because this treatise spans two large volumes, it is not something you can easily transport or read in a single sitting. Because of its size, the treatise is a type you would likely not see outside of a law library. Copies, however, may be obtained from your local law library in print or through the subscription online legal database Westlaw.

4. Litigation Under the Federal Open Government Laws[89]

Often referred to as the "*EPIC FOIA Litigation Manual*," this manual approaches FOIA from a public interest, requestor-centric view, in contrast to the other resources listed above. Similarly to other works, it also discusses other open government transparency laws, however, its main focus is on FOIA. This hybrid work offers both insightful overviews and case note annotations that discuss the statutory exemptions and FOIA process in great depth, with specific cites to relevant statute sections, breaking down the law into granular-level discussion points. This book was last updated in 2010, so it does not cover more recent FOIA events and amendments.

B. *Online Resources*

Online, there are a plethora of FOIA guides and resources available from an almost infinite number of websites. As a result, each online FOIA resource should be considered with caution. Researchers should pay close attention

87. James T. O'Reilly, Federal Information Disclosure (1st ed. 2016).

88. Gina Stevens, Cong. Research Serv., The Freedom of Information Act and Nondisclosure Provisions in Other Federal Laws (2010), https://www.fas.org/sgp/crs/secrecy/R41406.pdf.

89. Electronic Privacy Information Center (EPIC), Litigation Under the Federal Open Government Laws 2010 (Harry A. Hammitt et al. eds., 2010).

to the objectivity and perspective of its authors. Some entities are focused on obtaining information, while others are dedicated to protecting information. For example, advocacy organizations like journalist groups are focused on obtaining information, regardless of the exemptions and their purposes, while DOJ and other federal agencies embrace the exemptions and often take a less transparent approach.

1. Journalist Organizations

One of the major groups utilizing FOIA and advocating for government transparency is the journalism profession. As a main conduit of information between the government and the public, journalists are always seeking government information, and find that much of it is not proactively delivered to the public at large. As a result, many journalist groups have created websites, guides, and advocacy organizations dedicated to improving FOIA transparency and functionality.

a. George Washington University's National Security Archive FOIA Resources[90]

The National Security Archive is an organization created by journalists and scholars to "check" government secrecy.[91] Along with collecting and archiving declassified documents and FOIA records related to national security, the organization, which is based at the university's Gelman Library, provides excellent guides to help the public make FOIA requests and engage in FOIA litigation. The archive has pages specifically dedicated to all aspects of FOIA including pages titled "How to File a FOIA Request" and "Making FOIA Work for You." The site also contains the text and legislative history of FOIA, and audits and reports about how agencies are complying with their FOIA obligations. The archive also publishes an e-book *Effective FOIA Requesting for Everyone: A National Security Archive Guide*.[92]

Although the organization has a national security focus, its FOIA resources apply to environmental information requests as well. Also, it is worth mentioning that many transportation and infrastructure projects (dams, airports, waterways, etc.) have environmental components that

90. National Security Archive, FOIA, http://nsarchive.gwu.edu/nsa/foia.html (last visited Dec. 2, 2016).
91. *Id.*
92. Effective FOIA Requesting for Everyone, *supra* note 58.

impact national security. Additionally, issues like disposal of nuclear waste,[93] lead ammunition from firing ranges impacting water quality,[94] and noise pollution from submarines injuring wildlife intersect both environmental and national security topics.[95]

b. MuckRock.com

Muckrock bills itself as "a collaborative news site that gives you the tools to hold the government accountable."[96] This website was founded by Michael Morisy, a journalist who wanted to create a centralized FOIA request and response repository. MuckRock has become a FOIA-focused community with many FOIA and information access resources. It is more collaborative and interactive than many other informational and research websites, and you can place FOIA requests through MuckRock (for a fee) and also view other FOIA requests and responses. Perusing MuckRock is also a great way to see the language and format of FOIA requests and the agencies' common responses to FOIA requests. (The website also contains state agency requests.)

MuckRock also publishes news stories about information obtained through FOIA requests it places, and it conducts large FOIA projects as well. One project especially helpful for FOIA users is called "Requester's Voice: FOIA Advice, Tips, and Tricks," which contains hyperlinks to all sorts of FOIA experts' advice on various FOIA issues and topics.[97] For information seekers looking for FOIA help, the website has a FOIA question and answer forum where you can ask "public record pros" for advice about FOIA issues you are having, whether they be with particular agencies and requests or larger FOIA process questions.[98]

93. *See, e.g.*, Nevada v. Department of Energy, 517 F. Supp. 2d 1245 (D. Nev. 2007).
94. *See, e.g.*, Milner v. Department of the Navy, 562 U.S. 562 (2011).
95. *See, e.g.*, Natural Resources Defense Council v. National Marine Fisheries Serv., 409 F. Supp. 2d 379 (S.D.N.Y. 2006).
96. Muckrock.com, *About*, https://www.muckrock.com/ (last visited Dec. 2, 2016).
97. Muckrock.com, *Requester's Voice: FOIA Advice, Tips, and Tricks*, https://www.muckrock.com/project/requesters-voice-foia-advice-tips-and-tricks-22/ (last visited Dec. 2, 2016).
98. Muckrock.com, *Questions and Answers*, https://www.muckrock.com/questions/ (last visited Dec. 2, 2016).

c. Reporters Committee for Freedom of the Press

This group publishes: (1) the regularly updated *Federal Open Government Guide*[99] that discusses FOIA and its benchmark cases; and (2) the *Federal FOIA Appeals Guide*[100] that covers the administrative appeals process and also contains a section-by-section guide for each exemption and considerations like FOIA litigation and attorneys fees. Both are available online and downloadable in PDF format.

This organization has also created an online FOIA letter generator. Through their iFOIA project,[101] users can compose and track FOIA requests, free of charge for agencies that use e-mail for FOIA requests (most agencies). If you run into FOIA litigation issues, the Reporters Committee for Freedom of the Press also provides a free legal telephone hotline.[102]

d. Agency Resources

The best place to start when looking for FOIA information is to visit agencies' websites. Websites will contain tips and guidance, FOIA administrator contact information, links to agency-specific FOIA resources, and even submission forms. Moreover, each agency has its own FOIA office and procedures, and FOIA manuals that usually fall under the proactive disclosure requirements under §552(a)(2) of the transparency law.[103]

■ EPA Resources. According to EPA's FOIA website,[104] EPA FOIA requests can be submitted via mail or courier to EPA's Headquarters in Washington, D.C., or to any one of the 10 Regional FOIA Officers.[105] In addition, for the last four years, EPA has subscribed to the online FOIA submission and tracking website FOIAonline.[106] Formerly known as FOIA Module, FOIAonline is designed to provide "up-to-the-minute reports on FOIA processing."[107]

99. Reporters Comm. for Freedom of the Press, Federal Open Government Guide (2009), http://rcfp.org/federal-open-government-guide (last visited Dec. 1, 2016).

100. Reporters Comm. for Freedom of the Press, Federal FOIA Appeals Guide (2012), http://www.rcfp.org/federal-foia-appeals-guide (last visited Dec. 1, 2016).

101. iFOIA.org, https://www.ifoia.org/#!/ (last visited Dec. 2, 2016).

102. Reporters Comm. for Freedom of the Press, *Legal Defense Hotline*, http://www.rcfp.org/legal-hotline (last visited Dec. 1, 2016) (noting free legal help can be reached at 800-336-4243 or 703-807-2100).

103. 5 U.S.C. §552(a)(2).

104. U.S. EPA, *Freedom of Information Act (FOIA)*, https://www.epa.gov/foia (last visited Dec. 2, 2016).

105. U.S. EPA, *The FOIA Request Process*, https://www.epa.gov/foia/foia-request-process (last visited Dec. 2, 2016).

106. *See About FOIAonline*, *supra* note 26.

107. FOIAonline, *Frequently Asked Questions*, https://foiaonline.regulations.gov/foia/action/public/home/faqs (last visited Dec. 2, 2016) (scroll down to "What is the difference between FOIAonline and FOIA.gov?").

The portal itself is simple and gives researchers the option to "Search FOIA Requests," "Make a FOIA Request," or "Create an Account."[108]

To search FOIA requests that have been previously submitted by other researchers, select the "Search" tab. Here, users can search all submitted FOIA requests by "tracking number, description, requester . . . , organization, and record text," by type and by agency.[109] Once a search is run, results can be filtered by tracking number, type, phase, requester, requester organization, date submitted, and date due. As a limitation, the search feature does not currently allow for a keyword search of any of the responsive documents, although those documents do exist.

When making a FOIA request, users are required to select an agency (or a sub-agency), list their contact information, and provide a description of the type of information and/or documents to be requested. Researchers are also able to put a monetary cap on processing fees, request a fee waiver, or expedited processing. After a request has been submitted, users can track the status of their submission. Status alerts are assigned one of the following categories: submitted; evaluations; assignment; processing; and closed.[110]

Also of interest to researchers is that FOIAonline has the ability to generate comprehensive reports organized by type, agency, and time period.[111] Reports can include statistics on the number of requests, exemptions, and appeals, as well as information on the length of processing time, fee waivers, costs, backlogs, and more.[112] Results are displayed as bar and table graphs and can be exported in readable and printable formats. As a research note, the earliest date to search both FOIA requests and reports is from Oct. 1, 2012.[113]

e. Other Agency Resources

Sometime, federal agencies themselves do a good job of presenting FOIA best practices and procedures to the public. Although all agencies have

108. *See About FOIAonline*, *supra* note 26.

109. FOIAonline, *Search*, https://foiaonline.regulations.gov/foia/action/public/search (last visited Dec. 2, 2016).

110. For more information, see FOIAonline, FOIA Public Users Guide (2015), https://foiaonline.regulations.gov/resources/public/FOIAonline%20Public%20User%20Guide.pdf; Taryn L. Rucinski. *A Fresh Look at FOIAonline*, Bridge 2 Excellence Blog (B2E) (Special Libr. Ass'n NY Chap.) (July 7, 2014), http://slanypublications.org/2014/07/07/a-fresh-look-at-foia-librarians-and-foiaonline/.

111. FOIAonline, *Reports*, https://foiaonline.regulations.gov/foia/action/public/report (last visited Dec. 2, 2016).

112. *Id.* (from the dropdown menu).

113. *Id.* (by clicking the link for "Information you can search varies by agency").

their own FOIA offices, administrators, and policies, there is no requirement that these contacts and tools be shared in a particular way. Thus, some agencies have very little information about how to access their records, while others have in-depth guides and user manuals. As an example, here are some environmental information-focused FOIA materials provided on federal agency websites:

- Council on Environmental Quality's *FOIA Handbook*.[114] This slightly dated handbook is designed to provide the public with information about how to submit a FOIA request to the Council on Environmental Quality (CEQ) and to explain some of the key provisions in the CEQ's particular FOIA regulations.[115]
- DOJ's Environmental and Natural Resources Division FOIA Page.[116] This webpage provides special information about DOJ FOIA requirements and how the agency's requests are processed. This is also a good reminder that not all environmental records will be located at EPA. DOJ deals with environmental law enforcement, while other agencies work with other environmental and procedural issues.
- The U.S. Army Environmental Command.[117] This agency's FOIA pages include an explanation of "Common FOIA Mistakes" to avoid, including requesting records from the wrong office and incomplete or vague requests.
- Foia.gov.[118] This website, hosted by DOJ, contains video tutorials on various aspects of FOIA, as well as all sorts of FOIA statistics and reports. While this website is a rather broad look at FOIA, it is useful to someone who wants basic information about the statute and how it works.

f. Sample FOIA Requests Online

Many government transparency advocacy organizations have posted FOIA form letters online. Some are topical (specific to the FBI or to national

114. Council on Envtl. Quality, FOIA Handbook (2009), https://www.whitehouse.gov/sites/default/files/microsites/ceq/foia_handbook_2-27-12.pdf.
115. 40 C.F.R. §§1515 et seq.
116. U.S. DOJ, *Environment and Natural Resources Division FOIA*, http://www.justice.gov/enrd/enrd-foia (last updated Dec. 1, 2016).
117. U.S. Army Envtl. Command, *Common FOIA Mistakes*, http://www.aec.army.mil/ContactUs/FOIARequests/FOIAMistakes.aspx (last visited Mar. 22, 2016).
118. U.S. DOJ, *What Is FOIA*, https://www.foia.gov/report-makerequest.html (last visited Dec. 1, 2016).

security requests, etc.) while others are geared toward certain types of requesters (journalists, etc.) but overall, they tend to be good starting points for crafting FOIA request letters. A simple Google search for "sample FOIA request" retrieves a bevy of FOIA letter and FOIA appeal letter forms. Also, looking at the correspondence on previous requests like those on MuckRock and FOIAonline can help you word and craft FOIA requests. As federal agencies improve their online FOIA reading rooms, many are adding links to FOIA request response documents and FOIA requests, themselves.

VII. Conclusion

While the back-and-forth battle for records between the government and its citizens is somewhat eternal, and pervades every record type and topic, recently, environmental FOIA requests have gotten especially combative. A recent court opinion admonished EPA for withholding information pursued by climate change-denier groups,[119] and there have been several FOIA requests for e-mail records to the same agency designed to bureaucratically shame the environmental agency and bog down EPA's FOIA system.[120] Environmental groups have also been accused of harassing scientists after filing FOIA requests for documents to check for links between scientific studies and corporations that pollute.[121]

Despite the contentious nature of FOIA-related environmental politics, the ultimate lesson of FOIA is not to be afraid to approach the government with information requests using FOIA processes. Government transparency benefits everyone and is vital to environmental welfare and public health. As Rachel Carson wrote in *Silent Spring*, "the obligation to endure gives us the right to know."[122]

---

119. *See* Landmark Legal Found. v. EPA, 959 F. Supp. 2d 175, 182, 43 ELR 20199 (D.D.C. 2013).
120. *See, e.g.*, Competitive Enter. Inst. v. EPA, 12 F. Supp. 3d 100, 44 ELR 20026 (D.D.C. 2014).
121. Hank Campbell, *Environmental Groups Continue Their Harassment of Scientists*, Science 2.0, Aug. 6, 2015, http://www.science20.com/science_20/environmental_groups_continue_their_harassment_of_scientists-156785.
122. Rachel Carson, Silent Spring 13 (1962) (quoting French writer and biologist Jean Rostand).

Chapter 6: Federal Transparency Laws Beyond FOIA

Law in Action

In 2016, Southern Ohio RiverWatchers, a local environmental organization, learned that the Bureau of Land Management (BLM) was considering opening up 50,000 acres of greenspace in Wayne National Forest to oil and gas development. Soon after learning about this development, RiverWatchers received a tip that Landowners for Energy Access and Safe Exploration (LEASE), an organization whose membership consists of representatives from major natural gas extraction companies, had met with BLM officials numerous times in order to sway the agency to create more opportunities for natural gas extraction in the National Forest. RiverWatchers is currently trying to determine how to get more information about these meetings so that they can seek an injunction preventing the BLM from opening the land until an environmental impact statement can be prepared in accordance with the National Environmental Protection Act.[1]

A constellation of laws and rules exist at the federal level designating what information must be made available to the general public. Aside from the Freedom of Information Act (FOIA)—the "heavy hitter" of federal transparency laws, which this book discusses in depth in the previous chapter, Chapter 5—several other federal laws also provide different types of transparency, and conversely, secrecy, for various types of government information. There are several major categories of transparency laws in the federal statutory scheme:

1. Federal open meeting laws: These laws require public access to certain types of federal meetings, both by providing notice that meetings will occur and providing minutes and other information from the meetings after they convene.

2. Laws Governing the Management of Agency Information: These laws govern the maintenance, storage, and quality of certain types of fed-

1. For more information, see RiverWatchers, 216.68.102.178/riverwatchers/default.asp (last visited Dec. 3, 2016); Jackie Stewart, *Strong Support for Leasing in Wayne National Forest at Marietta BLM Meeting, Energy in Depth Ohio* (Nov. 19, 2015), http://energyindepth.org/ohio/strong-support-for-leasing-in-wayne-national-forest-at-marietta-blm-meeting/.

eral government records, requiring that certain records be archived, preventing arbitrary and secretive reccords disposal, and decreasing the amount of reporting required to create some types of records. These laws also create oversight for the quality of information in the records generated and compiled by the federal government.

3. Laws Restricting Access to Government Information: This set of laws shields information from public access. Some of these laws embargo records for a limited period of time, and others restrict the release of certain types of information from certain people (like personally identifiable information, for example), while guaranteeing access to others.

Generally, these laws share a common theme: government transparency. Most of the laws were created with the express purpose of providing access to information about the inner-workings and goings-on of the federal government, from its meetings to the activities of its highest officials. These laws try to ensure that important documents are maintained for future information seekers and that the data included in these records is verifiable and scientifically sound. Even the laws that restrict information access include provisions limiting their restrictions, whether it is in the form of mandatory time limits on embargoes or classifications, or requiring information disclosure to certain stakeholders such as Members of Congress or individuals discussed in the records. The federal government is engaged in a constant effort to balance government transparency with the need to withhold certain information from the public to ensure national safety, honor personal privacy, or protect economic and other enterprises. While some transparency advocates argue that a completely open government is necessary for true democracy, there are compelling arguments for keeping some sensitive documents out of the public view.

This chapter will focus on these types of federal transparency and classification provisions as access points, or restrictions from access, to environmental information. Similar laws exist in each state.[2]

I. Federal Meetings Laws

Environmental decisions are often made informally at meetings between stakeholders and federal agencies or advisory committees. There are two major laws that require disclosure of certain types of information from

2. For a useful guide on various federal transparency laws, see P. Stephen Gidiere III, The Federal Information Manual: How the Government Collects, Manages, and Discloses Information Under FOIA and Other Statutes (2d ed. 2013) [hereinafter Gidiere III Second Edition].

federal meetings to be made available to the public: the Government in the Sunshine Act (The Sunshine Act)[3] and the Federal Advisory Committee Act (FACA).[4] Environmental information seekers can use these laws to discover what goes on in meetings between agencies and various lobbying and interest groups. The laws require access to lists of meeting attendees, meeting minutes and transcripts, and scheduled meetings, which can be used to notify interested members of the public about meetings in case they wish to attend.[5] These laws also give information seekers access to reports, transcripts, working papers, and other materials prepared for or by federal advisory committees.[6]

A. *The Sunshine Act*

The Sunshine Act[7] is the major open meetings law for federal agencies. The law is meant to guarantee that "every portion" of every agency meeting be available for public observation.[8] It requires agencies to notify the public about agency meetings and to improve the availability of meeting records. The Sunshine Act is intended to provide more than mere superficial meeting minutes or "background papers tangentially related to the official agenda," and to go deeper, "requiring important decisions to be made openly."[9] As FOIA is limited in scope to agency records, the Sunshine Act's requirements only affect agency meetings. Thus, the meeting notices and transparency are only required of meetings held by agencies under the statute's definitions of "agency" and "meeting."[10]

1. Agency

Under the Sunshine Act, an agency is defined as an entity that is (1) subject to FOIA and (2) headed by a body of two or more members, a majority of whom are selected by the president with the advice and consent of the Senate.[11] This distinction is important because it means that the Sunshine Act does not apply to agencies headed by a single secretary, like the U.S. Environmental Protection Agency (EPA). Nevertheless, many of EPA's sub-

3. Pub. L. No. 94-409, 90 Stat. 1241 (1976).
4. Pub. L. No. 92-463, 86 Stat. 770 (1972).
5. 5 U.S.C. §552(b) (2014).
6. *Id.* app. 2.
7. Pub. L. No. 94-409, 90 Stat. 1241.
8. 5 U.S.C. §552b(b).
9. S. Rep. No. 94-354, at 5 (1975).
10. 5 U.S.C. §552b(a)(1)-(2).
11. *Id.* §552b(a)(1).

agencies and related environmental organizations do fall under the Sunshine Act's provisions. For example, the Chemical Safety and Hazard Investigation Board (CSB), an independent federal agency that investigates chemical accidents, violated the Sunshine Act by failing to notify the public that it would move to terminate five ongoing investigations at one of its meetings in 2015.[12]

To qualify as an agency under the Sunshine Act, the entity must be headed by a collegial body or it must be a subdivision of a collegial body.[13] Also, the FACA and the Sunshine Act are mutually exclusive—an entity may be an agency or an advisory committee, but it cannot be both.[14] A list of agencies that Congress intended to be covered by the Sunshine Act is in the statute's legislative history.[15]

Under the Sunshine Act, Congress, the federal courts, governments of U.S. territories, courts martial and military commissions, or military authority exercised in the field in times of war are not considered agencies.[16] Also, if a collegial body's only purpose is to "advise and assist the President," it is not an agency under the Sunshine Act.[17] Finally, although the Sunshine Act applies to agency subdivisions, it does not apply to boards or committees staffed by outsiders who are not members of the "collegial body."

2. Meeting

The Sunshine Act only covers meetings where there is a quorum, or the proper number of individual agency members to take action on any deliberations that occur during the course of the meeting. The Act also only applies when deliberations deal with conducting or disposing of official agency business.[18] Thus, the law does not include gatherings held only to plan for future meetings or where no official actions or motions take place.

---

12. Office of Inspector Gen., U.S. EPA, At a Glance (Sept. 30, 2015) (15-P-0304), www.epa.gov/sites/production/files/2015-10/documents/20150930-15-p-0304_glance.pdf.
13. 5 U.S.C. §552b(a)(1)(2014). In the FACA, a "collegial body" is composed of two or more individual members, a majority of whom are appointed to their position in the body by the president with the advice and consent of the Senate, and any subdivision thereof authorized to act on behalf of the agency.
14. S. Rep. No. 94-354, at 14 (1975).
15. H.R. Rep. No. 94-880, at 11-15 (1976), *reprinted in* 1976 U.S.C.C.A.N. 2183, 2193-96.
16. None of these entities are considered "agencies" under the definition provided in 5 U.S.C. §552b(a)(1). Further, if an entity is not subject to FOIA, then it is not subject to the Sunshine Act. Rushforth v. Council of Econ. Advisors, 762 F.2d 1038, 1044 (D.C. Cir. 1985).
17. *Rushforth*, 762 F.2d at 1044 (holding that the sole purpose of the Council of Economic Advisors is to advise and assist the president, so it is outside of the scope of the Sunshine Act).
18. 5 U.S.C. §552b(a)(2).

3. Access Provisions Under the Sunshine Act

The Sunshine Act contains 10 exemptions from the open meeting requirements. Meetings can be closed when the agenda features confidential and classified matters related to national defense or foreign policy, matters exempted by statute, trade secret disclosures, criminal accusations, official censure or other personal privacy issues. In addition, meetings covering topics that may interfere with law enforcement proceedings, endanger financial institutions, or discuss an agency's involvement in a civil action or particular case of formal agency adjudication are exempt from the Sunshine Act.[19]

If only part of a meeting falls under one of the 10 enumerated exemptions, then the agency must segregate the exempt discussion materials from any non-exempt materials. In order to close a meeting, an agency must follow the provisions set out in §552b(d)(1), taking a vote to determine whether the meeting will be closed for falling under one of the 10 exemptions. Yet even if an exemption applies to a prospective meeting, the agency may still be required to hold the meeting in an open session if it finds that public interest requires the meeting be open.[20]

The Sunshine Act does not create a right of public participation at meetings, in contrast to FACA, which is discussed later in this chapter.[21] The Sunshine Act merely provides meeting notice and transparency measures.

To comply with the Act's openness requirement, an agency must publicly announce the time, place, and subject matter of the meeting, and whether the meeting will be open or closed at least one week prior to the meeting date.[22] The agency must submit this information to the *Federal Register* for publication immediately following the public announcement. Courts have enjoined, or stopped meetings from happening, for failure to provide adequate notice of the meeting.[23]

In practice, the intent of the law's public announcement provisions is unclear, and, as a result, different agencies interpret them in different ways. Some agencies post notices at their office or record special voice mail systems that journalists and the public can call, or they maintain mailing lists to notify interested persons by mail.[24] On the other hand, some agencies do the

19. *Id.* §552b(c)(1)-(7).
20. Common Cause v. Nuclear Regulatory Comm'n, 674 F.2d 921, 929 (D.C. Cir. 1981).
21. We the People Inc. of the U.S. v. Nuclear Regulatory Comm'n, 746 F. Supp. 213, 217 (D.D.C. 1990).
22. 5 U.S.C. §552b(e)(1).
23. Coalition for Legal Servs. v. Legal Servs. Corp., 597 F. Supp. 198, 200 (D.D.C. 1984).
24. Reporters Comm. for Freedom of the Press, *The Government in the Sunshine Act*, www.rcfp.org/federal-open-government-guide/federal-open-meetings-laws/government-sunshine-act (last visited Dec. 5, 2016).

bare minimum—publishing notices in the *Federal Register*—and consider their Sunshine Act obligations met. Here is an example a CSB Sunshine Act notice in the *Federal Register.*

Figure 6.1 CSB Sunshine Act Meeting Notice

Federal Register / Vol. 80, No. 81 / Tuesday, April 28, 2015 / Notices

'e receive may be
'w.regulations.gov/#
'HIS-2013-0047 or in
vhich is located in
SDA South Building,
ependence Avenue
)C. Normal reading
m. to 4:30 p.m.,
iday, except
e someone is there to
ll (202) 799–7039

IATION CONTACT: Mr.
gram Analyst,
n, and Decision
IS, 4700 River Road
, MD 20737; (301)
A.Tadle@

)RMATION: On
'e published in the
) FR 5729–5731,
2013–0047) a
e that the U.S.
culture was holding
cultural coexistence,
ch was to advance
f agricultural
cuss how to make
ible for all
-day workshop,
March 12–13, 2015,
pportunity to learn
epresenting a wide
ith respect to

CHEMICAL SAFETY AND HAZARD INVESTIGATION BOARD

Sunshine Act Meeting

TIME AND DATE: May 6, 2015, 2014, 9:30 a.m.–1:00 p.m. EDT.

PLACE: U.S. Chemical Safety Board, 2175 K St. NW., 4th Floor Conference Room, Washington, DC 20037.

STATUS: Open to the public.

MATTERS TO BE CONSIDERED:

The Chemical Safety and Hazard Investigation Board (CSB) will convene a public meeting on May 6, 2015, starting a 9:30 a.m. at the CSB's headquarters located at 2175 K St. NW., 4th Floor Conference Room, Washington, DC 20037. At the public meeting, the Board will discuss and may vote on motions related to the following:

1. Proposed amendments to 40 CFR 1600 to provide for regular Sunshine Act meetings and to address timely voting on calendared notation item votes;
2. Proposed schedule for regular CSB public business meetings;
3. Notation Item 2015–07 relating to Board governance, the issuance of two Board Orders on Scoping and Investigations, respectively, and the administrative closure of three investigations (calendared on March 10, 2015); and the

Public Comment

Members of the public are invited to make brief statements to the Board concerning the agenda items outlined in this **Federal Register** notice. The time provided for public statements will depend upon the number of people who wish to speak. Speakers should assume that their presentations will be limited to five minutes or less, but commenters may submit written statements for the record.

Contact Person for Further Information

Hillary J. Cohen, Communications Manager, *hillary.cohen@csb.gov* or (202) 446–8094. General information about the CSB can be found on the agency Web site at: *www.csb.gov.*

Dated: April 23, 2015.

Mark Griffon,

Board Member.

[FR Doc. 2015–09913 Filed 4–24–15; 11:15 am]

BILLING CODE 6350–01–P

COMMISSION ON CIVIL RIGHTS

Notice of Public Meeting of the Illinois Advisory Committee for a Meeting To Review and Vote on Its Hate Crime Report

AGENCY: U.S. Commission on Civil Rights.

ACTION: Announcement of meeting.

Here is an example a CSB Sunshine Act notice in the *Federal Register. See* https://www.federalregister.gov/articles/2015/04/28/2015-09913/sunshine-act-meeting.

The Sunshine Act also requires that agencies prepare electronic recordings, transcripts or minutes of their meetings.[25] These materials must disclose the identity of each speaker[26] and must be promptly made available to the public.[27] Agencies are further required to keep complete sets of these items for at least two years, or at least one year after the agency concludes the proceeding for which the meeting was held, whichever is later.[28]

25. Pursuant to 5 U.S.C. §552b(f)(1) of the Sunshine Act.
26. *Id.* §552(f)(2).
27. *Id.*
28. *Id.*

Like FOIA and the FACA, the Sunshine Act is a transparency law that gives the public standing to sue in federal court when agencies violate its statutory requirements, including the ability to sue prior to a scheduled meeting or within 60 days after it occurs to enjoin meeting secrecy.[29] Individuals may also challenge rules and laws passed during meetings that violate the Act. Because most outsiders, even journalists, rarely learn of agency meetings beforehand, information from administrative meetings is often derived from meeting transcripts required by the Sunshine Act rather than information actually obtained from being present at an agency meeting.[30]

B. *The FACA*

The FACA[31] was created in 1972 to provide policies and oversight for the hundreds of federal advisory committees.[32] This oversight is provided by the General Services Administration (GSA), which reports to the president and Congress about the activities of advisory committees.[33] With FACA, the legislature sought to limit the proliferation and power of advisory committees and also increase their transparency.[34]

Members of Congress were concerned that creating advisory committees—the provisional, ad hoc groups outside of the federal executive agencies—would lead to poor oversight, lack of transparency, and lack of adequate control over the entities which advise the federal government.[35] Thus, the FACA imposes open meetings requirements on advisory com-

---

29. 5 U.S.C. §552b(h); Time Inc. v. U.S. Postal Serv., 667 F.2d 329, 330 (2d Cir. 1981).
30. Like the FACA, environmental information seekers have invoked the Sunshine Act. In 1980, when a nonprofit environmental group called Pacific Legal Foundation sued the Council on Environmental Quality, the agency tasked with overseeing National Environmental Protection Act processes, the court found that the Sunshine Act had been violated and that the agency had not properly given notice of its meetings and also had not properly closed the meetings according to the Sunshine Act's provisions. Pacific Legal Found. v. Council on Envtl. Quality, 636 F.2d 1259, 10 ELR 20919 (D.C. Cir. 1980). Similarly, in 1982, Common Cause, a government accountability non-profit, successfully sued to enjoin the Nuclear Regulatory Commission from withholding notice and materials from budget preparation meetings. Common Cause v. Nuclear Regulatory Comm'n, 674 F.2d 921 (D.C. Cir. 1982).
31. 5 U.S.C. app. 2.
32. Wendy R. Ginsberg, Cong. Research Serv., Federal Advisory Committees: An Overview (2009), www.fas.org/sgp/crs/misc/R40520.pdf. This report states that during fiscal year 2008, there were 917 active advisory committees governed by the FACA, with nearly 64,000 total members providing advice and recommendations to 50 federal agencies and racking up operating costs of $344.3 million. *Id.* at 1.
33. GSA, *Federal Advisory Committee Act (FACA) Management Overview*, https://www.gsa.gov/portal/content/104514 (last visited Jan. 5, 2017). In 1976, Executive Order No. 12024 delegated to the administrator of GSA all responsibilities of the president for implementing the FACA, and 5 U.S.C. app. 2, §7(a), states that the GSA is "responsible for all matters relating to advisory committees."
34. 118 Cong. Rec. 16296 (1972).
35. Some even refer to these advisory committees as the "fifth arm of government." *See id.*

mittees to assure that they are held accountable to the public and to avoid off-record dealings that skirt public transparency. This transparency is especially vital given that advisory committees are composed of private citizens, many with personal and commercial interest in the outcomes of government decisionmaking processes. The FACA defines an advisory committee as

> any committee, board, commission, council, conference, panel, task force, or other similar group, or any subcommittee or other subgroup thereof . . . which is (A) established by statute or reorganization plan, or (B) established or utilized by the President, or (C) established or utilized by one or more agencies, in the interest of obtaining advice or recommendations for the President or one or more agencies or officers of the federal government . . . .[36]

This definition is not all-encompassing and omits certain types of advisory entities from the FACA's scope. For example, groups that advise legislative or judicial branches but not the executive,[37] committees composed wholly of officers or employees of the federal government,[38] and committees providing advice not directed to governmental policy are all excluded from the FACA.[39] Additionally, courts have created FACA coverage loopholes, declining to extend the FACA's scope to advisory committees organized and operated by government contractors,[40] or to advisory committees that are not "amenable to strict management" by federal agencies,[41] or situations where people who are not members of government committees attend committee meetings and offer information and advice.[42]

For advisory committees within the FACA's scope, the law helps information seekers by requiring advisory committees to provide: (1) meeting records to the public[43] and annual comprehensive review records for advisory committees to Congress, which the GSA provides in the form of an accessible database[44]; and (2) public notice in the *Federal Register* 15 days prior to any

36. 5 U.S.C. app. 2, §3(2).
37. Wash. Legal Found. v. U.S. Sentencing Comm'n, 17 F.3d 1446, 1450, 25 ELR 21189 (D.C. Cir. 1994).
38. 5 U.S.C. app. 2, §3(2).
39. Judicial Watch v. Clinton, 76 F.3d 1232, 1233 (D.C. Cir. 1996).
40. Byrd v. EPA, 174 F.3d 239, 29 ELR 2150 (D.C. Cir. 1999) (holding that a peer review panel convened by an EPA contractor was not a FACA advisory committee).
41. Public Citizen v. Department of Justice, 491 U.S. 440 (1989).
42. In re Cheney, 406 F.3d 723 (D.C. Cir. 2005) (holding that National Energy Policy Development Group meetings, where the advisory committee composed of federal officials met with energy production companies and interest groups, were not within the FACA's scope).
43. 5 U.S.C. app. 2, §10(b).
44. The FACA originally required the GSA to file an annual report on federal advisory committees with Congress and the president on advisory committee activities pursuant to the FACA (5 U.S.C. app. 2, §6(c)), but in 1995, Congress passed the Federal Reports Elimination and Sunset Act of 1995, Pub.

of their meetings[45] and open access to meetings that are not closed.[46] Notice must include the proper name of the committees; the time, place, and purpose of the meeting; a summary of the agenda; and whether all or part of the meeting is open to the public or closed. If an advisory committee meeting is closed, the *Federal Register* entry must explain why, citing exemptions in the Sunshine Act that the committee relied upon for closure.[47]

I. Records Provisions Under the FACA

The FACA requires public access to records, reports, transcripts, minutes, appendices, working papers, drafts, studies, agendas, or other documents that were either made available to or prepared by each advisory committee unless a FOIA exemption applies to the documents.[48] This access is only required for as long as the committee exists. Put another way, once the committee is disbanded, so is FACA-guaranteed access to the documents. Agency access to advisory committee information varies. Generally, agencies post, at a minimum, links to overviews and contact information for each of their active advisory committees.[49]

The FACA's database[50] features advisory committee information including terms of service, corporate affiliations, fact sheets, reports, expenses, a charter, an executive order establishing the committee, and financial information. Although the FACA's database is not the most user-friendly or attractive in its current state, it is not without merit.

The best way to access the database to look for particular advisory committees or committee interest area is by clicking "Search," which takes you to a page with a search bar for committee names and a scrolling menu of topics, including "Environmental Issues." There are also menu options that allow you to view all of the advisory committees under each agency or access Annual Comprehensive Review (ACR) documents. Beware, however, that

---

L. No. 104-66, §3003, which eliminated the report requirement. The 1995 Act did not remove the FACA's requirement that an Annual Comprehensive Review (ACR) of committees take place, 5 U.S.C. app. 2, §7, so the data requirement has since been fulfilled by the GSA through the maintenance of the FACA's database.

45. 41 C.F.R. §101-6.1015(b)(1) (2016).
46. 5 U.S.C. app. 2, §§10(a)(1), (3); 41 C.F.R. §101-6.1021.
47. Bureau of Nat'l Affairs Inc. v. President's Council of Advisors on Sci. & Tech., No. 92-1088 (D.D.C. 1993) (stipulation of liability for failure to provide timely notice of advisory meeting minutes); 41 C.F.R. §§101-6.1021, .1023. The FACA invokes the 10 Sunshine Act Exemptions. *See* 5 U.S.C. app. 2, §10(d). (FACA); 5 U.S.C. §552b(c) (Sunshine).
48. 5 U.S.C. app. 2, §10(b).
49. For instance, the U.S. Fish and Wildlife Service (www.fws.gov/pdm/advcom.html) and EPA (www.epa.gov/faca) provide that basic advisory committee information via their websites.
50. GSA, *FACA Database*, www.facadatabase.gov/ (last visited Dec. 3, 2016).

the ACR documents are not current.[51] You can also download entire data sets for fiscal years past, but note that the documents are very large files, containing thousands of pages.

Search results on the FACA's database webpage vary. For instance, it is easy to retrieve information about "virtual meetings" hosted by EPA-related advisory committees in 2014, but clicking on the meeting links retrieves a list of basic information about the committee rather than meeting proceedings. So, while the FACA's database exists, overall, coverage is uneven and records are not always available despite the law's requirements. Sidney Shapiro, an administrative law scholar, suggests that Congress should require agencies to utilize their websites as FACA venues to provide transparency and solicit public participation in the selection of advisory committee members.[52] Although the websites are improving, there is no legislative requirement for online access to the FACA's portals by agency.

2. Advisory Committee Meeting Information Under the FACA

FACA requires advisory committees to provide notice of their meetings in the *Federal Register*.[53] These notices include the time, place, and purpose of a meeting, as well as a summary of the meeting's agenda. If any part of the meeting is closed to the public, the notice must list the rationale for keeping the meeting closed, including any Sunshine Act exemptions that may apply. At each meeting, advisory committees must also keep minutes and make them available to the public.[54] The minutes will contain a record of the persons present, a complete and accurate description of matters discussed and conclusions reached, and copies of all reports received, issued, or approved by the advisory committee.[55]

The FACA also authorizes interested persons to attend, appear before, or file statements with advisory committees.[56] While this provision does not allow members of the public to compel agencies to convene, it does allow the public to contribute to the discussions that occur.[57] Also, like the Sunshine Act, any person has standing to challenge denials of access to advisory com-

51. 2014 is the most recent year reported on that page.
52. *Activities of the Comm. on Oversight and Government Reform: Examining the Federal Advisory Committee Act (FACA)—Current Issues and Developments*, 110th Cong. 120 (2008) (testimony of Sidney A. Shapiro, Wake Forest School of Law).
53. 5 U.S.C. app. 2, §10(a)(2).
54. *Id.* app. 2, §10(c).
55. 41 C.F.R. §101-6.1025.
56. 5 U.S.C. app. 2, §10(a)(2).
57. Gates v. Schlesinger, 366 F. Supp. 797, 801 (D.D.C. 1973).

mittee meetings and records.[58] Although the FACA does not provide that right explicitly, courts have recognized a right of action under the Administrative Procedure Act (APA).[59]

The FACA is important for environmental information access, as many natural resource and environmental issues in the United States are dealt with in advisory committees.[60] From government determinations about dam-building and electrical projects to climate change and natural disasters, environmental decisions are informed by task forces or committees of scientific experts that report to the president and other high officials. Committees with names like the Clean Air Act Advisory Committee, Good Neighbor Environmental Board, and Pesticide Program Dialogue Committee advise the executive branch on undertakings as varied as environmental regulation development, operation activities, and the environmental impacts of government projects.[61]

II. Laws Governing the Management of Agency Information

The management of agency records directly impacts what kind of information is available to the public. After all, if the agency is not properly maintaining, storing, and ensuring the quality of certain records, those records will not be accessible to the public. Federal agencies obtain, curate, and create mountains of records. Several laws govern the organization, storage, security, preservation, and destruction of those records. Those include the Federal Records Act (FRA), Presidential Records Act (PRA), and Federal Information Quality Act or Data Quality Act (FIQA), and the Paperwork Reduction Act.

The management of agency information directly impacts information seekers. For instance, the length of time agencies must retain documents will affect whether the documents are available to the public, and laws requiring agencies to maintain certain documents directly impact how and if the agency will maintain those documents rather than disposing of them.[62] According

58. Public Citizen v. Department of Justice, 491 U.S. 440, 450-51 (1989).
59. Reporters Comm. for Freedom of the Press, Federal Open Government Guide 28 (10th ed. 2009), www.rcfp.org/rcfp/orders/docs/HOW2FOI.pdf.
60. G. William Domhoff, *The Corporate Community, Nonprofit Organizations, and Federal Advisory Committees: A Study in Linkages*, Who Rules America? (Apr. 2005), www2.ucsc.edu/whorulesamerica/power/federal_advisory_committees.html (last visited Dec. 5, 2016).
61. U.S. EPA, *ALL Federal Advisory Committees at EPA*, www.epa.gov/faca/all-federal-advisory committees-epa (last visited Dec. 5, 2016).
62. Gidiere III Second Edition, *supra* note 2. at 46.

to the Office of Management and Budget (OMB): "The Federal Government is the largest single producer, collector, consumer, and disseminator of information in the United States."[63] As such, government administrators recognize their duty to maintain federal records, data, and information as a "valuable national resource" available to provide the public with knowledge and ensure government accountability.[64] Thus, these laws have been created to protect useful information from disposal, to safeguard the records of presidential administrations, and to otherwise manage the flow of information through the federal government and to the public.

A. *The FRA*

The FRA[65] is the major law controlling the quantity and quality of records produced by the federal government and their preservation and disposal.[66] The National Archives and Records Administration (NARA)[67] is the federal agency that oversees the implementation of the FRA, which applies to the records of federal agencies, including the Government Accountability Office (GAO).[68] The PRA governs the record management processes of other non-agency executive offices including the Executive Office of the President and the National Security Council.[69] In 2014, the FRA was amended to expand the definition of federal records to include "all recorded information, regardless of form or characteristics," including electronic records. The 2014 amendments also put the NARA Archivist in charge of deciding whether recorded information is a federal record under the definition of the statute.[70]

In addition to encompassing many forms of records, the FRA's scope also reaches records submitted to the agency by outside parties, like government

63. OMB, Memorandum for Heads of Executive Departments and Agencies, Management of Federal Information Resources (Gov't Printing Office, Nov. 28, 2000) (Circular No. A-130 Revised, Transmittal Memorandum No. 4), www.whitehouse.gov/omb/circulars_a130_a130trans4/ (last visited Dec. 5, 2016).
64. *Id.*
65. 44 U.S.C. §§2101-2118 (Pub. L. No. 81-754).
66. *Id.* §2902(5) (2014). The FRA's provisions apply to most federal agencies, including EPA and some legislative branch entities. It does not apply to presidential records, which are governed by the PRA, or the records of Congress. Wendy Ginsberg, Cong. Research Serv., Common Questions About Federal Records and Related Agency Requirements 4 (Feb. 2, 2015), https://www.fas.org/sgp/crs/secrecy/R43072.pdf. There is also the Records Disposal Act, 44 U.S.C. §3301, which, when combined with the FRA and the PRA, are often collectively called the Federal Records Acts.
67. NARA, *Homepage*, www.archives.gov/ (last visited Dec. 5, 2016).
68. U.S. GAO, *Homepage*, www.gao.gov/ (last visited Dec. 5, 2016).
69. Armstrong v. Executive Office of the President, Office of Admin., 1 F.3d 1274, 1293 (D.C. Cir. 1993) (holding that National Security Council records fall under the PRA, which is distinct from and mutually exclusive from the FRA).
70. Pub. L. No. 113-187.

contractors, license applicants, registrants, etc.[71] Certain materials, however, are considered non-record materials, and are not covered by the FRA's maintenance and disposition requirements. Non-record materials include extra copies of documents kept only for reference, stocks of publications and processed documents, personal files, and personal papers.[72] In addition, working papers (drafts, rough notes, etc.) are only considered records under the FRA if they have been circulated for official purposes like comments or approval and if they contain unique, substantive information that adds to a proper understanding of the agency's formulation and execution of its basic duties.[73] Materials made or acquired by libraries and museums solely for reference or exhibition purposes also fall outside of the FRA's scope.[74] NARA's publication, *Disposition of Federal Records: A Records Management Handbook* also provides a list of items that do not fall under the FRA.[75]

Agencies must create procedures for organizing and storing records governed by the FRA to make sure that records are properly filed and properly identified and preserved for ease and accuracy of retrieval.[76] Records storage facilities must comply with NARA's standards of design and security.[77] The FRA also differentiates between permanent records (records of continuing value deemed appropriate for preservation) that must be retained by the agency or by NARA, and temporary records, which get destroyed or donated after a specified period of time.[78] These documents must be kept separate and NARA is the final determiner of whether records are permanent or temporary.[79]

When FRA records are no longer needed for the daily business of the agency, they must be destroyed or disposed of using NARA-compliant procedures.[80] This disposition can happen in a few ways: (1) disposal of temporary records by destruction or donation; (2) transfer of records to federal agency storage facilities or records centers; (3) transfer of records to the national archives for preservation; or (4) transfer of records from one federal agency to another.[81] Agencies' systems of preservation and disposal are organized to help end-users access records as needed. For instance, the U.S. Department

71. 36 C.F.R. §1222.10(b)(4) clarifies that the FRA covers records "regardless of their origin."
72. 44 U.S.C. §3301(a)(1)(B)(ii).
73. 36 C.F.R. §1222.12(c).
74. 44 U.S.C. §3301(a)(1)(B)(i).
75. NARA, Disposition of Federal Records: A Records Management Handbook (2000), www.archives.gov/records-mgmt/pdf/dfr-2000.pdf.
76. 44 U.S.C. §3302.
77. 36 C.F.R. §§1222.34(h), 1232.
78. 44 U.S.C. §3302(2).
79. 36 C.F.R. §1222.18(a); Gidiere III Second Edition, *supra* note 2, at 52.
80. 36 C.F.R. §1220.18; *see also* 44 U.S.C. §3105(1).
81. 44 U.S.C. §2901(5). Records Disposal rules are generally found in the Records Disposal Act of 1943.

of Energy (DOE) specifically separates environmental information documents in its record retention scheme because of their significant public interest aspects.[82] Records of activities that may affect the physical environment are often organized in project case files to enhance their retrievability. Those files are grouped by categories such as regulatory compliance, monitoring, and permitting to make them easier to find.[83]

Agencies develop record schedules, which are approved by NARA, that determine when documents will be disposed.[84] There are volumes of records schedules among the various agencies. For example, EPA has hundreds of records schedules that direct how long the Agency must keep its records and what to do with those records upon disposition.[85]

In contrast to the transparency laws, which allow citizens to sue over transparency obstacles, the FRA provides no private right of action to recover agency records wrongfully removed from government custody, and the U.S. Supreme Court has held that no private right of action should be implied under the law.[86] Instead, agency heads or the NARA archivist can pursue judicial action to recover or protect records through the attorney general if any records in their custody face actual, impending, or threatened unlawful removal, alteration or destruction.[87] Some lower courts have found that the public does have a limited right to file suit to recover records under the APA.[88] Nevertheless, the APA cause of action is acutely limited by the deference granted to agencies under the APA, and many courts have limited this type of FRA review.[89]

Also, keep in mind that although the FRA forces agencies to establish record keeping systems and maintain federal records, the statutory "key" for unlocking these records and releasing them to the public is FOIA. If you seek information about the systems or the underlying documents main-

82. U.S. DOE, *DOE Records Schedule for Environmental Records Introduction*, Aug. 25, 2003, energy.gov/sites/prod/files/cioprod/documents/Environmental.pdf (last visited Dec. 5, 2016).
83. *Id.*
84. 36 C.F.R. §1220.18 (defining the term "records schedule").
85. EPA maintains an updated list of record schedules at U.S. EPA, *EPA Records Schedule—Detailed Information*, www.epa.gov/records/epa-records-schedules-detailed-information (last visited Dec. 5, 2016).
86. Kissinger v. Reporters Comm. for Freedom of Press, 445 U.S. 136, 149-50 (1980).
87. 44 U.S.C. §3106(a).
88. Armstrong v. Bush, 924 F.2d 282, 291-93 (D.C. Cir. 1991) ("In sum, neither the statutory scheme nor the legislative history [of the FRA] evinces a congressional intent to preclude judicial review of the adequacy of the National Security Council's recordkeeping guidelines and directives.").
89. *See, e.g.*, Competitive Enter. Inst. v. EPA, 67 F. Supp. 3d 23, 32 (D.D.C. 2014) (finding that the Armstrong holding applies to the improper destruction or removal of records, and not to agencies' recordkeeping determinations).

tained therein, and that information has not proactively been published by the agency, you will have to file a FOIA request to obtain it.

B. The PRA

The PRA[90] requires the preservation of and public access to presidential records. The law was created in 1978 to provide public access to presidential and vice presidential records following the Watergate scandal, which demonstrated the need for protective measures to prevent the destruction of presidential materials.[91] The PRA governs official presidential records created or received after January 20, changing the records' legal ownership from private to public, and setting up requirements, via NARA, for managing those records. The president and vice president may only dispose of records no longer having "administrative, historical, informational, or evidentiary value"[92] after the NARA Archivist has examined them and permitted their disposal.

During the archiving process, the documents are embargoed from the public for five years. After that embargo ends, the incumbent president may impose another embargo for up to 12 years. The embargo's purpose is to protect presidents, vice presidents, and their staffs from the exposure of their thought processes during their Administration, in order to prevent the chilling of candid conversations and deliberations during presidential terms. During these embargoes, only the president, vice president, and special representatives selected by the president may access the collections of records. After the embargo period ends, NARA is responsible for maintaining presidential libraries as the records' final repositories. Once the presidential libraries are open to the public, the public may access the presidential records via FOIA.[93] Like FOIA, the PRA policies are crafted by each president via executive order at the outset of their time in office.[94]

---

90. 44 U.S.C. §§2201-2207.
91. Bruce P. Montgomery, *Presidential Materials: Politics and the Presidential Records Act*, 66 AM. ARCHIVIST 102-38 (2003), americanarchivist.org/doi/pdf/10.17723/aarc.66.1.8v80638064l1716t.
92. 44 U.S.C. §2203(c).
93. A discussion of the process for making a FOIA request can be found in the previous chapter, Chapter 5.
94. Presidential Records, Exec. Order No. 12667, 54 Fed. Reg. 3403 (Jan. 18, 1989) (Reagan); Further Implementation of the Presidential Records Act, Exec. Order No. 13233, 66 Fed. Reg. 56025 (Nov. 1, 2001) (G.W. Bush); Presidential Records, Exec. Order No. 14389, 74 Fed. Reg. 4669 (Jan. 21, 2009) (Obama: superseding Exec. Order No. 13233).

C. The FIQA or Data Quality Act

The FIQA was actually a mere two-sentence rider in a 2000 spending bill,[95] and it had no name in the actual legislation. The rider has since been called both the FIQA and/or the Data Quality Act.[96] The law requires the OMB to issue guidance to federal agencies to ensure the quality, objectivity, utility, and integrity of information that gets disseminated to the public. Agencies must create their own information quality guidelines and establish systems to let people seek information corrections when information fails to comply with OMB guidance.[97] Pursuant to the Act, agencies must provide annual reports to the OMB on the number, nature, and resolution of those public complaints.

Information quality is especially important for environmental information, where so much depends on scientific data and expert studies of varying quality. For example, EPA utilizes processes like peer review, to assure the quality of "influential" information, including scientific, financial, and statistical data disseminated in support of top agency actions.[98] The importance of accuracy and quality in environmental data is so high that the OMB created a special category for information about risks to human health, safety, and the environment. This category incorporates the data quality provisions of the Safe Drinking Water Act (SDWA) Amendments of 1996, which requires information to be based upon the best available science and mandates the presentation of this information to be "comprehensive, informative, and understandable."[99] Because there is little legislative history or judicial interpretation of the FIQA, OMB guidance interpreting the Act is the main source of instruction on the law.[100]

---

95. Pub. L. No. 106-554, §515:

> The Director of the Office of Management and Budget shall . . . with public and Federal agency involvement, issue guidelines . . . that provide policy and procedural guidance to Federal agencies for ensuring and maximizing the quality, objectivity, utility, and integrity of information (including statistical information) disseminated by Federal agencies . . . .

96. Thomas O. McGarity et al., Truth and Science Betrayed: The Case Against the Information Quality Act (2005), www.progressivereform.org/articles/iqa.pdf.
97. Curtis W. Copeland & Michael Simpson, Cong. Research Serv., The Information Quality Act: OMB's Guidance and Initial Implementation (2004), www.fas.org/sgp/crs/RL32532.pdf.
98. P. Stephen Gidiere III, The Federal Information Manual: How the Government Collects, Manages, and Discloses Information Under FOIA and Other Statutes 64 (2006) [hereinafter Gidiere III First Edition].
99. 42 U.S.C. §§300g-1(b)(3)(A) & (B).
100. *See* OMB, *Agency Information Quality Guidelines*, www.whitehouse.gov/omb/inforeg_agency_info_quality_links (last visited Dec. 5, 2016).

EPA has also adapted the SDWA information quality standards across all of its media and activities pursuant to OMB directives.[101] Despite setting these information quality goals, however, EPA often relies on information that does not meet the SDWA standards.

For instance, EPA and the U.S. Department of the Interior both only require peer review when available, not as a baseline standard.[102] Additionally, EPA is known for failing to respond to correction requests within OMB time limits, with some changes taking as long as five months to put into place.[103] So, it is unclear how much the Act and its resulting OMB guidance positively impact the quality of environmental information.

Environmental advocates fear that the Act and its OMB guidelines could empower regulated entities to slow or even stop new environmental standards by lodging time-consuming and burdensome information complaints.[104] Although the Congressional Research Service (CRS) claims this has not happened, they have noted that many of the requests to "improve" information have been placed by regulated entities, and some of those requests have resulted in the removal of information from public availability.[105]

D. *Paperwork Reduction Act*

The Paperwork Reduction Act is an information management law that reduces the amount of data and information collected and stored by the federal government.[106] The law was created to assuage private entities' complaints that federal paperwork and recordkeeping requirements were too burdensome.[107]

---

101. *See* 67 Fed. Reg. 8452, 8460 V(3)(b)(ii)(C) ("With regard to analysis of risks to human health, safety and the environment maintained or disseminated by the agencies, agencies shall either adopt or adapt the quality principles applied by Congress to risk information used and disseminated pursuant to the Safe Drinking Water Act Amendments of 1996. . . ."). *See also* GIDIERE III FIRST EDITION, *supra* note 98, at 65.
102. *Id.* at 66.
103. James W. Conrad Jr., *The Information Quality Act—Antiregulatory Costs of Mythic Proportions?*, 12 KAN. J.L. & PUB. POL'Y 521, 542-43 (2003).
104. COPELAND & SIMPSON, *supra* note 97.
105. *Id.* at 11. The concern that information complaints would stall environmental work has been realized in cases like *Competitive Enter. Inst. v. Office of Sci. & Tech. Policy*, No. 14-CV-01806 (APM), 2016 WL 544463 (D.D.C. Feb. 10, 2016). This FOIA lawsuit was filed after the Competitive Enterprise Institute, a conservative think tank known for denying climate change, used the FIQA to demand the Office of Science and Technology Policy correct a video about the polar vortex to remove the statement that "a growing body of evidence suggests that the kind of extreme cold being experienced by much of the United States as we speak is a pattern that we can expect to see with increasing frequency as global warming continues."
106. Paperwork Reduction Act of 1980, Pub. L. No. 96-511, 94 Stat. 2812 (codified at 44 U.S.C. §§3501-3521 (2014)) (with major amendments in 1986 and 1995).
107. HHS.gov, *Information Collection / Paperwork Reduction Act*, www.hhs.gov/ocio/policy/collection/ (last visited Dec. 5, 2016).

Thus, the law's purpose is to limit federal information collection activities for private entities. Prior to the passage of this law, agencies could request unlimited data points from regulated entities, including environmental polluters. The Supreme Court called the Paperwork Reduction Act a response to federal agencies' "seemingly insatiable appetite for data."[108]

From an environmental information perspective, this law may restrict agencies' ability to establish certain reporting requirements for polluters, and complicate pollution data collection processes by making agencies undergo an OMB review process before collecting information.[109] On the other hand, this law protects private entities from being overrun with bureaucratic federal processes, as data collection requirements can quickly overburden time and personnel resources.[110]

The law minimizes paperwork burdens by certifying that information that agencies collect is necessary for the agency functions and that the information has practical utility.[111] The law also requires agencies to get OMB approval for their information collection processes before imposing them on the public.[112] The law has been used to reduce the reporting requirements of entities governed by environmental laws, such as hazardous waste creators, handlers, and disposers under the Resource Conservation and Recovery Act.[113] By reducing the reporting burden, agencies also reduce the amount and quality of information available to the public about these governed entities' activities.

---

108. Dole v. United Steelworkers of Am., 494 U.S. 26, 20 ELR 20447 (1990) (discussing the history of the Paperwork Reduction Act of 1980).

109. While [a paperwork reduction] focus may be necessary in order to overcome institutional biases which prevent agencies from objectively determining their need for information, the process loses sight of the fact that reporting requirements can serve important public interests. Given this bias in the Paperwork Reduction Act, the APA's failure to afford protection to the public values found in non-regulation reporting requirements is especially disturbing and could seriously jeopardize the public interest which these various reporting requirements serve.

    William F. Funk, *The Paperwork Reduction Act: Paperwork Reduction Meets Administrative Law*, 24 Harv. J. on Legis. 1, 70 (1987) (pointing out that the Paperwork Reduction Act is focused on reducing reporting requirements and placing the OMB in charge of agency judgments over what data needs to be collected with the OMB clearance process directed against collection).

110. Robert L. Glicksman & Stephen B. Chapman, *Regulatory Reform and (Breach of) the Contract With America: Improving Environmental Policy or Destroying Environmental Protection?*, Kan. J.L. & Pub. Pol'y, Winter 1996, at 9, 27 (referring to the Paperwork Reduction Act in a footnote as "reduc[ing] the economic burdens associated with the blizzard of paperwork generated by compliance with environmental regulations").

111. 44 U.S.C. §3506(c)(2)(A)(i).

112. 44 U.S.C.A. §3503 (West) (establishes the Office of Information and Regulatory Affairs within the OMB to oversee federal agency information collection processes).

113. *See, e.g.*, U.S. EPA, *RCRA Burden Reduction Initiative, Wastes Information Resources*, http://www.epa.gov/wastes/inforesources/data/burdenreduction/index.htm (last updated Feb. 22, 2016).

The Paperwork Reduction Act governs anything within a "collection of information," which the statute defines broadly. According to the law, collections of information include any third-party disclosure to a federal agency of facts or opinions regardless of form or format, so long as the facts or opinions are solicited or obtained through answering identical questions imposed on 10 or more persons.[114] The Act excludes certain types of information, such as information from criminal, civil, or administrative investigations, criminal prosecutions and the disposal of criminal matters, or civil and administrative actions.[115] All information collections are given an OMB control number.

Figure 6.2 Information Collection Request (ICR)

Federal Register/Vol. 73, No. 17/Friday, January 25, 2008/Notices

1/15/2008.
umber: 20080117–0039.
ite: 5 p.m. Eastern Time
›bruary 5, 2008.
bers: ER08–446–000.
Kelson Energy III LLC.
Kelson Energy III, LLC
›lication for an order
for filing and for certain
anket approvals, FERC
Original Volume 1 etc.
1/14/2008.
umber: 20080117–0040.
ite: 5 p.m. Eastern Time
bruary 4, 2008.
desiring to intervene or to
of the above proceedings
cordance with Rules 211
Commission's Rules of
ocedure (18 CFR 385.211
n or before 5 p.m. Eastern
ecified comment date. It
y to separately intervene
ocket related to a
ng if you have previously
he same docket. Protests
red by the Commission
the appropriate action to
vill not serve to make
ties to the proceeding.
a motion to intervene or
rve a copy of that
he Applicant. In reference
ting a new proceeding,
or protests submitted on
omment deadline need
n persons other than the

with any FERC Online service, please e-mail *FERCOnlineSupport@ferc.gov*. or call (866) 208–3676 (toll free). For TTY, call (202) 502–8659.

Nathaniel J. Davis, Sr.,
Deputy Secretary.
[FR Doc. E8–1307 Filed 1–24–08; 8:45 am]
BILLING CODE 6717–01–P

ENVIRONMENTAL PROTECTION AGENCY

[EPA–HQ–RCRA–2007–1116, FRL–8520–9]

Agency Information Collection Activities; Proposed Collection; Comment Request; Facility Ground-Water Monitoring Requirements; EPA ICR No. 0959.13; OMB Control No. 2050–0033

AGENCY: Environmental Protection Agency (EPA).

ACTION: Notice.

SUMMARY: In compliance with the Paperwork Reduction Act (PRA) (44 U.S.C. 3501 *et seq.*), this document announces that EPA is planning to submit a request to renew an existing approved Information Collection Request (ICR) to the Office of Management and Budget (OMB). This ICR is scheduled to expire on May 31, 2008. Before submitting the ICR to OMB for review and approval, EPA is soliciting comments on specific aspects

www.regulations.gov, including any personal information provided, unless the comment includes information claimed to be Confidential Business Information (CBI) or other information whose disclosure is restricted by statute. Do not submit information that you consider to be CBI or otherwise protected through *www.regulations.gov* or e-mail. The *www.regulations.gov* Web site is an "anonymous access" system, which means EPA will not know your identity or contact information unless you provide it in the body of your comment. If you send an e-mail comment directly to EPA without going through *www.regulations.gov* your e-mail address will be automatically captured and included as part of the comment that is placed in the public docket and made available on the Internet. If you submit an electronic comment, EPA recommends that you include your name and other contact information in the body of your comment and with any disk or CD–ROM you submit. If EPA cannot read your comment due to technical difficulties and cannot contact you for clarification, EPA may not be able to consider your comment. Electronic files should avoid the use of special characters, any form of encryption, and be free of any defects or viruses. For additional information about EPA's public docket visit the EPA Docket Center homepage at *http://www.epa.gov/epahome/dockets.htm*.

FOR FURTHER INFORMATION CONTACT:

Proposed ICR Requests are also published in the *Federal Register* to notify stakeholders about the potential data collection scheme and provide for notice and comment. See https://www.gpo.gov/fdsys/pkg/FR-2008-01-25/pdf/E8-1312.pdf.

114. 44 U.S.C. §3502(3)(A).
115. *Id.* §3502(3)(B).

Absent a valid control number, one does not need to respond to an agency information collection response.[116] Using these control numbers can help information seekers find out more about particular data collection processes. For instance, Proposed Information Collection Requests (ICR) and ICR renewals are published in the *Federal Register*. Information seekers curious about what kinds of data are collected pursuant to various environmental laws, the purpose of data collection, and other ICR background details can obtain additional information and participate in notice and comment on proposed ICR schemes.

III. Laws Requiring Finding Aids for Researchers

The 1996 Electronic Freedom of Information Act (E-FOIA) Amendments[117] and the Paperwork Reduction Act require agencies to create inventories, finding aids, and indexes for "major information systems" (information systems that are important to an agency mission or play a significant role in the administration of agency programs).[118] While, in theory, such finding aids would greatly assist information seekers in locating important agency materials, in practice, the OMB directives allow agencies to provide incomplete inventories. Thus, while this statutory requirement is codified as law, these types of agency research tools are not usually available.[119]

IV. Laws Restricting Access to Information

A number of laws remove entire swaths of information from public access. These provisions usually protect information that the federal government considers sensitive in nature or in content, or whose release could jeopardize national security.

A. *The Records Classification System*

One of the ways the federal government restricts access to information is through the U.S. records classification system.[120] The system for classifying

116. 7 C.F.R. §1437.404.
117. Pub. L. No. 104-231, §4, 110 Stat. 3048 (1996).
118. *See* OMB, Circular A-130, *supra* note 63.
119. Lawsuits have been filed arguing that the OMB directives should be rescinded, Public Citizen, *Office of Management and Budget Rescinds Memorandum on Freedom of Information Act Challenged by Public Citizen*, Apr. 24, 1998, http://www.citizen.org/pressroom/pressroomredirect.cfm?ID=318.
120. This classification system has been managed through executive order since 1940. The current executive order concerning classified information is Exec. Order No. 13526, signed by President Barak Obama on Dec. 29, 2009. *See* NARA, *Frequently Asked Questions on Identifying and Handling Classified Records*

government information is not statutorily defined, and the current executive order policy on classification creates three classification levels.[121] Those levels are:

1. top secret (information whose release would cause exceptionally grave damage to national security);
2. secret (information whose release could be expected to cause serious damage to national security); and
3. confidential (information whose release could be expected to cause damage to national security).

The government estimates that "approximately one petabyte of classified records data accumulates every 18 months. One petabyte of information is equivalent to approximately 20 million four-drawer filing cabinets filled with text."[122]

In an environmental context, information is often classified when it involves sensitive activities or federal facilities. Examples of classified environmental information include information about nuclear facilities,[123] military bases, and infrastructure projects and dams.[124] This type of environmental information, while important for public participation in environmental decisionmaking, is often restricted in the interest of national security.

Unclassified materials can be obtained via FOIA and other federal information access methods.[125] Materials that can be classified may take any form, including photographs, maps, video, databases, hard drives, and, of course, paper (the format of most classified documents).

Information may not be labeled as classified merely because it may be embarrassing or because it hides illegal activities.[126] Ordinarily, classified information becomes declassified and available to the public after a certain

---

in Private Papers, https://www.archives.gov/isoo/faqs/identifying-handling-classified-records.html (last visited Dec. 5, 2016).

121. Exec. Order No. 13526, §1.2, 47 Fed. Reg. 14874 & 15557 (Apr. 2, 1982).

122. NPR Pub. Interest Declassification Bd., Transforming the Security Classification System 17 (2012), http://www.archives.gov/declassification/pidb/recommendations/transforming-classification.pdf.

123. Atomic Energy Act of 1954, Pub. L. No. 83-703, 68 Stat. 919 (provides for the classification of information related to the design of nuclear weapons.

124. *See, e.g.*, U.S. EPA, *Monitoring Federal Facility Compliance, Enforcement*, http://www.epa.gov/enforcement/monitoring-federal-facility-compliance (last updated Aug. 4, 2015).

125. *See infra* pt. VI.

126. New York Times Co. v. United States, 403 U.S. 713, 714 (1971) (enjoining the newspaper from printing a classified report).

predetermined number of years. For example, most information created prior to the 1940s is no longer classified.[127]

B. *Exceptions: Mandatory Declassification Review Requests*

Members of the public may request mandatory declassification review (MDR) under certain circumstances.[128] This is a common practice among information seekers who cannot get information through FOIA because the information falls within one of the classification categories. Reporters Committee for Freedom of the Press has a guide on how to make an MDR request.[129]

Agencies that have the authority to classify records must designate an official to receive MDR requests. These requests are similar to FOIA in that they should be in a letter format and they should describe the records at issue as specifically as possible and request that the agency declassify the records. Final agency determinations are made within 180 days, and if an agency has not responded within 120 days or does not declassify the records requested, you may file an appeal with the Interagency Security Classification Appeals Panel (ISCAP) within 60 days.

Only the president can reverse an ISCAP decision. If an agency decision is reversed, the agency head may petition the president by contacting the assistant to the president for National Security Affairs for a reversal.

Sometimes, information seekers rely on the MDR system instead of FOIA to obtain government information because MDRs lack mechanisms for charging copying or research fees and there are far fewer delays. Requestors should note, however, that there is no way for citizens to file lawsuits if records are not released through the MDR process as ISCAP is the only review available for these requests. Also, you cannot file a FOIA request for information already requested via MDR for two years after the review request.[130] If a requester submits a request but does not clarify whether it is a FOIA or MDR request (or invokes both access venues simultaneously), the

127. NARA, *Frequently Asked Questions on Identifying and Handling Classified Records in Private Papers*, http://www.archives.gov/isoo/faqs/identifying-handling-classified-records.html (last visited Dec. 5, 2016).
128. Section 3.5 of Executive Order No. 13526 states that classified information "shall be subject to a review for declassification by the originating agency if . . . [ ] the request for a review describes the document or material containing the information with sufficient specificity to enable the agency to locate it with a reasonable amount of effort."
129. Reporters Comm. for Freedom of the Press, *A FOIA Alternative: Mandatory Declassification Review*, http://www.rcfp.org/federal-open-government-guide/sidebars/foia-alternative-mandatory-declassification-review (last visited Feb. 22, 2016).
130. Exec. Order No. 13526, §3.5(g).

agency will make the requestor select one or the other, or else the agency will treat the request like a FOIA request.[131]

C. *Exceptions: FERC Critical Energy Infrastructure Information Requests*

In addition to outright classification, there are a variety of regulatory exceptions that provide or limit access to environmental information. An example of these special information access exceptions is the Federal Energy Regulatory Commission (FERC) policy on Critical Energy Infrastructure Information (CEII). The information access exception was created by a series of executive orders[132] and regulations.[133] It applies to information about proposed or existing critical infrastructure that:

1. relates details about the production, generation, transmission, or distribution of energy;
2. could be useful to a person planning an attack on critical infrastructure;
3. is exempt from mandatory disclosure under FOIA; or
4. gives strategic information beyond the location of the critical infrastructure.[134]

CEII exceptions were developed after the FOIA embargo on CEII information that followed the September 11th attacks.[135] FERC determined that FOIA's process "is not well suited for handling CEII requests."[136] Thus, the government imposed a more intensive background layer on CEII requests. While FOIA is blind to requesters' identities and their reason for requesting records, CEII request processes take this information into consideration. The CEII process is also more restrictive after requests have been satisfied. While under FOIA, request records given to one requestor must be given to all subsequent requestors, the CEII request system requires requestors to sign non-disclosure agreements, thereby restricting who can

131. NARA, *Mandatory Declassification Review Appeals*, http://www.archives.gov/declassification/iscap/mdr-appeals.html (last visited Dec. 5, 2016).
132. FERC, *Critical Energy Infrastructure Information (CEII) Regulations*, www.ferc.gov/legal//maj-ord-reg/land-docs/ceii-rule.asp (last visited Dec. 5, 2016).
133. 18 C.F.R. pts. 375 & 388 (2016).
134. FERC, *Critical Energy Infrastructure Information (CEII) Defined*, www.ferc.gov/legal/ceii-foia/ceii.asp (last visited Dec. 5, 2016).
135. Final Rule: Critical Energy Infrastructure Information, 68 Fed. Reg. 9857-9873 (Mar. 3, 2003).
136. *Id.* at 9857.

access the information and what they can use it for, effectively barring it from journalistic or any other widespread publication.[137]

FERC claims that by allowing this limited use, non-anonymous access to the public, they are "protecting energy facilities" nationwide.[138] Environmental information seekers, however, argue that this CEII process prevents the investigation and monitoring of energy facilities and also limits the publication or dissemination of potentially relevant environmental information to the public.[139] The Society of Environmental Journalists warned that obscuring public access to CEII prevents public awareness of infrastructure vulnerabilities, like pipeline leaks and failures to properly oversee the regulatory compliance of energy companies, endangering the environmental health and safety of U.S. communities.[140]

VI. The Privacy Act

The Privacy Act of 1974[141] regulates the collection, maintenance, use, and dissemination of personal information by federal agencies. Although the law's main goal is to protect citizens' privacy and govern the way agencies collect data, the law also requires federal agencies to disclose personally identifiable government records to citizens who request them and to amend personal records that are inaccurate, irrelevant, untimely, or incomplete.[142] Access is limited to personal records maintained in a system of records. Systems of records are those in which records can be retrieved by an individual's name, social security number, or some other personal identifier.[143]

In addition to allowing citizens to access their personally identifiable records, the Privacy Act also requires federal agencies to notify the public about the systems they develop and use to gather personally identifiable information. Notice is provided via systems of records notices published in the *Federal Register*.[144]

---

137. 18 C.F.R. §388.112.
138. *See* FERC, *CEII Defined*, *supra* note 134.
139. Comments of the Reporters Committee for Freedom of the Press & the Society of Environmental Journalists to FERC, Re: Public Access to Critical Energy Infrastructure Information (Nov. 13, 2002), http://www.rcfp.org/news/documents/20021113fercceiico.html.
140. *Id.*
141. The Privacy Act of 1974, Pub. L. No. 93-579, 88 Stat. 1896 (codified at 5 U.S.C. §552a (2014)).
142. Sarah J. Lamdan, *The Privacy Act of 1974*, *in* The SAGE Encyclopedia of Surveillance, Security, and Privacy (forthcoming 2018).
143. 5 U.S.C. §552a(a)(5) defines "system of records" and 5 U.S.C. §552a(b) limits the Privacy Act's scope records within systems of records.
144. *Id.* §552a(e)(4).

For the purposes of environmental information access, the Privacy Act usually plays a tangential role, as it applies only to personally identifiable information, or information dealing with people, rather than information about the environment. Depending on the situation, however, the Privacy Act may be able to provide an information seeker with useful glimpses of personal information gathered by the federal government and about systems of records where personal information is collected and stored by federal agencies.

The process for making a Privacy Act request for information is similar to that for making a FOIA request. Like a FOIA request, gather information about the specifics of your request so that the agency knows what you seek and where it may be. If you know which "system of records" the information is in, be sure to specify that in your request.

Because many agencies do not distinguish between the two transparency laws, it is recommended that a Privacy Act request letter to an agency cite both Acts (the Privacy Act and FOIA).[145] Also provide relevant dates, key words, and most importantly, the self-identifying information that will help agency administrators sort through the system of records (if the information is stored by social security number, provide that information).

One special aspect of Privacy Act requests is that you must also provide sufficient personal verification. Check the agency's Privacy Act policies to ensure you are giving the agency every bit of personal verification they need to fulfill the request. Best practice for Privacy Act requests would be to include the requestor's full name, address, a notarized signature, and as much identifying information as the requester can divulge. This personal information helps administrators find the correct records and ensures they are giving out personal records to the correct person.[146] An attorney requesting records on behalf of a client should submit a notarized authorization from the client.

Unlike FOIA, the Privacy Act does not contain time limits. The OMB guidelines state, however, that agencies "should" acknowledge Privacy Act requests within 10 business days to let the requestor know, what, if any, records access will be granted. The guidelines also suggest that agencies should provide that access within 30 business days, and let the requestor know if there is good cause for further delay past that 30-day guideline.[147]

145. Electronic Privacy Information Center (EPIC), Litigation Under the Federal Open Government Laws 2010, 440 (Harry A. Hammitt et al. eds., 2010).

146. *Id.*

147. *Id.*

Also unlike FOIA, agencies cannot charge fees for searching and can only charge copying fees.[148]

Because the Privacy Act is meant to protect the personal information of U.S. citizens, the law does not apply to information about non-citizens, and if you request information about other people, you must prove that you are authorized to receive their information for them. Also, because the Privacy Act is limited to information in systems of records, you cannot get all personally identifiable information using the law. The Privacy Act does provide civil and criminal remedies against agencies that violate the law. Finally, unlike FOIA, the Privacy Act does not require an administrative appeals process, and plaintiffs can proceed to court without appealing to the agency first.[149]

148. 5 U.S.C. §552a(f)(5).
149. EPIC, *supra* note 145, at 440.

Chapter 7: Rulemaking Documents

Law in Action

The federal U.S. Department of Transportation's Pipeline and Hazardous Materials Safety Administration (PHMSA) recently issued a pre-publication version of proposed rules governing natural gas pipelines. This 549-page notice of proposed rulemaking describes how the proposed rule would broaden the agency's authority, extending its jurisdiction over the nation's natural gas pipeline system and ending exemptions for pipelines created before 1970, greatly impacting the nation's natural gas creation and transport practices. The document caught the interest of many stakeholders, whose attorneys quickly went to work sifting through the pages to parse out the highlights and analyze their implications. College junior Sarah Smythe blogs about environmental law issues. She wants to follow the rulemaking process so she can blog about the agency's decisionmaking on the natural gas pipelines, and would like to know more about who the agency meets with and whether the agency is open to public participation in the decisionmaking process. She would also like to know more about how the agency decided to create the proposed rule in the first place.[1]

I. Introduction

The passage of regulations, general statements issued by agencies that have the force and effect of law,[2] is one of the most environmentally consequential actions taken by the federal government. Administrative agencies are part of the executive branch, the branch of government charged with carrying out and enforcing laws. Agencies are governmental authorities separate from courts and legislative bodies that "affect[ ] the rights of private parties" through various forms of official action including adjudication, rulemaking, investigating, prosecuting, and negotiating.[3] Agencies are both

1. Keith Goldberg, New Gas Pipeline Safety Rules Expand PHMSA's Reach, Law360, Mar. 18, 2016, http://www.law360.com/articles/773341/new-gas-pipeline-safety-rules-expand-phmsa-s-reach.
2. Please note that the terms rule and regulation are used interchangeably. Maeve P. Carey, Cong. Research Serv., The Federal Rulemaking Process: An Overview 1 (2013), http://www.fas.org/sgp/crs/misc/RL32240.pdf.
3. Charles H. Koch Jr. & Richard Murphy, 1 Administrative Law and Practice §1:21 (3d ed. 2016) (quoting Kenneth Culp Davis, Administrative Law and Government 6 (2d ed. 1975)) [hereinafter Admin. L. & Prac.].

established and empowered to issue regulations by Congress[4] or sometimes by executive order.

Regulations are critical, filling in the important details of governmental aspirations. The actions of the legislative and judicial branches of government are overarching, announcing broad goals rather than particulars. Take, for example, the broad words of the Clean Air Act (CAA). In its first section, the statute declares its purpose, which is ambitious and largely rhetorical:

> [T]he growth in the amount and complexity of air pollution brought about by urbanization, industrial development, and the increasing use of motor vehicles, has resulted in mounting dangers to the public health and welfare, including injury to agricultural crops and livestock, damage to and the deterioration of property, and hazards to air and ground transportation and "air pollution prevention [. . .] and air pollution control at its source is the primary responsibility of States and local governments", so we must "protect and enhance the quality of the Nation's air resources so as to promote the public health and welfare and the productive capacity of its population. . . ."[5]

This declaration lacks specific details about how these goals will be achieved. The actual steps that the nation must take to prevent air pollution are actually laid out by federal agencies, and in this case, the U.S. Environmental Protection Agency (EPA). EPA establishes the actual national ambient air quality standards (NAAQS) that convert the rhetoric of the CAA into action through rulemaking.[6] The federal agency creates the rules that put the statute into play.

Thus, the rulemaking process results in the regulations that inform business and individuals, with great specificity, about what types of environmental activities they can engage in.[7] These rules, with their specific directions on the mundane aspects of daily life (for example, how fast you can drive a car, or how wide a walkway must be to support wheelchair access) effect the daily lives of all U.S. citizens. It is fitting that EPA calls writing regulations "one of its most significant tools."[8] Polluters are only permitted to emit pollutants according to the rules made by agencies like EPA during rulemaking processes. Regulations are where the specific scope of conservation efforts

---

4. *Id.*
5. 5 U.S.C. §553 (2014).
6. Here is an example of an NAAQS rulemaking from March 2015: Proposed Rule: Fine Particulate Matter National Ambient Air Quality Standards: State Implementation Plan Requirements, 80 Fed. Reg. 15340 (Mar. 23, 2015), http://www.gpo.gov/fdsys/pkg/FR-2015-03-23/pdf/2015-06138.pdf.
7. For a very brief overview of the EPA's rulemaking process, see U.S. EPA, *The Basics of the Regulatory Process*, http://www2.epa.gov/laws-regulations/basics-regulatory-process (last visited Dec. 10, 2016).
8. *Id.*

are determined, pollution remediation activities are commanded, and the enforcement of environmental laws is dictated.

The basics of the rulemaking process are described in greater detail in Chapter 3. This chapter covers the documents involved in the rulemaking process, which are often referred to as the rulemaking record.[9] The rulemaking record is especially important as most federal environmental laws require federal agencies to create regulatory programs, processes, and projects to grapple with conservation and pollution issues.

How do agencies decide what to regulate? When agencies are granted the authority to issue regulations from laws (enabling statutes) or, less often, by the president via executive order that authority also establishes parameters and limits the scope of what the agencies can regulate. In other words, agencies cannot undertake rulemaking activities outside of their statutory (or presidential) authority. Additionally, federal agencies create regulations according to Administrative Procedure Act (APA) requirements.[10] There are two types of rulemaking processes permitted under the APA: formal and informal rulemaking. Formal rulemaking is a more time-consuming, hearing-like process, and not used nearly as frequently.[11] Thus, the focus of this chapter will be the far more frequently utilized informal rulemaking process and the information and records associated with informal rulemaking.

II. Informal Rulemaking: The Basics[12]

Informal rulemaking incorporates public participation and transparency. As one legal scholar summarizes: "[I]n short, the agency must declare what it is thinking about regulating, give others a chance to say what they think about the agency's thoughts, and wrap things up by justifying

9. *See* Corrosion Proof Fittings v. EPA, 947 F.2d 1201, 1213 n.12, 22 ELR 20037 (5th Cir. 1991), opinion clarified (Nov. 15, 1991).

> The term "rulemaking record" means (A) the rule being reviewed; (B) all commentary received in response to the (EPA) Administrator's notice of proposed rulemaking, and the Administrator's own published statement of the effects of exposure of the substance on health and the environment, the benefits of the substance for various uses and the availability of substitutes for such uses, and "the reasonably ascertainable economic consequences of the rule" on the national economy, small business, technological innovation, the environment, and public health; (C) transcripts of hearings on promulgation of the rule; (D) written submissions of interested parties; and (E) any other information the Administrator deems relevant.

10. 5 U.S.C. ch. 5.

11. *See* Admin. L. & Prac., *supra* note 3, §2:33 (noting that "[f]ormal, trial-like rulemaking procedures exist under the federal APA but they are rare").

12. RegInfo.gov, *Reg Map*, http://www.reginfo.gov/public/reginfo/Regmap/index.jsp (last visited May 12, 2016). The Reg Map was created by OIRA to provide a visual, graphical representation of the rulemaking process.

the agency's ultimate regulatory choice."[13] To facilitate public participation, agencies publish multiple informative records during the rulemaking process. The informal rulemaking process generates many documents useful to researchers who want to learn about environmental regulations and administrative environmental policy.

The informal rulemaking process has several well-defined steps. These steps allow for public participation and the synthesis of public comments in the final rule ultimately culminating in a new regulation that considers public input. The Office of Information and Regulatory Affairs (OIRA), part of the Office of Management and Budget (OMB) under the Executive Office of the President, describes the rulemaking process this way:

> Once an agency decides that a regulatory action is necessary or appropriate, it develops and typically publishes a proposed rule in the *Federal Register*, soliciting comments from the public on the regulatory proposal. After the agency considers this public feedback and makes changes where appropriate, it then publishes a final rule in the *Federal Register* with a specific date upon which the rule becomes effective and enforceable. In issuing a final rule, the agency must describe and respond to the public comments it received.[14]

Documents like comment notes, testimony, agency research, and other types of information, are collected by the issuing agency throughout each step of this process. Additionally, at the end of each step, the agency notifies the public about the rulemaking progress through the *Federal Register*. Because the *Federal Register* is the publication of notice for most steps of the agency rulemaking process, it is useful to begin with an understanding of the *Federal Register* and its contents.

A. *The* Federal Register

The *Federal Register* is known as "The Daily Journal of the United States Government."[15] From a research perspective, the *Federal Register* is the most important tool for searching for information about rulemaking activities. It publishes documents generated by the executive branch of the federal government, including rule notices, proposed rules, final rules, general notices (non-rulemaking documents applicable to the public like notices of public meetings, hearings, investigations, and environmental impact state-

13. Richard Murphy, *Enhancing the Role of Public Interest Organizations in Rulemaking Via Pre-Notice Transparency*, 47 Wake Forest L. Rev. 681 (2012).
14. RegInfo.gov, *Frequently Asked Questions*, http://www.reginfo.gov/public/jsp/Utilities/faq.jsp (last visited May 12, 2016).
15. Federal Register.gov, https://www.federalregister.gov/ (last visited May 12, 2016).

ments), and presidential documents. The *Federal Register* is published daily, except on weekends and federal holidays, since 1936.[16] Publishing in the *Federal Register* serves as constructive notice to the public for the executive. Once a year, the *U.S. Government Manual*, which provides detailed information about the federal agencies, including lists of agency officials and agency histories and statements of purpose, is published as a special edition of the *Federal Register*.

You can purchase a subscription to the *Federal Register*, but it is far more efficient and affordable to simply use the online versions available at federalregister.gov or FDsys (the Government Printing Office (GPO) website), each of which has archived versions of the *Federal Register* going back to 1994. Additionally, most public access law libraries will have a set, as will all federal depository libraries.[17] Finally, the *Federal Register* is available on commercial legal research systems including Lexis (since July 1980), Westlaw (since January 1981) and HeinOnline (full coverage since the inception of the *Federal Register* in 1936).[18] Bloomberg Law also contains a database of *Federal Register* entries from 1999 to the present.[19]

While the *Federal Register* is an excellent source for regulatory information, its entries are often dense with jargon and complex language, and the annual volume is gigantic in size—each year's compilation of daily reports numbers in the tens of thousands of pages. Thus, without a citation, specific date, or identification number, the *Federal Register* can be quite difficult to search. In fact, one of Pres. Richard Nixon's staffers declared the *Federal Register* to be "absolutely incomprehensible to the average consumer."[20] If you cannot locate a specific *Federal Register* citation, conducting a key word search in a *Federal Register* database on a legal research product like Lexis or Westlaw, or some other research system utilizing advanced search algorithms, may be the best source for yielding useful *Federal Register* search

16. *See* Office of Fed. Register, A Brief History Commemorating the 70th Anniversary of the Publication of the First Issue of the Federal Register March 14, 1936-March 14, 2006 (2006), https://www.archives.gov/federal-register/the-federal-register/history.pdf.

17. Any library that is part of the Federal Depository Library Program receives an array of government documents (published by the GPO) free of charge. In exchange for receiving those documents, Federal Depository Library Program participants must make their government document collection freely available to the public. For more information about the Federal Depository Program as well as a list of participating libraries, see FDLP.gov, http://www.fdlp.gov/ (last visited May 12, 2016).

18. Westlaw and Lexis Advance versions of the *Federal Register* are available in their respective subscription services. After logging onto the website, either browse content types and click on "Federal Register," or type "Federal Register" into the main search bar at the top of the screen.

19. Bloomberg Law's version of the *Federal Register* can be accessed by Bloomberg Law subscribers by clicking on the "Legislative & Regulatory" tab at the top of the screen after logging onto the website, and selecting "Search Federal Register."

20. 36 Fed. Reg. 5203 (Mar. 18, 1971).

results. Every rulemaking process is the culmination of many *Federal Register* entries, and rules are frequently amended, updated, and altered, so there may be many *Federal Register* entry search results that include your search terms and address the same topics as the *Federal Register* entry that you seek.

With the volume and complexity of *Federal Register* entries, an excellent way to follow rulemaking processes in the *Federal Register* is by using regulation identification numbers (RINs) to search for particular rulemaking. A RIN is assigned by the Regulatory Information Service Center[21] to identify each regulatory action listed in the Unified Agenda of Federal Regulatory and Deregulatory Actions (discussed in the pre-rulemaking section). You can also follow the Agency Docket Number, which is the internal file number that an agency might assign to a rulemaking process, if an Agency Docket Number has been assigned by the agency.[22] Also, each daily publication of the *Federal Register* lists its contents at the beginning of the issue. The contents are listed by agency, and if an agency is not listed in the contents, then that agency has no *Federal Register* notices or publications for the day.

The federal government recognizes the research shortcomings inherent in the *Federal Register*, and has attempted to improve rulemaking clarity and transparency. For example, the "plain language" movement pushed by the Administrative Conference of the United States (ACUS)[23] led to a 2012 OIRA directive telling agencies to provide "straightforward executive summaries" when presenting long or complicated rules in the *Federal Register*. But, because this movement has failed to magically transform the *Federal Register* into a complete, easy-to-read-and-search record of the totality of the rulemaking process, this book will break apart and examine each step of the rulemaking process to identify what types of research materials may be harvested from those steps, beyond the *Federal Register* notices. Before we look at each step, here is an overview of the process as a whole[24]:

21. U.S. General Services Admin. (GSA), *Regulatory Information Service Center*, http://www.gsa.gov/portal/content/104622 (last visited Dec. 10, 2016).
22. Since 2005, the federal government, as a part of its eRulemaking initiative, has maintained a Federal Docket Management System (FDMS) to promote better rulemaking process transparency. There is an FAQ page that provides answers to questions about the FDMS, *Answers to Frequently Asked Questions*, https://www.fdms.gov/fdms-web-agency/component/loginInfo?page=faq#3 (last visited Aug. 10, 2016).
23. ACUS. gov., *Homepage*, https://www.acus.gov/ (last visited May 12, 2016) (quoting President Barack Obama: "ACUS is a public-private partnership designed to make the government work better.").
24. Office of the Fed. Register, *A Guide to the Rulemaking Process* 7 (n.d.), https://www.federalregister.gov/uploads/2011/01/the_rulemaking_process.pdf [hereinafter *Fed. Reg. Rulemaking Guide*].

Figure 7.1 Rulemaking Process

Overview of Rulemaking Process

The first bar = the steps of the process

The second bar = where each step is published

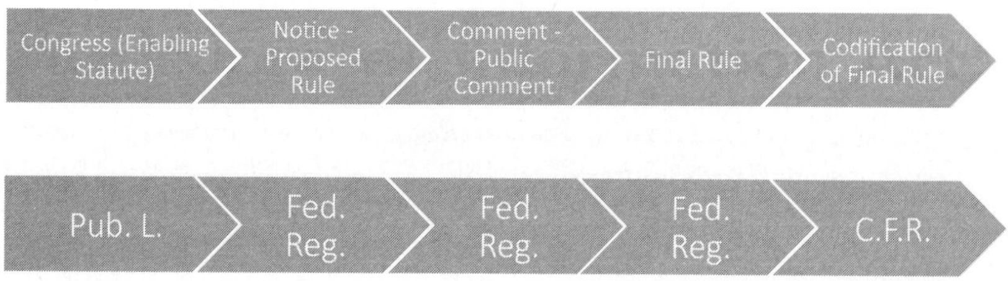

The Rulemaking Process as described by the APA, has four major steps:

- Notice of Proposed Rule
- Notice and Comment
- Final Rule
- Codification

OIRA has created Reg Map, an in-depth flow chart, to illustrate the informal rulemaking process in greater detail. As Reg Map demonstrates, the rulemaking process has many steps, and some of those steps occur before any rule is proposed and after a rule is finalized.[25] These grey areas of agency activity surrounding rulemaking, processes not governed by the APA, are often called pre-rulemaking and post-rulemaking, and each generates their own varied records and sources of information. They will be discussed in this chapter despite the fact that they are not officially part of the rulemaking process, because useful environmental information-filled documents are created and dispersed during pre- and post-rulemaking. The parts of the rulemaking process occurring outside of the APA's procedural scope are rather opaque from a government transparency perspective. One administrative law scholar describes notice-and-comment rulemaking as "Kabuki theater—a highly stylized process for displaying in a formal way the essence of something which in real life takes place in other venues."[26] It is troubling that these processes are so opaque given

25. *Reg Map*, *supra* note 12.
26. Donald Elliott, *Reinventing Rulemaking*, 41 Duke L.J. 1490, 1492 (1992).

that the pre-rulemaking period is when a lot of administrative decision-making happens.[27]

III. Informal Rulemaking Records

After the unofficial pre-rulemaking processes occur, the actual rulemaking process, as dictated by the APA, begins. Several Internet sources provide documents and information from the four phases of the informal rulemaking process, in addition to the *Federal Register*, which provides a searchable version of the "daily journal" and all of its rulemaking-related entries. FDsys.gov also provides the actual text of the *Federal Register*. These Internet sources provide additional background rulemaking information like public comments and other rulemaking-related material:

A. *Regulations.gov*

Regulations.gov is the federal government-wide rulemaking portal.[28] Regulations.gov is intended to open up the rulemaking process and its results to all people to create a more participatory rulemaking system.[29] The online portal provides mechanisms that allow visitors to comment on regulations online, as well as to view the comments of others.

Generally, Regulations.gov is a good place to view rules throughout the rulemaking process, as well as their associated dockets and comments. Regulations.gov is always improving, and more and more agencies are inputting rulemaking documents promptly on Regulations.gov as government services and access continue to move to online formats. Initially, not all agencies submitted their rulemaking documents to Regulations.gov, but now many do, including EPA. According to the U.S. General Services Administration (GSA), over 35 agencies and departments post rulemaking activities on Regulations.gov.[30] Some of the agencies that post materials on the website do not participate in the eRulemaking program, so while you can see the agency's rulemaking activities, you cannot provide comments or see other people's comments on Regulations.gov. The website FAQs suggest that for non-participating agencies, you contact the agency directly to learn more about the rulemaking process.[31] Regulations.gov is supported by the Federal

27. Jeffrey S. Lubbers, A Guide to Federal Agency Rulemaking 197 (5th ed. 2008).
28. Regulations.gov, *Homepage*, https://www.regulations.gov/#!home (last visited May 12, 2016).
29. Recommendation No. 2013-5: Social Media in Rulemaking, 78 Fed. Reg. 76269 (Dec. 17, 2013).
30. GSA, *Regulations.gov*, http://www.gsa.gov/portal/content/102264 (last visited May 12, 2016).
31. Regulations.gov, *Frequently Asked Questions*, https://www.regulations.gov/#!faqs (last visited May 12, 2016).

Docket Management System, which is the agency side of Regulations.gov, established in 2005.

In contrast to FDsys.gov and other websites that store *Federal Register* materials, Regulations.gov makes no assurance that it will archive the information on the website. As its goal is to enhance public participation, the website does not promise to forever store comment materials and rulemaking dockets, although you can often find information for already finalized rules on the website. If you are looking for comments for a final rule and they are no longer available on Regulations.gov, you can obtain the comments by placing a Freedom of Information Act (FOIA) request with the agency and get additional information by contacting the person listed at the end of the Final Rule (published in the *Federal Register*).

B. *Agency Websites*

Agency websites are a good place to catch up and keep current on a rulemaking process. As agencies come into compliance with website and public access requirements, their websites continue to gain more well-organized, accessible, up-to-date content about proposed and final rules, and agency news generally. Most agencies' websites contain links to regulatory information, which provide online access to proposed rulemaking, established *Code of Federal Regulations* (CFR) sections promulgated by the agency, and recent Unified Agendas published by the agency.

In particular, EPA's website contains an online feature called the Regulatory Development and Retrospective Review Tracker (Reg DaRRT).[32] In contrast to the *Federal Register*, Reg DaRRT is not the official publication of notice for rulemaking activities, but it is a resource that provides information about EPA rules in all phases of the rulemaking process: rulemakings that have not yet been proposed, rulemakings open for public comment, rulemakings where EPA is working on a final rule, and rules that have recently been finalized. Reg DaRRT even provides retrospective reviews of already finalized rules to determine whether rules should be modified, streamlined, expanded, or repealed.[33]

You can search Reg DaRRT records a number of ways: by phase (pre-proposal, proposal, open for comments, end of development, and even withdrawn rulemakings), by topic, by effect (by what the rule would effect, in

---

32. U.S. EPA, *Regulatory Development and Retrospective Review Tracker*, https://yosemite.epa.gov/opei/RuleGate.nsf/ (last updated Apr. 1, 2016).

33. U.S. EPA, *About Reg DaRRT*, https://yosemite.epa.gov/opei/RuleGate.nsf/content/about.html?opendocument (Apr. 1, 2016).

categories running the gamut from federalism, international, small business, to environmental justice), and "priority rulemaking" (rulemakings that are new, most viewed, etc.). The records contain RIN, docket numbers, current phase of rulemaking, and associated rulemaking materials, as well as links to contact EPA about the particular rulemaking. Although Reg DaRRT is EPA-specific and not the same as the *Federal Register*, it does interact with *Federal Register* websites, sending visitors to Regulations.gov to comment on rulemaking and to read Proposed and Final Rules, as well as other items in EPA dockets.[34]

IV. Information Access by Rulemaking Phase

Beyond the general access points for rulemaking information, each specific step of the rulemaking process creates particular rulemaking documents and other assorted content that are helpful for information seekers who want to learn more about certain rules. The *Federal Register* also includes a significant amount of supplemental information about each rulemaking, including background information about the purpose and rationale for particular rules.[35]

Although the APA lays out the process for rulemaking, the rulemaking process is not rigid—like the legislative process, there is flexibility, and sometimes rulemakings do not go all the way through the process in a linear fashion. For instance, it is not uncommon for proposed rules to be withdrawn or reconsidered after public comments are collected. So while each step in the rulemaking process is statutorily described, the actual process may not be so straightforward.

A. *Proposed Rule*

The public may first learn of a proposed rulemaking in several ways: advanced notice of proposed rulemaking (ANPRM); notice of negotiated rulemaking; inquiry or intent; or just a plain old notice of proposed rulemaking (NPRM).[36] Assuming an agency has proper rulemaking authority, and that it has decided to issue a regulation instead of a more informal and nonbind-

---

34. *Id.*

35. Note that the National Archives and Records Administration (NARA) has created a guide for agencies about how to draft *Federal Register* documents (including those in the rulemaking process), which very specifically sets out what each type of entry should contain. NARA, Federal Register Document Drafting Handbook (Oct. 1998 rev.), http://www.archives.gov/federal-register/write/handbook/ddh.pdf [hereinafter NARA Fed. Reg. Handbook].

36. *Fed. Reg. Rulemaking Guide, supra* note 24, at 3.

ing type of guidance document, the agency will commence the rulemaking process in one of these ways:

1. ANPRM

An ANPRM is published in the *Federal Register* during preliminary stages of rulemaking in order for the agency to get more information. An ANPRM is described by the Office of the *Federal Register* as "a formal invitation to participate in shaping the proposed rule," putting the notice-and-comment process in motion.[37]

2. Negotiated Rulemaking

Some agencies formulate proposed rules through negotiated rulemaking (often called reg neg), which is a style of rulemaking that was formally established through the Negotiated Rulemaking Act of 1990,[38] and is now codified in the APA.[39] Although reg neg is not used most of the time, environmental regulations were some of the first to be created through this process, and EPA continues to be the most frequent reg neg process user.[40] The reg neg process allows an agency head to create a rulemaking committee to negotiate a rule so long as the procedure would be in the public interest.[41] If the agency decides to gather a negotiated rulemaking committee, it must announce its intent by publishing notice in the *Federal Register* and providing a 30-day comment period.[42] These notices are often labeled "Intent to Establish a Negotiated Rulemaking Committee" or with some similar phrase that includes the term "Negotiated Rulemaking." The agency then receives comments on the notice, and may respond to comments by changing the membership or duties of the committee.

---

37. *Id.*
38. Negotiated Rulemaking Act of 1990, Pub. L. No. 101-648, 104 Stat. 4970.
39. 5 U.S.C. §§561-570a.
40. Cary Coglianese, *Assessing Consensus: The Promise and Performance of Negotiated Rulemaking*, 46 DUKE L.J. 1255, 1263 (1997) (noting "[i]n 1983, the Federal Aviation Administration (FAA) initiated the first formal negotiated rulemaking. A few other agencies followed the FAA in experimenting with the alternative procedure, most prominently the EPA."). The author continues: "[The] EPA has attempted and completed the most negotiated rulemakings." *Id.* at 1273.
41. 5 U.S.C. §563.
42. *Id.* §564. This section also provides details about what the publication must include, which includes descriptions of subject and scope of the rules to be developed, interests likely to be significantly affected by the rule, and a list of people proposed to represent those interests.

When the committee meets, it must comply with the FACA,[43] making its records and meeting notices available to the public.[44] After the meetings conclude, the committee may publish a report on its deliberations, its determinations, or its failure to meet a consensus. The report may then become the basis for a proposed rule, and follow the usual rulemaking process. The agency's deliberations are a large part of the rulemaking decisionmaking process, but the reg neg is not binding on the agency until it is adopted as a final rule in the rulemaking process. ACUS' guide *Negotiated Rulemaking Sourcebook* contains more information on this style of rulemaking.[45]

3. Proposed Rulemaking

A proposed rule or Notice of Proposed Rulemaking (NPRM) gets published in the *Federal Register*. An NPRM is "the official document that announces and explains the agency's plan to address a problem or accomplish a goal."[46]

Proposed rule entries in the *Federal Register* contain all sorts of information in addition to the words of the proposed regulation. The National Archives and Records Administration (NARA) advises that each proposed rule should include a billing code, headings, a preamble, a list of subjects, words of issuance, regulatory text, and a signature block. The heading section should contain the issuing department and sub-agency names, the CFR section affected by the proposed rule, the Agency Docket Number, if there is one, the RIN, and the subject heading.

The proposed rule's Preamble section in the *Federal Register* is full of useful background information. Preambles include summaries detailing what action is being taken, why the action is necessary, and the intended effect of the action, comment deadline dates, addresses and contact information for commenting and participating in the rulemaking process.

Another useful section in the proposed rulemaking is the Supplementary Information section. The Supplementary Information section describes the history of the rulemaking proceeding, including a detailed account of the background information. It answers questions like "What law or directive auhorizes the rulemaking?", "What existing regulations address the problem?", and "What problem does the rulemaking address?"[47] The Supplemen-

43. *Id.* §§565(a)(1), 566(g).
44. The FACA is discussed in Chapter 6.
45. David M. Pritzger & Deborah S. Dalton, ACUS, Negotiated Rulemaking Sourcebook (1995), https://www.acus.gov/sites/default/files/documents/Reg%20Neg%20Sourcebook%20Chap%201_0.pdf.
46. *Fed. Reg. Rulemaking Guide*, *supra* note 24, at 4.
47. NARA Fed. Reg. Handbook, *supra* note 35, at 1-12, 1-13

tary Information section is where the agency discusses a rule's merits relying on important data and other information, detailing its choices and reasoning. Agencies also provide legal authority for issuing the rule in the Supplementary Information section.[48]

In the proposed rulemaking, the agency also provides a full text version of the proposed regulation. If it is unclear how the text amends the current CFR, then the agency must describe the proposed action in narrative form.[49]

B. *Notice-and-Comment Period*

After the proposed rule is published, and the public has notice of the proposed rule via the *Federal Register*, the agency collects comments from the public on the proposed rule. Ordinarily, comments are taken in written form (or electronically by e-mail or Regulations.gov). The APA does not require agencies to publish comments to proposed rules in the *Federal Register* verbatim, but agencies will respond to major criticism communicated in public comments in their final rule publications. You can also read many of the comments for many proposed rules from 2002 and later (but not all) on Regulations.gov, or otherwise through agency webpages, agency online rulemaking dockets, or by contacting the agency.

Records of comments vary by agency.[50] Some agencies are required to hold rulemaking hearings, and others may hold public meetings to "collect more information or to help affected groups get a better understanding of the proposed rule," so there may be meeting records that contain commentary from the public on proposed rules.[51] Also, agencies increasingly use webcasts and live streaming video functions to broaden the audience for rulemaking-related meetings, so agency video archives may contain commentary on proposed rules. Records of these meetings, recorded or not, may fall under open meetings laws that require accessibility (see Chapter 9 for more details about what you can get through these open meetings laws).

Be aware that there may be more than one comment period for a particular rulemaking. For instance, some agencies add an additional reply period

48. *Fed. Reg. Rulemaking Guide*, *supra* note 24, at 4.
49. *Id.*
50. EPA reports that it receives as many as seven million comments annually on Regulations.gov, for example, and those comments are stored in the related rulemaking dockets. U.S. EPA, *EPA Docket Center*, https://www.epa.gov/dockets (last updated Dec. 30, 2015). EPA accepts written comments (if you send audio or video comments, they must be accompanied by a written comment), and even has a guide to best practices online, U.S. EPA, *Commenting on EPA Dockets*, https://www.epa.gov/dockets/commenting-epa-dockets#tips (last updated July 14, 2016).
51. *Id.* at 5.

after they hold a hearing for a proposed rule.[52] Also, sometimes comment periods are extended or re-opened when an agency is unsatisfied that it has enough high quality comments or when public comments provide compelling reasons for adding more time. Also, agencies may find that commenters raise new issues in their comments that may lead to a new series of proposed rules in the *Federal Register*, spawning new rulemaking activities from the original rulemaking process.[53] Information from the public can impact rules even after this public participation part of rulemaking has ended. In one U.S. Court of Appeals for the Ninth Circuit case, it was revealed that EPA added over 6,000 pages to the rulemaking record *after* the close of the final public comment period.[54] The record for informal rulemaking remains open even after the comment period, and agencies may add to scientific studies and other information it obtains, so long as it is made available to the public.[55] Changes to the proposed rule can still occur at this point in the process, after the public comments have been collected.

C. *Final Rule*

At the end of the rulemaking process, an agency may issue a final rule based on the totality of the rulemaking record, "consisting of the comments, scientific data, expert opinions, and facts accumulated during the pre-rule and proposed rule stages."[56]

Final rule publications in the *Federal Register* contain a spectrum of final actions that have "general applicability and legal effect.[57] *Federal Register* entries that impact the CFR may be interim rules, final rules, or entries that change the date a final rule goes into effect, suspend a previously published rule, or withdraw a rule not yet in effect. Other *Federal Register* entries that affect rules already in effect include petitions for reconsideration, and actions that don't amend the CFR but affect an agency nonetheless like policy statements, rule interpretations and clarifications, and waivers of agency rules.

The format and information in the *Federal Register* final rule entries is similar to those in proposed rules. Basic information about the rulemaking is included, like billing codes, agency names, CFR sections affected, docket and RINs, as well as preambles and lists of subjects. These information points are in both final and proposed rulemaking entries. Although the

52. *Id.*
53. *Id.*
54. Rybachek v. EPA, 904 F.2d 1276, 20 ELR 20973 (1990).
55. Charles H. Koch Jr. & Richard Murphy, 1 Administrative Law and Practice §4:44 (2010).
56. *Fed. Reg. Rulemaking Guide*, *supra* note 24, at 4.
57. NARA Fed. Reg. Handbook, *supra* note 35, at 2-3.

section headings are the same for the final and proposed rules, the words in the sections will be different in the proposed and final versions of each rule, because the final rule, especially in the Supplementary Information section, discusses the history of the rulemaking, including a summary of the comments collected by the agency and the agencies' responses to those comments.[58] Especially in older rules for which comments are not available online, this summary of comments may be the only insight we have into how the public felt about a particular regulation. Recently, ACUS recognized that comments and interactions between the public and the agency during informal rulemaking are part of the Administrative Record that should be preserved and indexed for possible judicial review purposes.[59] If you want more in-depth information about deliberations that led to a final rule, using the contact information in the final rule publication to contact the administrators may help.[60]

Final rules also provide the rule's effective date, or the date that the regulatory amendment becomes effective and replaces or changes current CFR sections. Some final rule postings in the *Federal Register* contain a compliance or applicability date. Compliance or applicability dates are the dates that the people affected by the regulatory change must start following the new rule. The Office of the *Federal Register* compiles effective dates into a Table of Effective Dates and Time Periods that appears in the Reader Aids section in the first issue of the *Federal Register* each month.

A rulemaking process does not always end with a final rule publication. An agency may decide to continue the rulemaking process, modifying the proposed rule in light of the rulemaking record. If the agency's changes to the original proposed rule are substantial, the agency may publish a "Supplemental Proposed Rule."[61] Also, an agency may decide to terminate the rulemaking process "if the rulemaking record contains persuasive new data or policy arguments, or poses difficult questions or criticisms," ending the process without ever issuing a final rule.

Sometimes agencies issue final rules without publishing a proposed rule.[62] Under the APA, agencies can finalize rules without first publishing pro-

58. *Id.* at 1-2.
59. Memorandum from Stephanie Tatham, ACUS Staff Counsel to the ACUS Committee on Judicial Review Re: Draft Recommendation—Administrative Record Project (Mar. 13, 2015), https://www.acus.gov/sites/default/files/documents/COJR%20_%20Draft%20Recommendation%20_%20Administrative%20Record%20_%2003152013.pdf.
60. FOIA is also a good option for obtaining the administrative record, as the documents fall squarely under the scope of FOIA's "agency records" coverage.
61. *Fed. Reg. Rulemaking Guide*, *supra* note 24, at 6.
62. *Id.*

posed rules if notice and comment would be impracticable, unnecessary, or contrary to the public interest.[63] Some examples of instances where the APA may permit agencies to create a rule without publishing a proposed rule include emergencies where immediate actions are needed to avert public safety threats; when the rules do not address a substantive issue; and when the agency has no discretion to propose a rule because Congress has already directed a specific regulatory outcome in a law. Whatever the rationale for not publishing a proposed rule, the agency must explain its reasoning in the preamble of the final rule published in the *Federal Register*.[64] Agencies can also issue rules by "actual notice" by directly notifying all affected persons and entities, but this is rare because very few regulations are so narrow in scope that all affected parties can be identified.

Rules that circumvent notice-and-comment periods often appear in the *Federal Register* as "Interim Final Rules" (final rules where the agency has found good cause to skip the proposed rule step) and "Direct Final Rules" (final rules where the agency decides that a proposed rule is unnecessary because the rule only deals with routine or uncontroversial matters).

D. *Codification*

After a Final Rule is published in the *Federal Register*, it is then integrated into the CFR, which is the compilation of all final rules sorted and codified by subject matter.[65] Rules that are immediately effective are integrated into the "Electronic Code of Federal Regulations" or e-CFR.[66] The e-CFR is unofficial, but authoritative editorial compilation published by the Office of the *Federal Register* and GPO, and it is the best place to view the most up-to-date version of the federal regulatory code. The update status of the e-CFR is in red on top of its homepage. If a final rule will not become effective immediately, it appears in an amendment file and linked from the main e-CFR database.

An official version of the CFR is available on FDsys.gov,[67] as are PDFs of the print version. The CFR in print is assembled from the material published in the e-CFR and published each year on a staggered, quarterly basis. The approximately 200 volumes of the CFR are published annually by title:

63. 5 U.S.C. §553(b)(B).
64. *Id.*
65. *Fed. Reg. Rulemaking Guide, supra* note 24, at 9.
66. GPO, *Electronic Code of Federal Regulations*, http://162.140.57.127/cgi-bin/ECFR?page=browse (last visited May 12, 2016).
67. GPO, *Code of Federal Regulations*, https://www.gpo.gov/fdsys/browse/collectionCfr.action?collectionCode=CFR (last visited Apr. 18, 2016).

- Titles 1—16 as of January 1
- Titles 17—27 as of April 1
- Titles 28—41 as of July 1
- Titles 42—50 as of October 1[68]

The code contains 50 subject matter titles, which are similar to, but do not necessarily match up with, those of the U.S. Code. This small but significant difference between the U.S. Code and the CFR can be confusing when researching on electronic databases like Lexis and Westlaw, where statutes look very similar to regulations. Be sure that when you are doing federal regulatory work, you are looking at the CFR and not the U.S. Code or some other type of legal code.

In the CFR, environmental regulations are scattered across several titles—for instance, Title 40 contains regulations dealing with environmental protection, Title 50 regulations address wildlife and fisheries, Title 43 contains public lands regulations, and Title 42 contains public health regulations. Depending on variables such as the era of a regulation's creation, the agency enabled to promulgate the regulation, and the subject matter of the regulations, you may find relevant environmental regulations in various titles of the CFR.

Remember: although the CFR contains the official version of the final rule, the *Federal Register* contains the regulatory history of the rule, and many more details about the rule. Thus, from a research standpoint, the *Federal Register* is an integral part of any search for regulatory environmental information. The CFR distills and codifies the final rules, but the actual rulemaking process plays out in the *Federal Register* entries associated with each rulemaking, which can be tracked in real time or researched after-the-fact via RIN or docket number.

V. Information Beyond the APA Information Rulemaking Process: Pre- and Post-Rulemaking Records

Although a great deal of information about a rule can be found by researching the steps of the rulemaking process from proposed rule to codification of a final rule, there is also information in materials created and gathered before

68. NARA, *About the CFR*, http://www.archives.gov/federal-register/cfr/about.html (last visited Mar. 12, 2016).

the official APA rulemaking process commences and after it concludes. This section will identify deliberations and processes outside of the rulemaking process for which there may be informational records, and discuss how to find those records.

A. *Pre-Rulemaking Records*

Pre-rulemaking generally includes all of the activity prior to rulemaking that triggers the start of the rule creation process. Although not a part of the APA rulemaking process, a lot of information is created and gathered during this formative period prior to the publication of an NPRM.

Catalysts for rulemaking vary, and include congressional and judicial activities, OMB directives and agency recommendations, public petitions, and advocacy. The Office of the *Federal Register* describes pre-rulemaking activities as "unstructured processes and informal conversations,"[69] which means from an information-gathering standpoint, that there is not a formal, organized way to collect and access pre-rulemaking records. The pre-rule-making process can include information and documents arising from:

- Congressional or judicial directives
- Another agency's action or nongovernmental organization's (NGO's) action or initiative
- Public petitions
- Advisory committee proposals
- Rulemaking agendas

Sources for pre-rulemaking information are often third parties outside of the agency system, comments submitted during prior rulemaking, and other contact with various parties (especially with regulated entities).[70] Due to the unofficial, ungoverned nature of pre-rulemaking activities, access to pre-rulemaking records is determined largely by who is "in the know." Lobbyists, large corporations with in-house counsel, and other entities with staffs dedicated to following regulatory activity have better notice of pre-rulemaking documents, as pre-rulemaking and rulemaking activities rarely garner

---

69. *Fed. Reg. Rulemaking Guide*, *supra* note 24, at 3.
70. Blair P. Bremberg, *Pre-Rulemaking Regulatory Development Activities and Sources as Variables in the Rulemaking Fairness Calculus: Taking a Soft Look at the Ex-APA Side of Environmental Policy Rulemakings*, 6 J. Min. L. & Pol'y 1, 4-5 (1991).

national attention.[71] Keeping current with agency activities on environmental processes and projects will make it easier to know about pre-rulemaking activities as they occur.

Even if you are aware of pre-rulemaking activities and wish to obtain them via the FOIA process, FOIA's Exemption 5, which protects inter- and intra-agency memorandums and letters from public disclosure, may apply to pre-rulemaking records.[72] But, note that Exemption 5 usually doesn't pass judicial muster unless the pre-rulemaking concern pending presidential or Cabinet-level review decisions.[73] For more information on FOIA and its exemptions, see Chapter 5.

There are several public notice requirements for pre-rulemaking information. Executive Order No. 12291 requires agencies to publish regulatory impact analyses (RIA) for proposed major rules.[74] RIAs must determine benefits and costs of proposed regulations and consider other alternatives to the proposed rules. Additionally, the Regulatory Flexibility Act requires agencies to publish agency rulemaking agendas twice a year; in October and in April.[75] Nevertheless, these public notice requirements do not provide the public with source material or underlying documents for proposed rules, so they do not necessarily offer much insight on pre-rulemaking activities or reveal how or why proposed rules were drafted and determined.

Open meetings laws and policies may be a good way to access pre-rulemaking documents. For instance, FACA[76] requires procedural notice for certain types of advisory meetings. FACA notice requirements, however, are "seemingly avoided at all costs in most pre-rulemaking settings," and frequently circumvented,[77] and thus the statute is often an ineffective tool for accessing information about pre-rulemaking activities.[78] In the current FACA scheme, agencies post varying amounts of information about their pre-rulemaking advisory committees. Some agencies, like the Consumer Prod-

---

71. Wendy E. Wagner, *Administrative Law, Filter Failure, and Information Capture*, 59 Duke L.J. 1321, 1344-45 (2010) (noting "[i]n a proposed rule, the agency will offer a proposal for addressing these many technical questions. Because the agency's proposal depends on mastering many of the issues of concern to industry, these plans are typically developed with heavy industry involvement years before the proposed rule is actually published and formally shared with the public.").
72. National Labor Relations Bd. v. Sears, Roebuck & Co., 421 U.S. 132, 149 (1975).
73. Wolfe v. Department of Health & Human Servs., 839 F.2d 768 (D.C. Cir. 1988).
74. Federal Regulation, Exec. Order No. 12291 (Feb. 17, 1981). The term "major rules" was amended to "significant regulatory action" in a later executive order: Regulatory Planning and Review, Exec. Order No. 12866 (Sept. 30, 1993), which replaced Executive Order No. 12866.
75. Regulatory Flexibility Act of 1980, Pub. L. No. 96-354, 94 Stat. 1164 (codified at 5 U.S.C. §§601-612 (2014)).
76. *See* Chapter 5.
77. Bremberg, *supra* note 70, at 33.
78. *Id.* at 45.

uct Safety Committee (CPSC), have official policies of openness, posting Internet-accessible calendars and records of their pre-rulemaking activities, including planned meetings without outside authorities.[79] Others, however, are less forthcoming with information about advisory sessions.[80]

1. Directives

Congressional and judicial pressure can spur rulemaking activities in federal agencies. Most directly, enabling statutes drafted by the legislature may require agencies to engage in rulemaking. These statutes, and their associated legislative history, are excellent sources for the rationales and purposes for various rulemaking undertakings. Looking through committee reports, hearing testimony, and examining other congressional documents can provide environmental background information for a federal regulation. Enabling statutes and related underlying legislative history materials may reveal the environmental problem that regulations were promulgated to solve, or may include testimony from scientific and environmental groups concerned with the issue that the regulations address.

a. Congressional Directives

Sometimes, congressional hearings and reports beyond the legislative process recommend or urge agencies to undertake rulemaking.[81] These reports and hearings can also be found in the *Congressional Record*[82] and in other congressional documents.

In other instances, informal contact between agency officials and Congress members can result in rulemaking action. For example, discussions between EPA officials and former Sen. Robert Byrd (D-WV) and his staff regarding CAA standards for carbon dioxide shaped that rulemaking process.[83] This type of information—arising from informal meetings or discussions between agency officials and legislative staff—may be harder to come by, as it will not appear in the *Congressional Record*. Nevertheless, if the meetings between legislators and agency officials are within the proper parameters, they may fall under open meetings laws and the FACA.

---

79. 16 C.F.R. §1011.4(a) (2016). CPSC also decided not to adopt a rule exempting meetings about early, exploratory issues from public disclosure. 46 Fed. Reg. 38322 (July 24, 1981).
80. See more about the FACA in Chapter 5.
81. Lubbers, *supra* note 27, at 197.
82. GPO, *Congressional Record*, https://www.gpo.gov/fdsys/browse/collection.action?collectionCode=CREC (last visited, Apr. 16, 2016).
83. Sierra Club v. Costle, 657 F.2d 298, 11 ELR 20455 (1981).

b. Judicial Directives

The judicial role in the rulemaking process has also grown in recent decades, where parties have sued the government for failing to comply with rulemaking deadlines or failing to follow through with rulemaking obligations.[84] Agencies have also been ordered to begin rulemaking by court mandates or litigation settlement agreements. For example, in a 2001 case, *Save the Manatee Club v. Ballard*,[85] a public interest group sued the government over its failure to protect the endangered Florida manatee. The parties settled, and part of the settlement agreement required the U.S. Fish and Wildlife Service to propose and promulgate new manatee refuges and sanctuaries in Florida.[86] In another settlement agreement between the environmental group, Save Our Cumberland Mountains Inc. and the U.S. Department of the Interior, the agreement required the agency to take rulemaking action, within 180 days, to change their permit blocking program.[87] Looking to court opinions and underlying docket materials in these cases may provide additional information about rulemaking and about the activities and events ancillary or leading to the rulemaking process.

Public response to a court decision, or the court's criticism of the agency in a decision, may also lead to rulemaking actions. One scholar claims that "[i]ndeed, on occasion, the mere filing of a lawsuit may lead to agency action." In one case, the Occupational Safety and Health Administration (OSHA) issued an emergency rule on certain pesticides a month after a public interest group brought an action in federal court to compel issuance of the standard.[88] In these instances, looking at the plaintiff's complaint may provide a great amount of information related to the regulation and why it was drafted.[89]

2. Third-Party Actions and Initiatives

Sometimes one agency will urge another agency to promulgate regulations. For example, the OMB issues "prompt letters" to agencies urging them to act on issues that the OMB considers a priority. OMB prompt

---

84. Lubbers, *supra* note 27, at 198.
85. Save the Manatee Club v. Ballard, 215 F. Supp. 2d 88 (D.D.C. 2002).
86. *Id.* at 89.
87. Settlement Agreement Between Save Our Cumberland Mountains Inc., and Manuel Lujan Jr., Secretary, U.S. Dep't of the Interior (Jan. 24, 1990) (codified at 104 Stat. 1927 (1990).
88. Lubbers, *supra* note 27, at 199 (referencing Benjamin W. Mintz, OSHA: History, Law, and Policy 99 n.5 (1984)).
89. Docket searches (covered in Chapter 3 of this book), would be the best way to find plaintiff complaints on rulemaking activities.

letters are available on the OMB's OIRA website[90] and often urge EPA to take regulatory action to promulgate federal standards. For instance, a 2004 OMB prompt letter urges EPA to fulfill its oversight duties and create updated pathogen criteria standards in accordance with the BEACH Act of 2000.[91]

In addition to the OMB, other agencies may conduct studies or compile research that motivates the agency to begin rulemaking. Agency specialists are often an important source for information relevant to environmental issues, and especially information related to scientific and engineering developments for best available technologies and other specifics that must be developed by a specialist.

Agencies are not the only entities that can initiate rulemaking. Occasionally, NGOs create voluntary standards, and agencies cooperate with these organizations and incorporate those standards into their rulemaking considerations.[92] In fact, the National Technology Transfer and Advancement Act of 1995 requires federal agencies to consult with voluntary, private-sector consensus standards bodies when consultations would be in the public interest and compatible with agency missions and goals.[93] Agencies may consult with appropriate non-governmental entities and benefit from the studies and research of specialists.[94]

In addition to standards and studies that inform rulemaking, sometimes agencies create directives like regulatory outreach programs to maintain contact with their constituencies. Office of Surface Mining Reclamation and Enforcement set up five types of outreach activities after continually having rules remanded by OIRA in the post-rulemaking review process.[95] The agency's outreach activities include advice on comments on priority setting and planning, background briefings and working groups, and negotiated rulemaking.[96]

---

90. OIRA, *OIRA Prompt Letters*, http://www.reginfo.gov/public/jsp/EO/promptLetters.jsp (last visited Aug. 10, 2016).
91. OMB, *OMB Prompt Letter*, http://www.reginfo.gov/public/prompt/epa_beach-act-2000.pdf. Beaches Environmental Assessment and Coastal Health Act of 2000, Pub. L. No. 106-457, 114 Stat. 870.
92. Lubbers, *supra* note 27, at 205 (referencing ACUS recommendation 78-4, *available at* https://www.acus.gov/sites/default/files/documents/78-4.pdf).
93. 15 U.S.C. §§3701-3722 (2014).
94. For more information about how to access specialists' documents, see Chapter 8 in this book.
95. Bremberg, *supra* note 70, at 36.
96. *Id.*

3. Public Petitions

The APA mandates that every agency give interested parties the right to petition for the issuance, amendment, or repeal of a rule.[97] Although there is no statute requiring agencies to create in-depth procedures for dealing with rulemaking petitions, the APA does require agencies to respond to petitions within a reasonable time.[98] EPA posts its rulemaking petitions online,[99] but does not indicate whether it denied or granted the petitions, or whether the petitions led to any sort of regulatory action.

Agencies receiving petitions for rulemaking from the public may announce the petition in the *Federal Register* and solicit comments on the issue.[100] Notice of rulemaking petitions in the *Federal Register* is optional, not mandatory, so the *Federal Register* may be a good place to look for rulemaking petitions, but there are no formal guarantees that they will be there.

Depending on whether agencies have procedures for dealing with rulemaking petitions, they may be difficult to find without resorting to FOIA requests or contacting the party who petitioned for the rulemaking. If an organization filed the petition with the agency, the petition may be posted on the organization's website.

4. Advisory Committee Proposals

Some regulatory statutes only allow agencies to create certain types of rules, after consulting with an advisory committee or panel before initiating the rulemaking process.[101] For instance, the Consumer Public Safety Commission cannot propose rules related to the risk of cancer or birth defects without first getting a report from a Chronic Hazard Advisory Panel.[102] Other statutes authorize the use of advisory committees prior to the initiation of rulemaking, but do not make them mandatory.

These advisory committees are usually chartered under the FACA.[103] Thus, the FACA is usually an ideal conduit for information related to advisory committee activities leading to rulemaking. Otherwise, turn to FOIA

97. 5 U.S.C. §553(a).
98. *Id.* §555(b). ACUS has also drafted agency recommendations for dealing with and responding to rulemaking petitions, *Petitions for Rulemaking*, Recommendation no. 2014-6 (Dec. 9, 2014), https://www.acus.gov/recommendation/petitions-rulemaking-0.
99. U.S. EPA, *Petitions for Rulemaking*, http://www2.epa.gov/aboutepa/petitions-rulemaking (last updated Dec. 15, 2015).
100. *Fed. Reg. Rulemaking Guide*, *supra* note 24, at 3.
101. LUBBERS, *supra* note 27, at 206.
102. *Id.* §2080.
103. LUBBERS, *supra* note 27, at 206 n.41.

to obtain the documents you need from pre-rulemaking exchanges between agencies and advisory organizations.

5. Rulemaking Agendas

Agencies are required to publish regulatory plans once a year in the fall and an Agenda of Regulatory and Deregulatory Actions in both the spring and fall. The regulatory plan and the regulatory agenda are often referred to collectively as the Unified Agenda. The Unified Agenda is where agencies announce future rulemaking activities and update the public on pending and completed regulatory actions. These plans and agendas are good sources of rulemaking forecasts. Rulemaking agendas are required documents, but they are non-binding, and agencies' activities can diverge from the agendas they publish.

The Unified Agenda is available on OIRA's RegInfo.gov website[104] and also Regulations.gov.[105] Agencies also publish most of this material in the *Federal Register*, and it should be available on their agency websites. FDsys.gov also contains a searchable Unified Agenda collection, although it is not easy to find on the website.[106]

B. Post-Rulemaking Review Process

Post-rulemaking activities occur in all three branches of the federal government. Congressional review processes, executive activities like those proscribed by the Paperwork Reduction Act[107] and Information Quality Act,[108] implement certain post-rulemaking activities. Judicial activities like interpretive court opinions, litigation settlements, and orders can also lead an agency to change or adjust a rule, or provide guidance on how the rule will be applied, as can policy interpretations by agencies in relation to the rule. All of these post-rulemaking activities create additional materials that may be useful to an environmental information seeker looking into the purpose,

---

104. OIRA, *Current Unified Agenda of Regulatory and Deregulatory Actions*, http://www.reginfo.gov/public/do/eAgendaMain (last visited May 16, 2016).
105. Regulations.gov, *Regulatory Agenda*, https://resources.regulations.gov/public/component/main?__dmf-ClientId=1461169545392&__dmfTzoff=240 (last visited May 16, 2016).
106. FDsys.gov, *Additional Resources*, https://www.gpo.gov/fdsys/browse/collection.action?collectionCode=GPO&browsePath=Unified+Agenda&isCollapsed=false&leafLevelBrowse=false&ycord=0 (last visited May 16, 2016).
107. The Paperwork Reduction Act of 1980, Pub. L. No. 95-511, 94 Stat. 2812, established the OIRA, which is charged under Exec. Order No. 12866 (Sept. 30, 1993) with reviewing final regulations.
108. This review is also carried out by OIRA.

rationale, and function of environmental rules.[109] Post-rulemaking activities can have major impacts on the rulemaking process, sometimes even derailing a rule before its implementation.

1. Congressional Review Under the Congressional Review Act[110]

The Congressional Review Act (CRA) requires all federal agencies to submit their rules to both houses of Congress and to the Government Accountability Office (GAO) before they can become effective.[111] Major rules (similar to economically significant rules that undergo OIRA review) are automatically delayed for at least 60 days for this congressional review, while non-"major" rules can go into effect without delay with congressional review is an optional step.[112]

The rules submitted to Congress and the GAO must be accompanied by a concise general statement on the rule, copies of cost-benefit analyses, and any other information or statements required by executive order or regulatory oversight laws.[113] After an agency submits this information, the GAO will draft a report to the congressional committees with jurisdiction over the major rule within 15 calendar days.[114] These GAO reports are available online within 24 hours of their dissemination to the committees.[115]

Although many rules are sent to Congress and the GAO each week pursuant to the CRA,[116] until 2017, members of Congress rarely responded negatively (or at all) to agency rulemaking or the GAO reports on the rules. Even though the legislative branch has the power to make rules void by passing resolutions of disapproval,[117] only one rule had been disapproved until 2017.[118] 2017 saw the rise, in Congress, of the tactical use of CRA to cancel Obama-era environmental regulations.[119]

109. Bremberg, *supra* note 70, at 5-6.
110. 5 U.S.C. §§801-808 (2014).
111. *Id.* §801.
112. *Id.* §801(a)(2)(A).
113. *Id.* §801(a)(1).
114. *Id.* §801(a)(2)(A).
115. *Oversight Hearings on the Congressional Review Act Before the House Subcomm. on Commercial and Administrative Law of the Comm. on the Judiciary*, 104th Cong. 3-4 (1997) (statement of Robert P. Murphy, General Counsel, GAO).
116. Lubbers, *supra* note 27, at 190.
117. These resolutions of disapproval must be signed by the president to take effect. 5 U.S.C. §801(a)(2)(B).
118. Maeve P. Carey et al., Cong. Research Serv., The Congressional Review Act: Frequently Asked Questions 5 (2015), https://www.fas.org/sgp/crs/misc/R43992.pdf.
119. *See* H. Res. 91 (disapproving BLM regulations on energy extraction measures under the CRA); S.J. Res. 28 (disapproving EPA CAA regulations under the CRA). The CRA disapprovals and related records are available on Congress.gov.

2. Executive Review by OIRA

OIRA, part of OMB,[120] reviews "significant" rules.[121] Significant regulations include those that would be economically significant—having a $100 million or more impact on the economy or otherwise materially adversely affect the economy, public health or safety, competition, or jobs—and those that would create a serious inconsistency or interfere with another agencies' plans or actions, or raise novel legal or policy issues. The parameters for OMB review are rather narrow, for instance, of the 135 actions pending review on Aug. 9, 2013, only 22 were designated as "economically significant" including two EPA rules.[122]

During the rulemaking process, OIRA does things like:

- Reviews draft proposed and final regulations to make sure they comply with administration policies.[123]
- Ensures that agencies incorporate public comments, consider alternatives to the rulemaking, and analyze the costs and benefits before rule implementation.
- Determines which agency regulatory actions are "significant" and thus subject to interagency review.[124]
- Oversees agency implementation of the Information Quality Act, including agencies' peer review practices.[125]

Overall, OIRA review ensures that agencies carefully consider the consequences of rules (costs and benefits) before enacting them.[126] During OIRA's review process, OIRA meets with interested parties, including people from state or local governments, private businesses, or nonprofit

---

120. OIRA was established in the 1980 Paperwork Reduction Act (44 U.S.C. ch. 35).

121. Regulatory Planning and Review, Exec. Order No. 12866 (Sept. 30, 1993).

122. Leland E. Beck, *Economically Significant: A Threshold Snapshot of OMB's Regulatory Docket*, Fed. Reg. Advisor, Aug. 9, 2013, http://www.fedregsadvisor.com/2013/08/09/economically-significant-a-threshold-shapshot-of-ombs-regulatory-docket/.

123. Pursuant to Exec. Order No. 12866, in 2009, the director of OMB issued Memorandum M-09-13, "Guidance for Regulatory Review," which increased OIRA oversight of the regulatory process throughout the rulemaking (and other regulatory) processes.

124. OIRA "generally designates between 500-700 regulatory actions as significant each year." *See Frequently Asked Questions*, *supra* note 14.

125. *Id.*; *see also* Memorandum for the President's Management Council from John D. Graham, Administrator of the OIRA, Executive Branch Implementation of the Information Quality Act (Oct. 4, 2002), https://www.whitehouse.gov/sites/default/files/omb/assets/omb/inforeg/pmc_graham_100402.pdf.

126. For a more in-depth discussion of this process, see U.S. GAO, Rulemaking: OMB's Role in Reviews of Agencies' Draft Rules and the Transparency of Those Reviews (2003), http://www.gao.gov/new.items/d03929.pdf.

advocacy organizations. A log of these meetings is available on OIRA's website. The log includes the subject, date, and participants involved in the meeting.[127] Environmental regulations that go under review deal with various environmental issues, including emissions standards, effluent guidelines, and even radiological incidents.[128] Information from these meetings available on OIRA's website include a list of attendees and links to handouts from the meeting, both of which are valuable for understanding the reasoning behind the rule and its major constituents and stakeholders.

Sometimes, after reviewing draft legislation, OIRA may send a letter to the agency returning the rule for reconsideration. This letter is commonly known as a return letter. A return letter explains why the regulation is not up to par with OMB standards. Reasons range from rules not being compatible with the law, rules with inadequate analysis or that are not justified by the analysis, or rules that unnecessarily conflict with other executive branch agency regulations or efforts.[129]

3. OIRA Return Letters

Return letters are not outright rejections of the rule, but rather OIRA determinations that the rulemaking would benefit from additional consideration and review by the agency. Return letters are publicly accessible on OIRA's website.[130] When a rule is returned, the agency must bring the rule in line before the agency may publish the rule. This may include doing further analysis regarding alternatives to the rule, etc. Agencies may conduct further work on the draft and re-submit it for OMB consideration.

4. OIRA Prompt Letters

Also, in 2001, OIRA began drafting prompt letters to proactively suggest issues that agencies may address: where further regulation may be needed to fill gaps in current environmental, health, and safety protections; or to suggest where current regulation is no longer needed and should be modified or

---

127. The Executive Order No. 12866 Meetings calendar allows you to search meetings by agency. *See* RegInfo.gov, *EO 12866 Meetings*, http://www.reginfo.gov/public/do/eom12866Search (last visited May 12, 2016).

128. OIRA's regulatory dashboard contains a regulatory review search option that allows you to find regulations under OIRA review. RegInfo.gov, *Search of Regulatory Review*, http://www.reginfo.gov/public/do/eoAdvancedSearchMain (last visited May 12, 2016).

129. *See Frequently Asked Questions*, *supra* note 14.

130. RegInfo.gov, *OIRA Return Letters*, http://www.reginfo.gov/public/do/eoReturnLetters (last visited May 12, 2016) (noting that there are no more recent letters than 2011).

rescinded. To date, OIRA has issued 12 prompt letters, which can be found on Reginfo.gov.[131]

5. OIRA's Regulatory Dashboard

OIRA's website, RegInfo.gov, is a virtual warehouse of rulemaking documents, statistics, and information. More than many other agencies, OIRA is efficient and up to date on offering information about federal agency rulemaking.

OIRA's website contains basic explanatory information about OIRA and its regulatory review process, OIRA guidance documents, executive orders affecting OIRA, reports, and "all substantive communications with parties outside the Executive Branch concerning regulatory actions under review."[132] These outside communications include meeting logs and written material received from outside parties on rules under review. Documents exchanged between OIRA and the rulemaking agency are only available after the regulatory action is published in the *Federal Register*, and even then, only select documents are made public.

OIRA also provides graphs, lists, and links to relevant documents going through the review process.[133] Significant regulatory actions regularly displayed on RegInfo's regulatory dashboard include:

- Notice documents (documents announcing new programs or agency policies)
- ANPRM
- Proposed Rule
- Final Rule
- Interim Final Rule[134]
- Direct Final Rule[135]

---

131. RegInfo.gov, *OIRA Prompt Letters*, http://www.reginfo.gov/public/jsp/EO/promptLetters.jsp (last visited May 12, 2016) (noting that there are no more recent letters than 2006).
132. OIRA, *Q&As*, https://www.whitehouse.gov/omb/OIRA_QsandAs (last visited May 12, 2016).
133. OMB, *Office of Information and Regulatory Affairs*, https://www.whitehouse.gov/omb/oira (last visited May 12, 2016).
134. Statutory provisions sometimes allow agencies to publish a final rule that becomes effective soon after publication without going through the proposed rule stage. "Good cause" exception in the APA allows agencies to bypass notice and comment on a rule when it would be impracticable, unnecessary, or contrary to the public interest. This move typically allows for public comment after the rule is published so that the agency can consider the comments and revise accordingly.
135. Similar to interim final rules except that there is no comment period after publication, on the ground that they are uncontroversial. The dashboard simply categorizes these as "Final Rules."

OIRA also creates annual reports, including a summary on the costs and benefits of significant regulatory actions, in its annual report to Congress, which is also available on the OIRA website.[136]

Keep in mind when searching for post-rulemaking review documents from OIRA, however, that OIRA only designates a fraction (only about one in six rules or so) as economically significant rules worthy of review. Thus, more rules than not make it through rulemaking without any OIRA review.[137] Also, it has been suggested that OIRA review is not apolitical: one report showed that in a recent decade, OIRA altered 84% of EPA rule submissions, and delayed tougher ozone pollution standards for over a year.[138] Other reports show that OIRA delays have led to serious public health and safety issues.[139]

Tip: *If you click on a rule in the RegInfo dashboard and the content for the rule is blank, it means that the agency didn't provide descriptive information in its Unified Agenda entries, because that is where OIRA draws its abstract and description information for rules. Also, you cannot see the actual documents and text for regulatory actions undergoing OIRA review, as they are considered deliberative and not available for public release during review.*[140]

Tip: *RegInfo.gov is different from Regulations.gov, which provides a government-wide portal for the public to review and comment on published regulatory actions and view any supporting materials provided by the agency. In contrast, RegInfo provides information about the status of agency documents undergoing OIRA review.*

---

136. OMB, *OIRA Reports to Congress*, http://www.whitehouse.gov/omb/inforeg_regpol_reports_congress/ (last visited May 12, 2016).
137. Richard Willian & James Broughel, *OIRA Quality Control Is Missing for Most Regulations*, Mercatus Center, George Mason University (Oct. 1, 2014), http://mercatus.org/publication/oira-quality-control-missing-most-regulations.
138. N.Y. Times. Ed. Bd., *Stuck in Purgatory*, N.Y. Times, June 30, 2013, http://www.nytimes.com/2013/07/01/opinion/stuck-in-purgatory.html.
139. Scot J. Paltrow, *How a Small White House Agency Stalls Life-Saving Regulations*, Reuters, Oct. 29, 2015, http://www.reuters.com/investigates/special-report/usa-regulations-oira/.
140. OIRA deliberations are generally exempt under FOIA's Exemption 5. *See, e.g.*, Wolfe v. Department of Health & Human Servs., 839 F.2d 768, 773 (D.C. Cir. 1988) ("Congress adopted Exemption 5 because it recognized that the quality of administrative decisionmaking would be seriously undermined if agencies were forced to operate in a fishbowl."). This is echoed on the OIRA FAQ webpage, which states: "Draft documents under review are considered deliberative and are not available for public release during review." http://www.reginfo.gov/public/jsp/Utilities/faq.jsp.

OIRA has a public docket library that you can visit in person. You can also file a FOIA request with the OMB to obtain rulemaking materials not proactively made available in the agency's online or in-person reading rooms.[141]

C. Records Created After a Final Rule Is Issued

The regulatory process does not end after the Final Rule is issued. After the Final Rule, the regulatory process includes compliance, interpretation, and review phases.[142] Agencies often issue guidance documents to better explain the regulatory requirements to those affected by the rule. These guidance documents take many forms, from brief "no-comment letters" that answer individual entities' questions about the rules on a case-by-case basis to voluminous manuals distributed to regulated entities. Guidance materials may appear in all sorts of places—they may show up in the *Federal Register* as Interpretive Rules or policy updates, or they may be posted on agency websites. Agency guidance documents do not create new regulatory standards or requirements; rather, they explain and supplement these final rules.[143] There are no rules about format or notice for agency guidance, so they may appear in a variety of forms in a variety of places, making researching for agency guidance on a particular rulemaking a hit-or-miss task.

Environmental rulemaking guidance for rules promulgated by EPA follows OMB best practices, and EPA maintains a list of significant guidance documents online, sorted by topic.[144] The guidance documents include many question and answer manuals, and information on proper practices for environmental practices required by environmental regulations. Because there are so many environmental regulatory programs, the types of guidance vary greatly—for instance, Comprehensive Environmental Response, Compensation, and Liability Act regulation guidance provides recommended practices for disposing of contaminated materials[145] while the Clean Water Act program guidance is much different—describing how to assess ambient water

141. OMB, *Public Handbook for Gaining OMB Information*, http://clinton4.nara.gov/OMB/foia/foiares4.html (last visited May 12, 2016) (site no longer being updated).
142. National Ass'n of Farmworkers Orgs. v. Marshall, 628 F.2d 604, 621-22 (D.C. Cir. 1980) (holding that the agency can continue to develop and collect information about a rule even after it is finalized as a normal requirement of competent administration).
143. *Fed. Reg. Rulemaking Guide*, *supra* note 24, at 7.
144. U.S. EPA, *Significant Guidance Documents*, https://www.epa.gov/laws-regulations/significant-guidance-documents (last visited Dec. 10, 2016).
145. U.S. EPA, *Significant Guidance Documents: Land and Emergency Management*, https://www.epa.gov/laws-regulations/significant-guidance-documents-land-and-emergency-management (last visited Dec. 10, 2016).

quality for bacteria or operate a mitigation bank.[146] One thing all of the guidance documents have in common is their role in filling in the functional gaps of regulatory language, so if you want to learn more about how to properly follow environmental regulations, agency guidance is an excellent starting point for your research.

VI. Conclusion

As this chapter reveals, there is a lot going on during the rulemaking process, and some parts of the process are less transparent than others. Namely, activities not governed by the APA, the FACA, or the Sunshine Act are not subject to transparency and notice requirements, so records related to those activities are often hard to find. ACUS and other nonprofit organizations focused on government transparency have issued plenty of advice on how agencies can improve rulemaking transparency, but no official legislative or regulatory changes have been made to improve the quality of rulemaking transparency.[147]

Transparency gaps can make researching environmental issues difficult because rulemaking is especially important in environmental law.[148] Much of environmental law and policy happens in agencies at the regulatory level. Some environmental laws even have specific rules about rulemaking records, like the CAA, which requires EPA to create a docket when an NPRM is published and to place all written comments and documentary information on the docket as soon as possible.[149]

Understanding the rulemaking process and identifying its discreet parts and their functions within the process makes it easier for information seekers to locate rulemaking records and information. If you know what each step of the rulemaking process entails, you can determine what types of documents and records may have been generated during the steps.

The rulemaking process results in a rule that can only be set aside by the courts if the court finds that the agency acted in an arbitrary and capricious manner, abused its discretion, or failed to act in accordance with law.[150] The great deference given to agencies in the rulemaking process makes the trans-

146. U.S. EPA, *Significant Guidance Documents: Water*, https://www.epa.gov/laws-regulations/significant-guidance-documents-water (last visited Dec. 10, 2016).

147. Cary Coglianese et al., Transparency and Public Participation in the Rulemaking Process, A Nonpartisan Presidential Transition Task Force Report (2008), http://www.hks.harvard.edu/hepg/Papers/transparencyReport.pdf.

148. Lubbers, *supra* note 27, at 317.

149. 42 U.S.C. §7607(d)(2)-(4) (2014).

150. 5 U.S.C. §706(2) (2014).

parency of this process all the more important, as it is hard to successfully argue that a rule should be changed once it has been finalized.

Chapter 8: Grey Literature: Other Sources of Environmental Information

Law in Action

Chilton, Hardwick & Associates is a corporate law firm that represents large corporations, including Exxon, a prominent oil company. Exxon has just been sued for pollution from a refinery that affected a nearby school.[1] *The complaint relies heavily on data from a scientific study about oil plumes and groundwater from an academic institution. Janice Parson, a new associate at the firm, has been tasked with finding the study in the complaint and locating other information to counter that study. She's also been asked to find information about similar lawsuits, verdicts, and settlements against oil companies, as well as examples of the corporate filings and any public relations information that accompanied those suits. Janice isn't sure where to start but she knows that her ordinary search tools and strategies do not necessarily contain things like scientific studies, information about settlement costs, corporate filings, and press releases.*

I. Introduction

Grey literature encompasses a variety of documents. What makes grey literature unique is that it is not controlled by commercial publishers. Things like guidance materials and academic studies are examples of grey literature. Some types of grey literature are especially relevant to environmental information seekers. Corporate filings, congressional documents, company websites, and nonprofit and academic work may contain useful environmental data and information.

II. Other Government Records

In addition to the various environmental and transparency laws that exist at the state and federal level, there are several U.S. laws that require entities to provide information to the public, and to federal and state governmental

1. *See* Benjamin Weiser, *Exxon Settles $9 Billion Pollution Case in New Jersey for Far Less*, N.Y. Times, Feb. 28, 2015, at A15.

agencies. Some of these required disclosures may be relevant for environmental information seekers. There are also other materials created by all three branches of government that may contain useful information. In the executive branch, other administrative agencies beyond the U.S. Environmental Protection Agency (EPA) provide public access to environmental information. The legislative branch conducts research and gathers testimony on legislative proposals dealing with environmental issues. The judiciary compiles a multitude of court documents on each case docket, beyond the opinions that contain the final dispositions of the courts. All of these documents are potential resources for those doing environmental work, providing additional information, background, and data.

A. *U.S. Securities and Exchange Commission Filings: Corporate Disclosure*

Environmental information seekers often require information about companies. Companies are common polluters, creators of environmental technologies, parties in environmental litigation, and lobbyists who draft and impact environmental legislation. Knowing more about companies, and what they do, is useful for research into environmental matters.

The U.S. Securities and Exchange Commission (SEC), a federal financial agency, requires corporations that deal in securities (stocks, bonds, etc.) to provide regular reports about their activities, and to send special reports when they engage in certain activities. These reports are commonly called SEC filings, and they are available to the public to keep investors up to date on the inner workings of the companies they invest in. Information about corporations' environmental activities, and their ongoing litigation, is available in these corporate disclosure documents.

Similarly, state agencies also require corporate entities to register and file certain documents with state governments.[2] State laws also require corporations dealing in securities to register and obtain licenses. These state laws, called Blue Sky Laws after a U.S. Supreme Court case that described the laws as measures to prevent the negative effects of "speculative schemes which have no more basis than so many feet of 'blue sky'. . . ."[3] These filings are generally filed in states' department of state offices and are sometimes made available to the public in online databases.

2. These state government departments are places such as the department of state, the state attorney general's office, etc.
3. Hall v. Geiger-Jones Co., 242 U.S. 539, 550 (1917).

In addition to the general corporate information in SEC filings, SEC-promulgated regulations also require publicly traded corporations to disclose environmental information beyond the disclosures required by state and federal environmental laws. These environmental disclosures are designed to inform investors about potential effects of climate change and other environmental impacts on certain business considerations. These disclosures include descriptions of costs incurred by companies in complying with environmental laws, information about litigation related to environmental issues, and any environmental risk factors that have impacted their business or that may impact their business in the future.[4]

Specifically, the SEC has created three reporting requirements in its S-K regulation,[5] which describes the required disclosures in the regularly recurring reports filed by publicly traded corporations including the 10-K (annual report) and 8-K (reports filed when certain things happen like bankruptcies or CEO departures) filings. The three required environmental disclosures are:

- Description of Business.[6] These corporate overviews must include the costs of complying with state, federal, and local environmental laws and regulations. For instance, if the company must comply with laws limiting the discharge of materials into the air or water, or other rules designed to protect the environment, the costs of such compliance must be reported in the filings. Companies must share information about material estimated capital expenditures for environmental control compliance undertaken by the parent company or its subsidiaries. Even smaller reporting companies are required to disclose information about the costs they expend to comply with environmental laws.[7]

- Risk Factors.[8] Companies are required to discuss the most significant factors that make investing in their enterprise speculative or risky. These risk assessments include climate change and other environmental risks. Climate change considerations include things like capital expenditures for reducing emissions and purchasing "cap-and-trade" allowances for companies that cannot meet target reductions, as well as changing prices of goods and services affected by a heating planet. They also include meteorological phenomena related to climate change, such

4. Jennifer L. Machlin & Tomme R Young, Managing Environmental Risk: Real Estate and Buisness Transactions (Environmental Law Series) §§13:2-13:14, 18:34 (2015/2016).
5. 17 C.F.R. §§229.101 to .915 (2016).
6. S-K Item 101.
7. 17 C.F.R. §229.10(f).
8. S-K Item 503(c).

as changes in weather patterns, sea-level rise, permafrost melts, and temperature extremes. Related impacts that should be reported include changes to resources, like water availability and quality, or drought and flooding impact on agriculture.[9] The company must clearly state the risk and specify how the risk could affect the company.[10]

- Legal Proceedings.[11] Companies must also report material pending legal proceedings involving the company or any of its subsidiaries beyond the ordinary, routine litigation incidental to the business. If any liabilities could materially impact the company's financial or competitive position, aka if they could affect investors, the legal proceedings must be discussed in the companies' S-K filings. So cases where liability or sanctions could exceed 10% of the current assets of the company, or government enforcement proceedings costing over $100,000 (including EPA-initiated enforcement actions and citizen lawsuits or injunctions) would be included in this discussion.[12] Disclosure of these types of legal proceedings are required even if the proceedings have not been initiated, so if such a proceeding is even being considered or contemplated, the company must report it to investors.
- Management Discussion.[13] This section gives investors the opportunity to view the company as the management views it.[14] This discussion requires companies to disclose any known trends or uncertainties that have, or may have, a material impact, whether favorable or unfavorable, on net sales or revenues or income from continuing operations. This discussion must further include environmental contingencies that may impact these aspects of the operations.

Additionally, SEC guidance links these disclosure mandates with certain environmental laws. For instance, the SEC requires environmental disclosures pursuant to National Environmental Protection Agency,[15] and the

9. Commission Guidance Regarding Disclosure Related to Climate Change, Final Rule, 75 Fed. Reg. 6290, 6291 (Feb. 8, 2010).
10. This Congressional Research Services (CRS) report identifies, describes, and analyzes the SEC's guidance on the climate change risk disclosures, Gary Shorter, Cong. Research Servs., SEC Climate Change Disclosure Guidance: An Overview and Congressional Concerns (Aug. 26, 2013), https://www.fas.org/sgp/crs/misc/R42544.pdf.
11. S-K Item 103.
12. 75 Fed. Reg. at 6293-294.
13. S-K Item 303.
14. Concept Release on Management's Discussion and Analysis of Financial Condition and Operations, 52 Fed. Reg. 13715, 13717 (Apr. 24, 1987).
15. This was the original rationale for these S-K environmental provisions, Notice of Adoption of Amendments to Registration and Report Forms to Require Disclosure With Respect to Compliance With

SEC in 1993 issued guidance stating that companies should report their estimated allocation of potential liabilities at Superfund sites, where liability is joint and several.[16]

An SEC guidance release explains these environmental disclosures more thoroughly than the regulations, providing more details about what should be disclosed. There are other SEC guidance documents that also provide insight about how these disclosures may be written and what kinds of information they should contain. Secondary sources written by nongovernmental authors, like law firm blogs[17] and bulletins, as well as scholarly resources like law review articles explain the inner workings of environmental SEC filing disclosures and discuss the SEC's intent for creating environmental and climate change disclosure requirements, etc.[18]

<table>
<tr><th>Item</th><th>Provision</th><th>Guidance</th></tr>
<tr><td>Item 101
Description of Business</td><td>17 C.F.R. §229.101</td><td rowspan="4">17 C.F.R. Parts 211, 231, and 241
Commission Guidance Regarding Disclosure Related to Climate Change, Final Rule[a]</td></tr>
<tr><td>Item 103
Legal Proceedings</td><td>17 C.F.R. §229.103</td></tr>
<tr><td>Item 503(c)
Risk Factors</td><td>17 C.F.R. §229.503</td></tr>
<tr><td>Item 303
Management's Discussion & Analysis (MD&A)</td><td>17 C.F.R. §229.303</td></tr>
</table>

a. Available at http://www.sec.gov/rules/interp/2010/33-9106fr.pdf.

Beyond these specific environmental law disclosures, other sections of SEC filings contain useful information like addresses, litigation updates, lists of chains of command and financial data for the corporations. All of this information can be useful if you are trying to find out things like: what a particular company does, where the company is located and has operations, and who is involved in the company's board or leadership.[19]

It is much more difficult to find information on private companies, or companies that do not have publicly traded securities, because they do not

---

Environmental Requirements and Other Matters, Securities Act Release No. 5386, Exchange Act Release No. 10,116, 1 SEC Docket 1 (Apr. 20, 1973).

16. SEC Staff Accounting Bulletin No. 92, 58 Fed. Reg. 32843 (June 14, 1993).

17. *See, e.g.*, Allen & Overy, *Environmental Disclosure and Corporate Environmental Transparency*, http://www.allenovery.com/expertise/practices/environment/Pages/environmental-disclosure.aspx (last visited Dec. 15, 2016).

18. *See, e.g.*, Mark Latham, *Environmental Liabilities and the Federal Securities Laws: A Proposal for Improved Disclosure of Climate Change-Related Risks*, 39 Envtl. L. 647, 728 (2009).

19. These data points in SEC filings are discussed further in Section IV of this chapter.

have these filing requirements. Information seekers, however, can often find some information on private companies if they sift through several assorted online resources and piece together data points—there is no single corporate dossier or set of agency filings that will provide comprehensive information about private companies. Look to the companies' websites, and to state corporation and company registers to find whatever limited information is available about private companies. Often, company webpages will have an "about us" page or a "history" page that may provide research leads. You can also search for news stories about a company, or search through court docket databases,[20] to learn about key events and litigation involving the company.

B. *Locating SEC Filings*

These records can be found on a company-by-company basis through the SEC's EDGAR database. While notoriously difficult to search, documents on EDGAR,[21] SEC filings can also be found in a variety of fee-based subscription databases. SEC filings for a particular company can often also be found on the investor relations pages on company websites, which are good places to look if you know which company's SEC filings you need, but not as helpful if you are looking for a type of company instead of a specific company; or if you need sample filing language from more than one company.

It must be noted, however, that although the information provided by the companies is required by statute and not voluntary, the environmental disclosures that companies do provide in their SEC filings are often very brief and superficial. In fact, most companies fail to meet SEC climate-related disclosure requirements. A 2013 study found that 40% of companies on Standard & Poor's 500 Index failed to include any climate-related disclosure in their annual 10-K filings.[22]

C. *Other Government Agency Documents*

Cass Sunstein, administrator of the White House Office of Information and Regulatory Affairs, describes the executive branch as "the most knowledge-

20. Searching for court dockets is discussed in Chapter 3.
21. EDGAR, *Company Filings*, https://www.sec.gov/edgar/searchedgar/companysearch.html (last visited Dec. 15, 2016).
22. Barry R. Burr, *Ceres: SEC Needs to Better Enforce Climate Change Disclosure Requirements*, Pensions & Invest., Feb. 7, 2014, http://www.pionline.com/article/20140207/ONLINE/140209893/ceres-sec-needs-to-better-enforce-climate-change-disclosure-requirements.

able branch" of government[23] because the executive branch is so well stocked with policy and scientific experts, gathering huge swaths of information from public and private entities. Although this book focuses on agencies that work intensively on issues affecting the environment, like EPA, environmental information is not limited to only a few government agencies.

In addition to the SEC, most federal and state agencies engage in activities that touch on environmental issues even if their mission is not explicitly related to the environment. The Federal Aviation Administration (FAA) and U.S. Department of Transportation (DOT) engage in anti-fuel and noise pollution activities, as well as the building of infrastructure; the U.S. Food and Drug Administration (FDA) regulates food safety issues including those involved with farming and ranching practices; and the U.S. Department of Justice enforces criminal sanctions within environmental laws. There are few agencies that do not generate environmental information. Beyond rulemaking and administrative adjudication proceedings, the agencies create guidance documents: guides; pamphlets; letters; and opinions that instruct the public on how to follow agencies' regulatory obligations. Guidance is often available on agency websites, but some items may be harder to find, requiring deeper research and records requests.

If you are looking for particular agency adjudication records, you can look on the agency's website which usually has a link to its administrative law judges' webpages that contain filings, procedures, and orders.[24] The Administrative Conference of the United States (ACUS), in cooperation with Stanford University has also created a website for their Adjudication Research Project that contains information and links to every agency and their related administrative tribunals.[25] For example, if you wanted to see all the types of adjudications done by the U.S. Department of Health and Human Services, you could click on that agency[26] and see the tribunals and adjudicatory schemes for the agency, with links for additional information about each one.[27]

---

23. *See* Cass R. Sunstein, *The Most Knowledgeable Branch*, 164 U. Pa. L. Rev. 1607 (2016), http://scholarship.law.upenn.edu/cgi/viewcontent.cgi?article=9547&context=penn_law_review.
24. *See* U.S. EPA, *Filings, Procedures, Orders and Decisions of EPA's Administrative Law Judges*, https://www.epa.gov/alj (last updated Dec. 15, 2016).
25. *See* Stanford Univ., *Federal Administrative Adjudication*, https://acus.law.stanford.edu/ (last visited Dec. 15, 2016) (a joint project of ACUS and Stanford University law school).
26. *See* Stanford Univ., *Department of Health and Human Services*, https://acus.law.stanford.edu/agency/department-health-and-human-services (last visited Dec. 15, 2016).
27. Under "agency scheme," it shows you 11 types of adjudication, and you can click onto each one to get more information about it. For example, there is more information for health insurance benefits under Medicare, including information on what exactly is adjudicated, the governing *Code of Federal Regulations'* sections, and how and where to appeal. *Id.*

Generally, the best place to look for an agency's records, including those related to administrative adjudications, are on their websites, but there is no uniformity in the way agency websites are created and organized, and there are few overarching requirements on what agency websites must contain. As a result, the types and amounts of environmental information you find on agency websites will vary. Generally, the more local an agency, the less information will be available on the website—for instance, federal agency websites tend to be well stocked with information and frequently updated, whereas state and municipal agency websites often have less information, and are less frequently updated. This is largely because local agencies have less staff and funding to maintain websites and fill their online archives with records. This is not always the case as there are some state and local agency websites that sometimes contain regularly updated databases and archives of information.

Beyond agency websites, there are also specialized databases that contain agency documents and information. There are topic-specific subscription databases, like Thomson Reuters Checkpoint[28] which focuses on tax issues and contain IRS records, including guidance and adjudication documents. There are also more general agency subscription resources, like Bloomberg BNA,[29] a service that historically reported on agency activities and created regular news updates about federal agencies, and continues to create those news reports as well as provide agency records in its databases. These topic-specific and regulatory-focused databases contain a wealth of information, but they can also be costly. If you would like to look at a specialized database service, it is a good idea to check out the holdings of your local public and academic libraries, as many may hold subscriptions.

D. *Congressional Documents*

Aside from the executive branch, federal, state, and municipal congressional bodies also engage in environmental work, creating and passing environmental laws and legislation that affects the environment. These legislative bodies create records related to their law making, and also the other congressional activities, including investigations and hearings. As administrative documents are often found on agency webpages, legislative committee's

---

28. Thomson Reuters, *Checkpoint*, https://tax.thomsonreuters.com/products/brands/checkpoint/ (last visited Dec. 15, 2016), (formerly RIA).
29. Bloomberg BNA, *Homepage*, https://www.bna.com (last visited Sept. 29, 2016) (formerly Bureau of National Affairs (BNA)).

webpages, and the webpages of individual legislators, are also good starting points for finding documents and information about legislative activities.

I. Congressional Hearings and Reports

Congressional hearings are meetings or sessions of the U.S. Senate, U.S. House of Representatives, or a joint or special committee of Congress. These hearings are ordinarily open to the public, and transcripts of the hearing proceedings are often available online. The focus of congressional hearings differs depending on the purpose of the meeting, however, they are largely exploratory in nature, helping inform congressional committees about proposed legislation or congressional investigations. A hearing may also be held to evaluate or oversee the activities of a particular government department or the implementation of a federal law. The federal government provides video coverage of many of these congressional hearings and meetings.[30] Additional video coverage of congressional activities, including activity on the floor and other committee events, are also available online, sometimes with live streaming.[31] The time between a hearing and its publication varies greatly. The U.S. Government Publishing Office (GPO) reports that a hearing is usually published anywhere from two months to two years after it is held.[32]

Depending on the committee, the GPO may provide access to the hearing testimony on its FDsys.gov Congressional Hearings page.[33] Selected hearings from 1986 to the present are available on the page. If a hearing is not on the page, then it is not currently available electronically through the GPO.

The GPO also provides online access to congressional reports going back to 1995.[34] Congressional reports are drafted in congressional committees, and they deal with proposed legislation and issues under investigation. There are two types of congressional reports: House and Senate reports are the reports of congressional committees concerning proposed legislation or investigation findings. Senate Executive Reports are from the Committee on Foreign Relations, and they deal with treaties between the United States and

30. Congress.gov, *House Committee Hearings and Meetings Video*, https://www.congress.gov/committees/video (last visited Dec. 15, 2016).
31. Congress.gov, *Audio and Visual Coverage*, https://www.congress.gov/resources/display/content/Audio+and+Visual+Coverage (last visited Dec. 15, 2016).
32. GPO, *Congressional Hearings*, https://www.gpo.gov/fdsys/browse/collection.action?collectionCode=CHRG (last visited Dec. 15, 2016).
33. *Id.*
34. GPO, *Congressional Reports*, https://www.gpo.gov/fdsys/browse/collection.action?collectionCode=CRPT (last visited Dec. 15, 2016).

foreign nations that have been submitted to the Senate for ratification. Senate Executive Reports may also be reports of Senate committees regarding executive nominations.

There are also congressional Committee Prints available through the GPO's FDsys.gov databases.[35] Committee Prints are congressional committee publications on topics related to the legislative and research activities of the committees. They often contain historical and statistical information, as well as legislative analysis. There are various categories of Committee Prints, including: draft reports and bills; directories; statistical materials; investigative reports; historical reports; situational studies; confidential staff reports; hearings; and legislative analyses. The FDsys.gov collection of Committee Prints goes back to 1975.

Finally, FDsys.gov has a "Browse Government Publications" page that allows visitors to search by congressional committee, collection, date, or author.[36] This page contains links to many organized collections of legislative and congressional documents in html and PDF formats. Beyond the FDsys.gov website, the fee-based subscription database ProQuest Congressional and its add-on ProQuest Legislative Insight provides access to legislative documents created by Congress.[37] These reports, hearing documents, committee prints, and other materials are especially useful when prominent environmental investigations and proposed environmental legislation are in play. For instance, the House Committee on Oversight and Government Reform conducted its own investigation of the Flint, Michigan, water crisis and the city's preparation and response to the water pollution threats. The committee's website contains a page dedicated to the Flint investigation, with links to the hearing materials and press releases.[38]

2. Congressional Research Service Reports

In addition to the Congressional Record discussed in Chapter 6, CRS reports are in-depth sources of information from the legislative branch of the federal government. The legislative branch contains its own think tank, the Congressional Research Service (CRS) that provides policy

35. GPO, *Congressional Committee Prints*, https://www.gpo.gov/fdsys/browse/committeetab.action (last visited Dec. 15, 2016).

36. GPO, *Browse Government Publications*, https://www.gpo.gov/fdsys/browse/committeetab.action (last visited Dec. 15, 2016).

37. ProQuest, *ProQuest Legislative Insight*, http://www.proquest.com/products-services/legislativeinsight.html (last visited Dec. 15, 2016).

38. House Comm. on Oversight & Gov't Reform, *Flint*, http://democrats.oversight.house.gov/investigations/flint (last visited Dec. 15, 2016).

and legal analysis to congressional committees.[39] To these ends, the CRS produces reports designed to help legislators make informed decisions. The CRS is not required to publish CRS reports of any type or on any topic, so there is no guarantee that a CRS report is available for your topic of interest.

As there is no right of publication of CRS reports, the CRS reports are not uniformly distributed to libraries or placed online for public access, so they can be difficult to find. Although transparency activists and information specialists advocate for making these reports publicly available, there is no central public access depository for the reports, and they must be sought in piecemeal fashion, using various search tactics to root out a specific report, like finding the proverbial needle in the haystack.

CRS reports can be searched for by report title or number, author names, or even the date of publication. Because multiple reports may be written on the same topic or statute, searching for a report on a broad topic, like the "Clean Water Act," is not usually specific enough to find a particular report.[40] Once you have some basic information about the report you need, you can proceed to several search venues.

The first place to search for CRS reports is through a general search engine, as many of the reports are available on the Internet. Although they are not part of a systemic government release scheme, they are also not proprietary materials, as they are created by a government entity, and are thus in the public domain. So many reports are available online for free. There are several online depositories that are attempting to build collections of CRS reports. Some of these are curated by special topics, while others are more general attempts to provide CRS reports to the public.

Demand Progress, an Internet activism organization, and the Congressional Data Coalition, have launched an effort to collect all CRS reports in one online repository, Everycrsreport.com.[41] This database includes 8,255 CRS reports, and it continues to grow. Some other online CRS report repositories include academic library websites like the University of North Texas online

39. Library of Cong., *Congressional Research Service Careers*, https://www.loc.gov/crsinfo/ (last visited Dec. 15, 2016).

40. A quick Google search reveals that there are at least a dozen CRS reports discussing the Clean Water Act. *See also* PEN-E, *Congressional Research Service Reports for Environmental Law* (Aug. 20, 2012), http://paceeenvironmentalnotes.blogspot.com/2012/08/congressional-research-service-reports.html.

41. EveryCRSReport, *Recently Published Reports*, https://www.everycrsreport.com/ (last visited Dec. 15, 2016).

CRS Reports Collection,[42] the Thurgood Marshall Law Library's database,[43] and Stanford's Archive-It collection, which pulls together various online CRS report services into a single place on the Internet.[44]

CRS collections organized by topic include the U.S. Department of State Foreign Press Center report collection, which goes back to 1999,[45] and a collection maintained by the Federation of American Scientists, which attempts to collect as many non-confidential CRS reports as possible.[46] Of special interest to those seeking environmental information is the CRS report collection from the National Agricultural Law Center, which arranges its CRS reports by environmental topic.[47] Several CRS report collections also specialize in security and military issues, including GlobalSecurity.org[48] and a website called Information Warfare which contains CRS reports from 2001-2004.[49]

There are also a number of subscription databases containing CRS reports that require a fee to access. For example ProQuest has an entire Congressional Research Digital Collection containing CRS reports.[50] Like the other paid databases, however, you may find that a local library has licenses to provide public access to these subscriptions, so you may be able to access them that way.

You can also contact members of Congress to request CRS reports. It is helpful to contact a congressperson who sat on a committee that considered legislation relevant to the report's subject.

---

42. UNT Digital Libr., *Congressional Research Service Reports*, http://digital.library.unt.edu/explore/collections/CRSR/ (last visited Dec. 15, 2016).
43. Thurgood Marshall L. Libr., *CRS Reports*, http://www.law.umaryland.edu/Marshall/crsreports/ (last visited Dec. 15, 2016).
44. Social Sciences Resource Group, Stanford Univ., *Archive-It: Congressional Research Service (CRS) Reports*, https://archive-it.org/collections/1078 (last visited Dec. 15, 2016). Several other efforts to provide access to CRS reports, including the Open CRS project by the Center for Democracy & Technology, and the Penny Hill Press CRS report service, have been discontinued.
45. U.S. Dep't of State, *Congressional Research Service Reports (CRS) and Issue Briefs*, http://fpc.state.gov/c18185.htm (last visited Dec. 15, 2016).
46. Federation of American Scientists (FAS), *Congressional Research Service [CRS] Reports*, http://www.fas.org/sgp/crs/ (last visited Dec. 15, 2016).
47. National Agric. L. Ctr., *Congressional Research Service Reports*, http://nationalaglawcenter.org/crs/ (last visited Dec. 15, 2016) (part of the University of Arkansas).
48. GlobalSecurity.org, *Congressional Research Service [CRS] Reports*, http://www.globalsecurity.org/military/library/report/crs/ (last visited Dec. 15, 2016).
49. Bureau of Public Affairs, U.S. Dep't of State, *IWS: The Information Warfare Site, Congressional Research Service (CRS) Reports and Issue Briefs*, http://www.iwar.org.uk/news-archive/crs/ (last visited Dec. 15, 2016).
50. ProQuest, *ProQuest Congressional Research Digital Collection*, http://www.proquest.com/libraries/academic/databases/ProQuest-Congressional-Research-Digital-Collection.html (last visited Dec. 15, 2016).

3. Court Information: Litigation/Settlement Information

Chapter 3 explains how to find court dockets and court reporters containing published final opinions. In addition to these major sources of court documents, there are resources that can provide limited information about litigation and settlements that is not provided in dockets. While it is nearly impossible to retrieve information about court deliberation and party preparation for trial, as attorney work product and judicial deliberations are protected by various disclosure privileges, information about verdicts and settlements is sometimes made available through specialized online databases. Information about settlements and verdicts can help people considering environmental litigation understand their chances of recovery, and estimate the awards or damages they may face as they proceed with the case. In an environmental law scheme that allows citizens to file suits for the enforcement of environmental laws and in a world that recognizes liability for injuries caused by pollution (toxic tort litigation, etc.), considering the potential costs and awards of litigation can be a key factor in determining whether to pursue environmental claims or assess the costs of defending against an environmental law claim.

Most of the databases containing verdict and settlement data and information require subscription fees, but there are a few options for free access to some of this case information. Law journals like *the Journal of Empirical Legal Studies* contain studies on trends in jury awards and settlements.[51] Additionally, there are several grass-root, private organizations that strive to present as much information about state court verdicts online as they can. The National Association of State Jury Verdict Publishers[52] (NASJVP) publishes jury verdict summaries from state courts, taking civil litigation information from attorneys and then writing summaries of case verdicts for use by attorneys and insurers for case evaluation. Zarin's Jury Verdict Review & Analysis[53] is a similar effort by attorneys to present verdict information to the public. Without a central repository for verdict and settlement data, information reflecting litigation rewards is available in several places, but there is no "main source" for verdict and settlement data.

51. *See, e.g.*, Seth A. Seabury et al., *Forty Years of Civil Jury Verdicts*, 1 J. Empirical Legal Studies 1 (2004), http://content.csbs.utah.edu/~dlevin/jud_pro/articles/jury%20awards%20over%2040%20 years.pdf.
52. JuryVerdicts.com, *Homepage*, http://juryverdicts.com/ (last visited Dec. 15, 2016).
53. Zarin's Jury Verdict Review & Analysis, *Homepage*, http://www.jvra.com/verdict_trak/ (last visited Dec. 15, 2016).

ALM, the publisher of the *National Law Journal* and many other legal periodicals, including the *New York Law Journal* hosts a subscription database called VerdictSearch,[54] which may be the most complete and up-to-date database of information about verdicts and settlements. Westlaw and Lexis also each have competing products that contain information about settlements and verdicts. Westlaw's Jury Verdicts & Settlements database is readily available on the homepage of its website, so long as you have a subscription. You can access a similar content type in Lexis by selecting the "Browse Sources" tab and narrowing the content type to "Jury Verdicts and Settlements." It appears that Lexis's verdicts and settlements come from ALM VerdictSearch databases. Some other paid database systems like Bloomberg BNA provide verdict and settlement analysis for certain types of cases.[55]

Finally, there are some print publications that may be useful to environmental information seekers looking for information about verdicts and settlements. Thomson Reuters, Westlaw's parent corporation, publishes *Dollar Verdicts: Personal Injury* as a regularly updated loose-leaf set that lists injuries alphabetically, and then provides relevant award amounts and case summaries for each type of injury.[56] Matthew Bender, a Lexis publisher, offers a similar, competing set called *What's It Worth? A Guide to Current Personal Injury Awards and Settlements*, with searchable, online eBook format options.[57] Additionally, *Causes of Action*, a popular series that provides overviews and practical information about every possible cause of action in U.S. jurisprudence may also provide some information about verdicts and settlements. This publication is available in most law libraries, and also on Westlaw with a subscription.

III. Individuals and Corporations: Researching Like a Reporter

Undertaking research about environmental issues (and really, most legal issues), sometimes requires information seekers to delve into research about people, places, and businesses. In order to know which court to file a complaint in, you have to know where a party to a potential lawsuit lives or has

54. ALM Media Properties, *VerdictSearch*, http://verdictsearch.com/state/all/ (last visited Dec. 15, 2016).
55. For instance, Bloomberg BNA has an employment discrimination verdicts and settlements navigator. *See* Bloomberg BNA, *Verdicts and Settlements Navigator*, http://www.bna.com/vands_lerc/ (last visited Dec. 15, 2016).
56. Thomson Reuters, Dollar Verdicts: Personal Injury (2008), http://legalsolutions.thomsonreuters.com/law-products/Treatises/Dollar-Verdicts-Personal-Injury/p/100028113.
57. LexisNexis Store, What's It Worth? (Matthew Bender 2016), http://www.lexisnexis.com/store/catalog/booktemplate/productdetail.jsp?prodId=57164#.

its center of business. To understand the magnitude of a company's activities, you sometimes must research its past record of activities or infractions. Finding fellow stakeholders who may care about advocating for the same thing you are interested in sometimes involves reaching out to people and entities you must discover on your own. There are many instances where research extends the boundaries of government documents and scholarly books and articles into the murky world of detective work and sleuthing. This type of research is similar to the research conducted by journalists and news reporters, which entails pulling together information about people, places, and things.

Researching like a reporter for environmental information may include tasks such as finding information about corporations, locating public records about people, researching the history of real property, gathering news stories and information about current events and many other types of research that help information seekers develop a clearer picture of the situation they wish to advocate for or against or spread awareness about. Here are some specific tips for gathering these different types of information:

- Corporate Overviews. Information about corporations and businesses beyond the environmental disclosures required in SEC filings is often necessary for environmental information seekers to proceed with their work. Business information can help you whether you are finding out what is going on in that factory sitting on the waterfront, who runs the local landfill company, or which entity is responsible for supplying power to a local electrical grid. Many environmental issues involve corporations and businesses.

  The best place to start when trying to find more information about a particular business entity is to go to their webpage (see Figure 8.1). Publicly traded corporations ordinarily have an "Investor Relations" tab or "Investors" link at the bottom of their webpage. (You may have to look in the small print on the margins of the webpage to find the link.) The investors' page will contain links to webcasts and earning report calls, as well as links to PDF filings and basic corporate information. SEC filings for publicly traded companies also contain basic company information including the names of the people, who run the company, where the company is located, and the company's purposes and goals.

  Private companies that do not trade stock publicly will not have links to investor information or SEC filings, as they likely do not have to file regularly with the SEC, but they probably have a link to an "About Us" or "Company History" webpage. That link may contain information

Figure 8.1 Exxon Mobil's Investor Relations Webpage

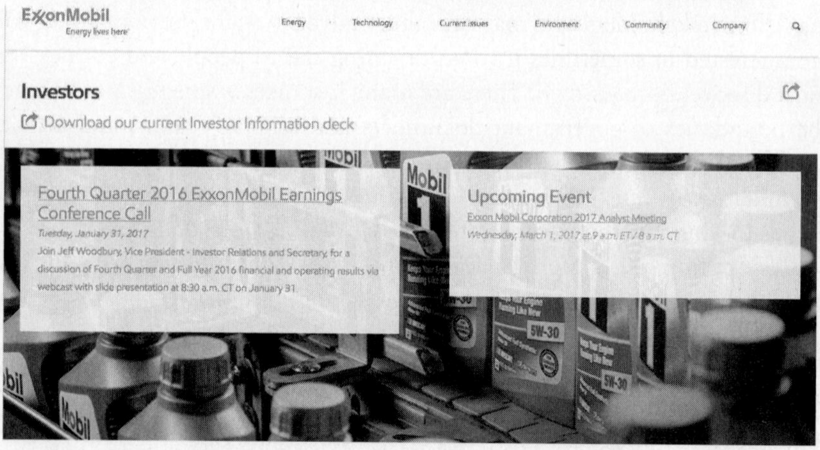

Exxon Mobil's investor relations page contains downloadable versions of SEC filings conference call transcripts and investor presentations, as well as information about the company's structure and upcoming events.

about the company's history, the owners, and the business purpose and model. All of that information can be used to conduct general Internet search engine, news, and social media searches. Private corporate research, however, is one of the most difficult types of research.

Beyond the companies' own webpages, there are online databases containing information about many companies. All publicly traded companies appear in these databases because the databases gather information from publicly available SEC filings, and some private companies may also have some information stored in these databases. There are also some free online databases that contain limited information like Google Finance[58] and Yahoo Finance,[59] which mostly contain financial data for companies' stocks, but most corporate information databases charge a fee to access the corporate data they collect. Companies like Bloomberg L.P.,[60] Dun & Bradstreet,[61] and Hoovers[62] have robust corporate information databases and information systems, but can be quite costly. Libraries at business schools or public business libraries may provide limited service to these products, free of charge.

58. Google, *Finance*, https://www.google.com/finance (last visited Dec. 15, 2016).
59. Yahoo!, *Finance*, https://finance.yahoo.com/ (last visited Dec. 15, 2016).
60. Bloomberg, *Homepage*, http://www.bloomberg.com/company/ (last visited Dec. 15, 2016).
61. Dun & Bradstreet, *Relationships & Growth*, http://www.dnb.com/ (last visited Dec. 15, 2016).
62. Hoovers, *Homepage*, http://www.hoovers.com/ (last visited Dec. 15, 2016).

Both private companies and publicly traded corporations have to register in the states where they do business. Most states make this registration information, which includes names, addresses for service of process, dates of incorporation, company name histories, and other information, available online. This data is usually locaated in a state's department or secretary of state office or an ancillary department of corporations, and the databases contain varying levels of information, depending on how much the state agency makes available. Also, some states charge money to obtain certain types of company registration information. There is no uniformity to what company information you will receive on a state-by-state basis.

- Property Information. Environmental information often involves buildings and property. Environmental issues focus on pollution in waterways; soil and groundwater contamination; animal and human habitat destruction; and point sources of air and noise pollution—activities that occur on and focus on parcels of real property and buildings. State and municipal agencies often have offices dedicated to collecting and storing information about parcels of properties, including deeds to land, records of building ownership, and information about building permits and licenses. Some of these documents may be stored online in agency databases, but ordinarily, obtaining these documents requires a trip to the state or municipal agency offices to retrieve prints of the original items. Remember, local agencies often do not have the robust online databases and services of larger federal, and even state, agencies.

  Federal and state projects like building highways, dredging waterways, and other undertakings that impact the environment often have publicly available permitting and hearing documents, licensing documents, and oftentimes even environmental impact statements and plans drafted under state and federal law requirements. See Chapter 2 for more information about how to retrieve environmental impact statements for projects involving federal agencies or federal agency cooperation or oversight.

- Public Records. Finding information about people online grows easier as people increase their online transactions, social media interactions, and other services that collect personally identifiable information. As more of our personally identifiable information flows through government and private databases, and as we engage more and more in online activities, we create larger online footprints. Typical Google searches for people will turn up results for their social media pages (Facebook, Twitter, LinkedIn, etc.), as well as any webpages posted by their workplaces ("About Us" pages or employee directories) and any news stories or other publicity that may have been published about the person (relay race times, awards for hobbies, wedding announcements, etc.). These leads are a good start for gathering information about people.

There are also pay-by-search websites, like Intelius.com,[63] or Spokeo.com[64] that will provide additional information about people, including their past residences, their phone numbers and addresses, and other records that may not be freely available on the Internet. Some subscription services, like LexisNexis Accurint[65] provide in-depth dossiers on people, but they charge premium fees.

If you are searching for a person's criminal record, depending on whether the crimes were prosecuted in federal or state court, you may be able to find information on either federal or state websites. New York State, for instance, provides criminal records search results for a $65.00 fee.[66] Federal criminal press releases and court documents, including complaints or judgments, may be available on the U.S. Attorneys' website.[67]

There are several useful tips for narrowing searches for people using people's names. It is sometimes hard to search for people by name, especially common names. For example, putting the phrase "Mark Smith" into a search bar will likely retrieve search results about many different people, as many people have that name in common. To ensure that you are getting results for the correct Mark Smith, include their middle name or middle initial, or even their state of residence. Searching for "Mark C. Smith or Mark Charles Smith" from Texas, or even more specifically, from Lubbock, Texas, is likely to bring about better results than a broader search for "Mark Smith."

Also, if you are searching in a database or search engine that allows for Boolean search terms, you may want to put the name in quotation marks (searching only for the precise phrase "Mark C. Smith" or "Mark Charles Smith") to eliminate results where the word Mark

---

63. Intelius, *Homepage*, https://www.intelius.com/ (last visited Dec. 15, 2016).
64. Spokeo, *Homepage*, http://www.spokeo.com/ (last visited Dec. 15, 2016).
65. LexisNexis, *Accurint*, http://www.accurint.com/ (last visited Dec. 15, 2016).
66. NY Courts.gov, *Criminal History Record Search*, https://www.nycourts.gov/apps/chrs/ (last visited Dec. 15, 2016) ($65 fee and mailed/hand delivered to New York State Office of Court Administration, 25 Beaver Street, Room 840. Results e-mailed in 24-48 hours).
67. U.S. DOJ, *Offices of the United States Attorneys*, https://www.justice.gov/usao (last visited Dec. 15, 2016).

is close to the word Smith, and you also may want to place words in the name within close proximity of each other if you do not know the middle name of the person you are searching about—Mark w/3 Smith will retrieve any results where the word Mark is within three words of Smith, which leaves room for any number of middle names to fall into those search results.

- News Stories. News stories are helpful not just as current awareness tools, keeping us up to date on environmental events, but as sources of information about people, places, and companies. You can run searches in news databases for people and company names, or names of a location, like Yosemite National Park, in order to learn more about newsworthy events that are currently happening or have happened in the past. Westlaw, Lexis, and even Bloomberg Law, with their Bureau of National Affairs, each have large news databases. There are also free databases, like Google News[68] that allow you to search in tens of thousands[69] of news sources with a single search, free of charge.

  When searching in news sources, pay close attention to where the news outlets are located, as local sources tend to have more detailed information about situations or people in that location. For instance, local Detroit newspapers have much more in-depth, continuous coverage about the Flint, Michigan, drinking water crisis and the people and companies involved than the *New York Times*. If you are looking for information about a certain person or company, it can be useful to seek out the local hometown newspaper and television websites where the people reside or where the businesses have their center.

  If you are looking for older news stories, you may need to visit actual libraries to use their microfiche or microform collections. Most public libraries maintain some form of archives for local news sources and newsletters. You can contact a reference librarian to find out where local news archives are held—it may be a central library in a larger municipality, and not one of the branches. Alternately, you may be able to pay fees to access a newspaper's historical archives.[70] You can

68. Google, *News*, https://news.google.com/ (last visited Dec. 15, 2016).
69. Over 75,000 news sources, according to a recent post. James Moorehead, "Putting a Spotlight on Local News Sources," *Google News Blog*, May 9, 2016, https://news.googleblog.com/2016/05/putting-spotlight-on-local-news-sources.html.
70. The *New York Times* Article Archive provides articles from 1851 to digital subscribers. *New York Times*, New York Times *Article Archive*, http://www.nytimes.com/ref/membercenter/nytarchive.html (last visited Dec. 15, 2016). Websites like Newspapers.com allow you to search an aggregated collection of historical newspaper articles for a subscription fee. Newspapers.com, *Homepage*, https://www.newspapers.com/ (last visited Dec. 15, 2016).

find out whether a newspaper has searchable archives by contacting the subscriptions department of the newspaper's offices. Google News Archives maintains an online archive of some scanned copies of many historic newspapers, so that is also worth searching if you are seeking historic news materials.[71]

Blogs are also excellent sources for news information. Blogs are often not vetted and fact checked like official news sources, so it is important to assess the veracity of the information offered on blogs critically, keeping in mind issues like the truthfulness and impartiality of the sources. From an information-gathering standpoint, however, blogs often contain up-to-date, hyper-local information. For instance, blogs about local New York City neighborhoods like Long Island City, Queens, and Brooklyn Heights, Brooklyn, provide detailed information about local building projects, superfund and brownfield sites, and development plans, often written by people who see these locales every day and delve deep into specifics on what is going on with those projects and issues.[72] Searching for a person, place, or thing in Google with the word "blog" or "blawg" (a common moniker for legal blogs) after your search terms may retrieve relevant blogs, and there are also some search platforms specifically focused on blog searches.[73]

It is also useful, when researching people, places, and businesses, to consider their social media presences as sources of information. Because social media platforms have pre-set information fields (word limits, content type requirements, etc.), the content within those platforms is sometimes limited. Also, social media entries are usually self-reported and unedited, more of a casual conversation than a formal report, so they are harder to fact-check and verify than other sources of information, like official government documents or edited newspaper stories. Most businesses have some social media presence, whether it is a Facebook page or a Twitter feed, and most people have some social media presence, as well, although it may be hidden behind privacy walls. Some websites, like Pipl.com and Cvgadget.com, aggregate search results from various social media sites in a single online location. These websites make it easier to search multiple social media platforms at once for a person or business, but they do not guarantee complete results containing all social media site data.

71. Google, *News: All Newspapers*, https://news.google.com/newspapers (last visited Dec. 15, 2016).
72. For instance, LICKTALK, *Homepage*, http://lictalk.com/ (last visited Dec. 15, 2016) and Brooklyn Heights Blog, *Homepage*, http://brooklynheightsblog.com/ (last visited Dec. 15, 2016).
73. BlogSearchEngine.org, *Homepage*, http://www.blogsearchengine.org/ (last visited Dec. 15, 2016).

When searching like a reporter on social media it is important to respect your ethical and professional obligations, especially if you are an attorney who is bound by rules of professional conduct. In any type of journalistic-styled research, legal practitioners must avoid using pretext or dishonesty to convince people to provide social media access. Lawyers must also keep their client's confidentiality in mind.

While you may access people's accounts and posts on social media sites like Facebook and Twitter, you may not lie about who you are, pretend, or otherwise trick anyone into providing you access to information. For instance, the New York City Bar Association's ethics opinion on obtaining evidence from social media websites clarifies that "[a] lawyer may not attempt to gain access to a social networking website under false pretenses, either directly or through an agent."[74] Generally, anything that the social media user makes publicly available is okay to use without violating ethical rules about pre-textual searches or getting information using false pretenses. In most cases, you can send friend requests without lying (using your real name and identity) and also without disclosing the reasons for making the friend request. If, however, you or anyone you know (like your coworker, family, or friends) tries to get evidence about a legal matter you are working on from social media using false pretenses, you have likely violated your ethical obligations as an attorney.

Another thing to be aware of, from an ethics standpoint, is violating client confidentiality. Some social media sites like LinkedIn tell people when you have accessed their online profiles, which may give people notice that something is going on related to litigation, or to your client. To be safe, if you are an attorney, it is best to use formal discovery devices (subpoenas, etc.) rather than engaging in "Google stalking," or doing undercover research on social media platforms.

IV. Searching Nongovernmental and Nonprofit Organizations

Although most of this book is dedicated to obtaining environmental information from federal and state governments, there are many nongovernmental sources of environmental information that can greatly enhance and shape effective decisionmaking about environmental issues. Sometimes, great

74. New York City Bar, *Committee Report: Formal Opinion 2010-02: Obtaining Evidence From Social Networking Websites*, Oct. 19, 2010, http://www.nycbar.org/member-and-career-services/committees/reports-listing/reports/detail/formal-opinion-2010-02-obtaining-evidence-from-social-networking-websites.

sources of environmental information are created for nonprofit organization campaigns and programs instead of governmental or academic sources. For example, Earthjustice, an environmental law nonprofit organization, provides an interactive map of U.S. coal ash contamination sites.[75]

The volumes of data and information gathered, organized, and often artfully presented by private entities should not be ignored when considering the spectrum of environmental information access. It can usually be found on organization websites. For instance, pro-environmental nonprofit organizations like Sierra Club, the Natural Resources Defense Council (NRDC), National Audubon Society, and the National Wildlife Federation all have very well-developed websites well-stocked with information.

Research for environmental information that comes from nonprofit organizations most often begins with a simple Google or other search engine search for the relevant search terms, like "climate change," or "fracking." In many instances, nonprofits' websites divide their information into particular advocacy campaigns,[76] and the information associated with each campaign is geared toward convincing others to join their campaign by either engaging in political activism (interacting with legislators, attending a public event) or to donate funds toward a certain cause. This means that the information may be accurate, but also may be biased or drafted to favor a particular point of view.

Thus, when looking at non-academic,[77] nongovernmental sources, it is important to remember the four factors by which to evaluate the sources you find: (1) identity of the author; (2) objectivity of the author; (3) accuracy of information; and (4) currency of information.[78] Many nonprofit organizations have a specific viewpoint (aka, pro-environmental conservation versus non-belief in climate change, etc.) and all of their online content is drafted with the goal of convincing viewers to join their campaigns. Some of this content may not be drafted by experts in the field, and may not utilize verifiable facts and figures. It may even contain inaccurate data. Also, the information might not be current—organizations that may be defunct and campaigns that may be long concluded may still have long-untouched webpages languishing on the Internet. Even environmental information created by the government should be judged this way by asking questions like:

---

75. Earthjustice, *Map Feature: Coal Ash Contaminated Sites & Hazard Dams*, http://earthjustice.org/features/map-coal-ash (last visited Dec. 15, 2016).
76. Some of NRDC's topics include "climate," "health," "energy," and "food." NRDC, *Homepage*, https://www.nrdc.org/ (last visited Dec. 15, 2016), while Sierra Club's campaigns are called things like "Move Beyond Fossil Fuels" and "Preserve Our Wild America." Sierra Club, *Homepage*, http://www.sierraclub.org/#! (last visited Dec. 15, 2016).
77. Aka sources that are not peer reviewed, cited, and researched.
78. The four factors are described in Chapter 4.

Was there bias in information collection? Are the agencies under corporate capture? Is there a political agenda or deadline to press? Critical literacy is especially necessary when looking at information often at the heart of controversy, which environmental information often is.

When considering nonprofit and other nongovernmental organizations, there are several key ways to check these four factors. The first place to check the objectivity and veracity of an organization's online content is to visit the "About Us" link on the organization's website. That link should provide you with basic information about the goals and purposes of the organization, including whether the organization is partisan, funded by any external sources, or affiliated with any academic institutions.

Beyond the organization's own website, you can learn more about the nongovernmental information sources by searching for news stories in Google News or some other news aggregator. News searches will provide you with a glimpse into what kind of activities and issues the organization is engaged in. For example, a search for The Nature Conservancy in Google News reveals news stories about counties asking the organization to pay for repairs on certain lands, and to take oversight of some federal lands, as well as local news updates about various events that the organization has planned. These stories can provide insight into the kind of organizational activities a nonprofit takes part in and its level of prominence or notoriety.

You can also use news searches and scholarship searches like Google Scholar to see who uses the organizations' information and data, and in what ways. If academics, who are subject to peer review, rely on an organization's data, it may indicate a higher level of veracity. You can also find different data sets with which to compare the organization's data, to see whether it seems in line with conventional wisdom. Additionally, you can search for online author biographies and updated versions of older documents in order to learn more about authors' expertise and objectivity and to make sure you are dealing with current information.

V. Access to Scientific and Academic Research

Politicians, and those influencing politics, often rely on academics to prepare studies to inform their decisionmaking activities. In one dramatic example of the relationship between scientific information and political interests, a recent exposé highlighted the intertwined nature of academic studies and political activity, revealing that Princeton and Penn State professors were willing to be paid by fossil fuel companies to draft papers promoting the benefits of carbon dioxide and the use of coal in developing countries, even though those concepts disagree with scientific realities.[79] For instance, another recent exposé revealed that Coca-Cola invested millions of dollars on an organization called Global Energy Balance Network that underwrites scientists' efforts and research showing that people are overly concerned with how much they eat and drink, and placing too much of the blame for their health issues on sugary drinks.[80]

Academic institutions are centers of scientific and social policy innovation, teeming with students and faculty conducting experiments, fieldwork, and research that lead to groundbreaking environmental concepts and progress. The halls of academia are brimming with environmental information. Some environmental information is the work of single professors or teams of students, but some academic centers house entire schools, think tanks, and organizations dedicated to environmental information development and dissemination. For example, Yale's Program on Climate Change Communication is composed of social scientists who study the causes and consequences of public opinion and behavior related to climate change, helping governments, media, companies, and other groups to communicate about climate change more effectively.[81] This academic organization has regular newsletters and radio programs containing environmental information.

Bias and objectivity tend to be lesser issues in academic environmental information, which must often undergo rigorous testing and peer reviews before publication, than in other types of nongovernmental environmental information. Although these information sources tend to lack corporate or advocacy agendas, it is useful to know who funded the studies and data collection. Academic studies and writing are often underwritten with grants and funding, and those sources can, in extreme cases (like the one discussed at the beginning of this section) influence the environmental information that is ultimately published.

There are several excellent resources for finding online academic research and data. Google Scholar is a free search engine focused on academic papers

79. Lawrence Carter & Maeve McClenaghan, "Exposed: Academics-for-Hire Agree Not to Disclose Fossil Fuel Funding," *Energydesk Blog*, Dec. 8, 2015, http://energydesk.greenpeace.org/2015/12/08/exposed-academics-for-hire/.

80. Anahad O'Connor, "Coca-Cola Funds Scientists Who Shift Blame for Obesity Away From Bad Diets," *N.Y. Times* Well Blog, Aug. 9, 2015, http://well.blogs.nytimes.com/2015/08/09/coca-cola-funds-scientists-who-shift-blame-for-obesity-away-from-bad-diets/.

81. Yale.edu, *Climate Change Communication*, http://climatecommunication.yale.edu/ (last visited Dec. 15, 2016).

and work, and there are many academic databases that are available for free in numerous libraries including JStor, EBSCO, and ProQuest. There are some open access online resources for scientific studies: PubMed[82] is the U.S. National Library of Medicine within the National Institutes of Health and it provides open access to many environmental studies and articles. The National Science Foundation (NSF's) awards database[83] contains thousands of papers and abstracts written under NSF grant awards.

VI. Searching for Grey Literature: How to Power Up Google Searches

Since grey literature is not centrally located online in any one depository, you will likely be relying heavily on Google and other similar search engines to find the items discussed in this chapter. When searching on the Internet, not all searches are created equal—there are ways to optimize your results by crafting better searches with more precise search terms and Boolean search tools. Google has a guide to its search operators.[84] Here are some specific ways to improve your search engine searches to create better Google searches:

Search Operators	Functions
-	Exclude results containing words/sites
*	Wildcard place holder
""	Put words in particular order
..	Indicate a range of numbers
#	Indicate media trend
Site:	Search within particular website
Filetype:	Search for particular type of file
OR	Search results will contain either term
define	Locate definition for search term

Google allows its users to use search operators, or words and symbols that can be added to searches to help narrow down specific types of results. Symbols can also trigger certain types of search specifications. Adding a dash symbol (-) before a word or website URL will exclude results containing those words or sites. To search for a phrase, or group of particular words in a particular order, put the phrase in quotation marks, like this: *"Flint*

82. PubMed.gov, *Homepage*, http://www.ncbi.nlm.nih.gov/pubmed (last visited Dec. 15, 2016).
83. NSF, *NSF Award Search*, http://www.nsf.gov/awardsearch/ (last visited Dec. 15, 2016).
84. Google, *Search Operators*, https://support.google.com/websearch/answer/2466433 (last visited Dec. 15, 2016).

water crisis." Use asterisks, the "wildcard" place holder, in place of words for unknown search terms, or terms that could be more than one thing. A search for *"global * change"* locates results for "global temperature change," "global climate change," and "global environmental change." To search for a range of numbers, place two periods between those numbers (..): *50..100* will find results containing any of the numbers from 50 through 100. If you are looking for a specific social media trend, you can use its hashtagged name in a Google search to retrieve more leads on the trend. A Google search for *#flintwatercrisis* retrieves a collection of news and social media-generated data on the topic.

There are also words that act as search operators. To search a particular website, add "site:" to your search without adding spaces between the colon and the web address. For example, to find a report on water quality from EPA's website, add *site:epa.gov* to your search. Similarly, if you add "info:" before the URL, you will retrieve information about the website, and "cache:" will retrieve the version of the web address the last time Google visited the site. Searching using the "filetype:" operator before a certain type of file name will retrieve only results in that format. Searching for *filetype:PDF* is especially useful for finding court documents. A search for *Exxon court tort filetype:PDF* retrieves a collection of litigation documents containing those terms. Similarly, to search for particular types of websites (for instance, to search just for government websites, or .gov sites), use the "site:" operator (*site:gov*, for example). Make sure not to add a space between the search operator and the website address in order to get the correct results.

Beyond the colon-separated operators, the "OR" operator also works in Google searches, so adding OR to your searches will retrieve search results that have either one term or another. A search for *wetland OR marsh* will provide results for online resources containing either of those terms. Type the word "define" before a word to obtain its definition. Finally, you can find unit conversions online by simply putting in the measurement and then typing the unit to which you would like to convert it. Entering the search *3ml to oz* retrieves a unit calculator that shows how many U.S. fluid ounces three millimeters equals.

Google has advance search options beyond its search operators.[85] Google's advanced search has similar features to most database advanced search options, allowing you to run searches for all words ("and" searches), exact words (phrase searches), any words ("or" searches), and searches that exclude certain words, as well as ranges of numbers. You may also narrow

85. Run a search for "Google Advanced Search" or simply go to https://www.google.com/advanced_search.

your searches by language, geographical region of publication, pages updated within a specified time, pages from a certain website or domain type (.edu, .org, .gov, etc.), and file type (PDF, Excel, Word, etc.). The advanced search page also allows you to find pages that are similar to a particular URL, search only in pages you have already visited (this is especially useful if you are trying to revisit a website and do not remember its URL), and customize your search settings for future searches.

There are also several search tools that are not contained in the general Google search bar or in the advances search options. Clicking on the "Search Tools" link at the top of the Google search page opens up options to limit your search results to a certain period of time or to isolate your results to those that contain your search terms verbatim.

Beyond these special search options, there are some general tips to keep in mind when conducting searches on any search engine or within any online database. One is to consider your search terms carefully. Before going online and typing in words, take a moment to consider which words are likely to appear in the results you seek. If you are looking for documents relating to a statute about groundwater, but the statutory term for groundwater in the jurisdiction is "subsoil water," be sure to search for "subsoil water" to find the related documents. If you are not sure what term a jurisdiction uses, add synonyms to your search. For instance, if you are looking for a state law about dumping waste, you may include the term "waste" and also similar terms like "trash," "litter," and "garbage" into your search. Especially when searching for law and policy materials, is important to be aware of the language used by the official sources of law.

In addition to using the proper terminology, remember to be specific when creating an online search query. If you are looking for information about local superfund sites, be sure you specify the correct geographic location for the data you seek. Finally, pay attention to search requirements for particular sites—search tools require varying degrees of sophistication and permit different Boolean search terms.

You can also look at old versions of websites. There are online "web archives" that catalog cached versions of websites. These historical webpage resources are useful if you want to look at an old webpage that is now defunct, or you want to find information that used to be online, but is no longer available, or to see what was or was not on a particular webpage at a certain point in time.[86] Google saves cached pages periodically, and the Internet Archive's Wayback Machine allows you to type in almost any URL and select cached versions from various points in time, as marked on a calendar.[87] Using these tools can help you find old agency guidance that is no longer linked on agency websites, information about organizations that may now be disbanded and old views of academic and nonprofit organizations' online presences. The archeology of the Internet is fairly new, but already contains a deep well of historical information, documents, and records.

VII. Conclusion

Grey literature, the hard-to-define collection of materials that are not official government records, and include a vast array of materials ranging from corporate documents, in-house journals and newsletters, technical reports, conference proceedings, and more, are sometimes overlooked by legal researchers.[88] These materials are full of environmental information, including scientific data, hyper-local events, and trade-specific knowledge. While the legal field has "struggled to develop a relationship" with grey literature,[89] environmental information seekers should explore grey literature resources, because they provide scientific, technical, and background facts, figures, and illustrations that enrich understandings of environmental issues and consequences.

86. Old versions of websites were very important after the relevant government information venues were shut down in the wake of the federal government sequestration in 2013. *See* WhiteHouse.gov, Impacts and Costs of the October 2013 Federal Government Shutdown (2013), https://www.whitehouse.gov/sites/default/files/omb/reports/impacts-and-costs-of-october-2013-federal-government-shutdown-report.pdf.
87. Internet Archive, *WayBack Machine*, https://archive.org/index.php (last visited Dec. 15, 2016).
88. Taryn L. Rucinski, *The Elephant in the Room: Toward a Definition of Grey Literature*, 107 L. Libr. J. 543, 544-45 (2015).
89. *Id.* at 545.

Chapter 9: State and Local Environmental Information Access

Law in Action

Ted Clark recently moved to Dimock, Pennsylvania, to teach at a local school. Ted had been warned about historical fracking incidents in Dimock that led to water contamination in local water wells, and even an explosion in one of the wells.[1] *While he was assured that the fracking issues have long passed, and that the water in his well is safe to drink, he would like to see some proof to ease his mind and help him determine whether he needs an alternate water source. He knows he can have his water tested, but he would also like to know what kinds of fluids go into the fracking chemicals that the nearby oil extraction projects inject into the ground. He would like to know whether that kind of information is available to the public in Pennsylvania, as he has heard fracking policies and requirements vary from state to state.*[2]

I. Introduction: Overview of State Environmental Transparency Schemes

As the federal Freedom of Information Act (FOIA) requires federal agencies to provide public access to their records, states have similar statutes to create public access points for government records. Like the federal government, state governments gather huge amounts of information.[3] Nevertheless, because state governments and their various governing bodies differ from those of the federal government, the types of information collected by states differ from those collected by the federal government. States not only collect more particularly local information, they also create and

1. Jad Mouawad & Clifford Krauss, *Dark Side of a Natural Gas Boom*, N.Y. Times, Dec. 7, 2009, at B1, http://www.nytimes.com/2009/12/08/business/energy-environment/08fracking.html.
2. Brad Plumer, *How States Regulate Fracking (in maps)*, Wash. Post, July 16, 2012.
3. An article discussing government data access in the state of California states: "Government agencies collect and archive a staggering amount of data every day on everything from traffic patterns to budget allocations, and usually that data, once it serves its immediate purpose, is stored away in databases and digital archives to collect virtual dust." Jason Barrett, "The Secret Password for California? Open Data," *Milken Institute Blog*, Mar. 20, 2015, http://www.milkeninstitute.org/blog/view/723.

enforce laws and policies under the police power granted to states by the U.S. Constitution's Tenth Amendment. Those police powers allow states to govern certain aspects of health, safety, and general welfare not delegated to the federal government, and different data sets are collected as those laws are carried out. Those types of laws go beyond criminal policing laws and include zoning, land use, building codes, discrimination, traffic, schooling, and sanitation laws. Thus, state records vary substantively from federal records in many instances.

The types of records that information seekers can get through state Freedom of Information (often called FOI) laws are different than the types they will get through the federal FOIA law. Thus, it is often useful to utilize state and local information access laws on top of the federal transparency laws. Also, state transparency laws often grant access to some state legislative records, which is something the federal FOIA does not do.

State FOI laws differ from the federal FOIA, and so do their exemptions. On a state level, common transparency law exemptions include records containing information like personally identifiable student information, acquisition of real property or water rights, trade secrets of public hospital boards, library records, medical treatment forms, reference letters, adoption records, and other more specific types of information often collected pursuant to laws established under states' police powers. The particularities of state FOI exemptions reflect the localized, micro nature of state laws and regulatory information collection in comparison with the much broader sweeping federal information collection scheme.

Not only are state FOI laws different, in some ways, from the federal FOIA, state FOI laws also vary from one another. Each state's FOI scheme and transparency law implementation methods are unique, and no two are exactly the same, although there are many parallels and themes between the states' assorted information access laws. For example:

- Some states have offices or departments within the state attorney general's office dedicated to educating the public and government entities about how to use and obey state records access laws. Sometimes, these attorney general's office departments are even tasked with hearing and mediating open meetings and open records act complaints. Other states have no such office, or have instead created statutes and regulations requiring other agencies in charge of finances, state contracts, or other tax-dollar-heavy operations to have their own transparency mechanisms and online portals. Some states' open meetings and open records laws establish ombudsmen, boards, committees, etc. to oversee and ensure the efficacy of FOI proceedings and processes in the states.

- Definitions of key terms in transparency laws vary from state to state. In some states, the definition of "meeting" covers any gathering of three or more members of a public body,[4] and in other states, meetings are only "meetings" under open records law if there is a quorum.[5] Some states create a wide scope of access for a variety of "governing bodies" and others adopt a stricter definition of agencies that must follow FOI and open meetings processes. Some states even go as far as to include nongovernmental entities that receive public funding or have public officials as members in the scope of their transparency laws.

- Several states combine open meetings and open records laws into a single statutory scheme, often called the state's Sunshine Law or FOIA, and others separate the two access points—access to meetings and access to agency records—into different parts of the state's code.

- Different states have different limitations on who can file FOI requests. Some states do not allow non-state citizens, incarcerated persons, or minors to make records requests. Similarly, some states require requesters to be citizens of the state, and some do not.

- Some states prohibit records requesters from using the materials they retrieve for certain activities, forbidding the use of FOI records for commercial, political, or litigation purposes.

- States have a variety of rules about what kinds of fees should be paid for FOI records, and what the limitations are on charging fees. While most states require that fees not exceed the minimal costs of providing copies of public records, some laws grant fee permissions for mailing costs and the personnel and overhead costs for providing copies. Many states also allow for fee waivers. Additionally, some states designate representative document custodians charged with upholding records access provisions within the agencies that fall under the scope of the state's FOI. Many state agencies, including environmental agencies, will have a special records access web page with online request mechanisms and e-mail contacts for FOIA requests.[6] In fact, the New York state Department

4. Like in Colorado's Sunshine Law.
5. Connecticut's law or the District of Columbia's law.
6. E.g., Department of Energy & Envtl. Prot., Ct.gov, *Freedom of Information Act (FOIA) Requests*, http://www.ct.gov/deep/cwp/view.asp?a=2688&Q=502686&deepNav_GID=1511 (last visited Dec. 18, 2016).

of Conservation created an e-mail service to send subscribers updates about various state conservation topics.[7]

- Most states have procedures set out for appeals to open records requests, with deadlines and particular courts or agencies with which to file the appeals. Some states even have administrative processes established for mediating appeals outside of a court setting.
- The sanctions for violating state transparency laws vary, and there are sometimes civil fines and even criminal sanctions attached to breaking a state's transparency laws.
- Some state FOI laws have special provisions for electronic records. E-mail is sometimes specifically listed as a type of record that falls under state transparency laws, and many laws explicitly state that any form of record, including electronic formats, fall under the scope of state transparency laws. Some open meetings laws also require online notice of meetings. Other state FOI laws require records requests to be fulfilled electronically when requested.

State legislatures often deviate from standards set by federal laws in their own legislation, making state laws with different contours than their federal counterparts, and state FOI laws are no exception to this practice. The scope of the transparency provisions in state information access laws are sometimes more broad or narrow than the federal information access laws, and they are also reactive to state issues. For instance, the Michigan legislature introduced a bill adding the state legislature and non-agency executive bodies, including the governor's office, to the scope of the state's records disclosure law in the wake of the Detroit water crisis and the related difficulties accessing the governor's e-mails regarding water safety.[8] So while state transparency laws are very similar to their federal counterparts, there are unique qualities to each state's open meetings and records statutes.

State level open government laws are common mechanisms for environmental information access: in 2016 in New Mexico, almost half of the state's 8,654 Inspection of Public Records Act (IPRA) requests were filed with

---

7. NYSDEC Email Subscription Service, https://public.govdelivery.com/accounts/NYSDEC/subscriber/new?preferences=true#tab1 (last visited Jan. 3, 2017).
8. Ted Roelofs, *Bipartisan Bill to Make Governor, Lawmakers Subject to Open Records Law in Doubt* (May 26, 2016), http://bridgemi.com/2016/05/bipartisan-bill-to-make-governor-lawmakers-subject-to-open-records-law-in-peril/.

either the Environment Department of the Department of Public Safety by environmental information seekers.[9]

In addition to state level environmental information access, local government (municipalities, villages, etc.) may have other rules and ordinances that supplement state laws and regulations.[10] On a municipal level, several major cities like New York[11] and San Francisco[12] have created frameworks for making government data publicly accessible. Various municipalities do various things—some do the bare minimum while others develop major transparency systems—for instance, Arlington, Virginia, now posts their responses to FOIA requests, along with the original requests, on an open access website,[13] and the town of Hempstead, New York, has an online portal for open meetings notices and access.[14]

II. Environmental Information Access in State Laws

Due to the variations in state law transparency schemes, there is little uniformity in what types of environmental information are available through transparency laws from state to state. Beyond federal environmental law provisions, each state and municipality has its own added layer of environmental laws requiring access to various bits and pieces of environmental information depending on the jurisdiction.

State laws do things like require state agencies to maintain data on and inform the public about air pollution "hot spots"[15] and collect data on the types and amounts of toxins used by certain companies.[16] Depending on geographical location and pollution type, a variety of information access points await the environmental information seeker. This chapter provides an extensive, though incomplete view of some of these laws and rules. There

9. New Mexico Foundation for Open Government, *FOG Conducts IPRA Audits of Counties and State Agencies* (May 17, 2016), http://nmfog.org/2016/05/fog-conducts-ipra-audit-of-counties-and-state-agencies/.
10. For instance, certain types of disclosure obligations often attach to local permitting ordinances like: Newburyport, Massachusetts, Code of Ordinances §XV-E, Materials for Review (requiring pollution impact statement for project permit); and Sonoma County, California, Code of Ordinances §26A-09-010 (general standards for mining permit and operations).
11. New York City's Department of Records and Information Services and its tech department have created a government publications portal containing thousands of New York City records, http://a860-gpp.nyc.gov/ (last visited Dec. 18, 2016).
12. The city of San Francisco has a city Sunshine Ordinance in Chapter 67 of its administrative code (http://www.sfbos.org/index.aspx?page=4459) and the city created an online portal for government records from its agencies (http://index.sfgov.org/#/home (last visited Dec. 18, 2016).
13. *See* News Release, Arlington Now Posting FOIA Responses Online (June 1, 2016), https://newsroom.arlingtonva.us/release/arlington-now-posting-foia-responses-online/.
14. Town of Hempstead, Long Island, New York, *Homepage*, https://toh.li/ (last visited Dec. 18, 2016).
15. California hot spots law, Health & Safety Code §44301 (West 2016).
16. Mass. Gen. Laws Ann. ch. 21I, §§1-23 (West 2016).

are laws governing very local aspects of pollution like sewage data reporting, dry cleaner chemical use, and other particular point source pollution types often found in municipalities. These laws require data reporting, and that data can often be retrieved from state and local environmental agencies. These state and local reporting requirements are too numerous to list in this chapter.

Because states are more localized than the federal government, they are better positioned to monitor polluters and the health and welfare of municipalities and communities. In fact, before 1970, environmental issues were seen as strictly local issues to be dealt with by states and municipalities. For instance, in 1960, President Dwight D. Eisenhower called water pollution a "uniquely local blight," declining to take further national action to clean up sewage in local waterways, claiming that "primary responsibility for solving the problem lies not with the Federal Government but rather must be assumed and exercised, as it has been, by state and local governments."[17]

The groundswell of environmental concerns in the 1970s (discussed in Chapter 2), led to a less state-driven and more federally focused environmental law scheme in the United States. The creation of the U.S. Environmental Protection Agency (EPA) and its many statutory and regulatory programs led to the current system, where the federal government oversees and conducts the majority of environmental programs in the nation, and baseline environmental standards are no longer determined by states.

Federal environmental governance, however, still recognizes the superior vantage point of state agencies to monitor polluters, and many federal environmental laws give states the ability to take on pollution controls and monitoring with federal oversight in a "cooperative federalism" framework where the federal government gives states primacy to carry out federal environmental provisions.[18] State programs are often the state iteration—implementation and enforcement—of federal environmental programs, and state environmental agencies are sometimes authorized to carry out and enforce federal law programs.[19] As a Texas environmental law primer explains: "Over the years, states have gradually assumed 'primacy' in more and more areas. They now have authority to implement and enforce the significant

17. 106 Cong. Rec. 16729, 16736 (1960); H.R. Rep. No. 86-346, at 1 (1960), *reprinted in* 1960 U.S.C.C.A.N. 1542, 1542-543.
18. 45 Texas Environmental Law §4:2 (2015/2016 ed.) (Texas Practice Series).
19. *Id.*

majority of the major federal environmental programs that can be delegated to them."[20]

This cooperative federalism, and the many parallel state and federal environmental programs that result from this shared environmental law scheme, lead to many state environmental information access laws that mirror their federal counterparts. For example, many states carry out clean air and water programs to meet the requirements of the federal Clean Air Act and the Clean Water Act. So oftentimes, environmental information will end up in state environmental agency files.

Additionally, many states have implemented environmental statutes that are like those created by the federal government, including "NEPA-like" provisions requiring environmental impact reports for proposed legislation, regulation, and government activities. For instance, California requires impact reports for projects proposed by its state agencies in the following NEPA requirements[21] and New York's State Environmental Quality Review Act SEQRA similarly requires New York's state and local agencies to conduct environmental assessments (and draft environmental impact statements) for proposed project with the goal of identifying and mitigating their environmental impacts.[22] Many states and localities also have versions of toxic "right-to-know" laws and ordinances that are very similar to the federal Emergency Planning and Community Right-To-Know Act (EPCRA) statutory scheme.[23]

In addition to state environmental information laws that parallel federal law schemes, some states go beyond the minimum thresholds created by federal laws. States create additional public notification requirements to inform the local populace about potentially dangerous environmental situations. Many states have laws like New York's Sewage Right-To-Know Act,[24] which requires public notification when any raw or partially treated sewage is discharged into state waterways. There are also state laws that require public notice for pesticide application.[25] States also create special statutes to facilitate and incentivize self-disclosure by environmentally regulated entities.[26]

---

20. *Id.*
21. Cal. Pub. Res. Code §§21100,1 (West 2016).
22. N.Y. Envtl. Conserv. Law §§8-0101 to -0117 (McKinney 2016).
23. For instance, New York City (tit. 24, ch. 7, 24-701 to -718) and Cleveland (Ordinance ch. 393) are both cities with Right-To-Know Ordinances.
24. *See* §17-0826-a.
25. Illinois Lawn Care Products Application and Notice Act, 415 Ill. Comp. Stat. 65/1 to /9 (2016).
26. *See, e.g.,* Ark. Code Ann. §§8-1-301-8-1-312. Arkansas's DEQ also has a "Self Disclosure Incentive Policy," https://www.adeq.state.ar.us/legal/pdfs/env_self_disclosure_incentive_policy.pdf.

States also create special environmental information access laws for environmental issues unique to their state. In Virginia, a state where many homes have historically used petroleum tanks for fuel, the state collects extra data on petroleum tank storage and has designated a fund for petroleum tank cleanup to minimize environmental harm from petroleum tanks.[27] The state's Department of Environmental Quality (DEQ) maintains a petroleum database that the public can download containing information about petroleum releases and registered tanks.[28]

These state and local laws are present in many state's statutory and local ordinance schemes, yet they are often difficult to locate and compile, as they tend to be scattered throughout a state or municipality's legislative code, and access to local codes online varies greatly throughout the United States.

III. Environmental Information Access: State-by-State

Here is a summary of state information access laws.[29] The basic open meetings laws and public records access laws for each state are provided in this chapter, as well as some additional statewide environmental information access provisions. Be aware that state laws are often created and amended in a faster, more streamlined manner than federal laws, so there may be even more recent laws and amendments to existing state laws that have not yet been incorporated into this state-by-state survey. Note also that most agencies have promulgated regulations that create agency-specific processes for complying with the state's FOI laws similar to the federal records access regulations of federal agencies.[30] Each of these statutes is further construed and explained in state agency regulations and guidance, as well as state court cases interpreting these laws. Such legal authority supplements the requirements of these statutes.

27. Carol Hazard, *State Fund Helps Pay to Mop Up Leaking Oil Tanks*, Richmond Times-Dispatch, Nov. 13, 2011.
28. Information on how to download the petroleum database is available here: Virginia DEQ, *Files & Forms*, http://www.deq.virginia.gov/Programs/LandProtectionRevitalization/PetroleumProgram/FilesForms.aspx (last visited Dec. 18, 2016).
29. Some online organizations also provide information about state FOI laws. The National Freedom of Information Coalition (NFOIC) provides links to each state's FOI laws. NFOIC, *State Freedom of Information Laws*, http://www.nfoic.org/state-freedom-of-information-laws (last visited Dec. 18, 2016). Each state's link contains statute citations for the relevant open meetings and public records acts, as well as a sample FOIA request for each state. It also lists the exemptions for each statute, which vary slightly from state to state. The Reporters Committee for Freedom of the Press has a useful *Open Government Guide* with even more detail on specific transparency statutes in each state: http://www.rcfp.org/open-government-guide (last visited Dec. 18, 2016).
30. For instance, the New York Department of Environmental Conservation's regulation, 6 NYCRR §§616.1 et seq., sets forth the specific information access rules for the agency.

Alabama

Alabama Open Meetings Act: Ala. Code §§36-25A-1 to -11.[31]
Alabama Public Records Law: Ala. Code §§36-12-40 to -42.

Alabama's Public Records law exempts records that could reasonably be expected to be detrimental to the public safety or welfare or against the public's "best interests." It also requires that for requests relating to "critical infrastructure," the infrastructure owner must be notified in writing of the request and be given time to comment on the request about any related public safety threats that could arise as a result of disclosing the records.[32]

Alaska

Alaska Open Meetings Act: Alaska Stat. §§44.62.310, .62.312 & .62.319.[33]
Alaska Public Records Act: Alaska Stat. §§40.25.110 to .25.295.

The Department of Commerce, Community, and Economic Development maintains a webpage containing primers and information about implementing the Open Meetings Act (OMA) in Alaska. The Alaska Public Records Act encourages public agencies to provide their information in "usable electronic formats" whenever feasible.[34]

Arizona

Arizona Open Meetings Law: Ariz. Rev. Stat. §§38-431 to -431.09.[35]
Arizona Public Records Law: Ariz. Rev. Stat. Ann. §§39-121 to -128.

Under the Arizona Open Meetings Law, if a court finds that a public officer intentionally evaded the statute, the person may be removed from office.[36] A recent court case determined that under Arizona Public Records Law, digital metadata attached to electronic files are also subject to disclosure.[37]

---

31. Alabama Open Meetings Act, *Homepage*, https://www.openmeetings.alabama.gov/generalpublic/publicdefault.aspx (last visited Dec. 18, 2016).
32. Ala. Code §36-12-40 (2016).
33. Alaska, Department of Commerce, Econ. & Community Dev., *Local Government Online*, https://www.commerce.alaska.gov/web/dcra/LocalGovernmentOnline/LocalGovernmentElectedOfficials/OpenMeetingsAct.aspx (last visited Dec. 18, 2016).
34. Alaska Stat. §40.25.115 (Michie 2016).
35. Open meetings law website: Arizona Attorney General, *Open Meeting Law*, https://www.azag.gov/sgo (last visited Dec. 18, 2016).
36. Ariz. Rev. Stat. Ann. §38-431.07 (West 2016).
37. Lake v. City of Phoenix, 222 Ariz. 547, 218 P.3d 1004 (2009).

Arkansas[38]

Arkansas Open Meetings Law: Ark. Code §25-19-106[39]
Arkansas Freedom of Information Act: Ark. Code §§25-19-101 to -19-110

Arkansas Open Meetings Law is part of its state FOIA. Violating any part of the Arkansas FOIA is a Class C misdemeanor.[40]

California

California Open Meetings Act[41] (Ralph M. Brown Act, Bagley-Keene OMA, Legislative Open Meetings Law): Cal. Gov't Code §§11120 to 11132 (Bagley-Keene), Cal. Gov't Code §§54950 to 54963 (Ralph M. Brown Act), Cal. Gov't Code §§9027 to 9031 (Legislative Open Meetings Law).
California Public Records Act: Cal. Gov't Code §§6250 to 6277

California has a constellation of three open meetings laws: (1) the Ralph M. Brown Act governs open meetings for local government bodies; (2) the Bagley-Keene OMA governs all statewide boards and commissions, including administrative agencies; and (3) the Legislative Open Meeting Law governs state and assembly committee, subcommittee, and conference committee meetings. The Bagley-Keene OMA governs state agencies, and the Ralph M. Brown Act parallels the state-level open meeting law, enforcing similar open meeting requirements on local government agencies.[42]

The California Public Records Act has a single, catchall exemption clause in §6255 of the law[43] that allows agencies to withhold records in any situation where the "public interest served by not disclosing the case clearly outweighs the public interest served by disclosure." Also, in addition to California's Public Records Act, the state adopted Proposition 59 in 2004, which adds an affirmative right to information in the state constitution, so that Article I, §3(b) of the constitution now reads: "The people have the right of access to

38. Arkansas' open meetings law and state FOIA are part of the same statutory scheme. The state's attorney general has a website dedicated to the law and publishes an annual FOIA handbook after every regular session of the General Assembly: Arkansas Attorney General, *FOIA*, http://arkansasag.gov/media-center/foia/ (last visited Dec. 18, 2016).
39. A statute within the larger FOIA statutory scheme, Ark. Code Ann. §§25-19-101 to -19-110 (Michie 2016).
40. *Id.* §25-19-104.
41. The California State Department of Justice has a website with links to handbooks on the Brown Act and Bagley-Keene Open Meeting Act.
42. Memorandum from Dorothea Johnson, Legal Affairs Division, California Department of Consumer Affair, to Executive Officers et al. (Jan. 1, 2015) (concerning the Bagley-Keene Open Meetings Act).
43. Cal. Gov't Code §6255 (West 2016).

information concerning the conduct of the people's business, and, therefore, the meetings of public bodies and the writings of public officials and agencies shall be open to public scrutiny."[44]

California's Air Toxics "Hot Spots" Information and Assessment Act of 1987[45] requires facilities that may release hazardous substances into the air to submit emissions inventory plans to local air quality management officials and report chemical releases into the air. The state's Air Resources Board must make all of the emissions data collected under the "Hot Spots" Program available to the public.

Colorado

Colorado Sunshine Law: Colo. Rev. Stat. §§24-6-401 to -6-402
Colorado Open Records Act: Colo. Rev. Stat. §§24-72-200.1 to -72-206

The Colorado Sunshine law proscribes open meeting requirements for any state public body where public business is discussed. Colorado's Open Records Act (CORA) has a long-standing electronic mail policy requirement for agencies, and since 1997, CORA has required state bodies to adopt written policies on monitoring electronic mail communications.[46]

Connecticut[47]

Connecticut Open Meetings Law: Conn. Gen. Stat. §§1-225 to -232 (this section is within the Connecticut FOIA statutory scheme)
Connecticut Freedom of Information Act: Conn. Gen. Stat. §§1-200 to -259

In Connecticut statutes, the open meetings provisions are located in the state's FOIA.[48]

44. Cal. Const. art. I, §3(b).
45. Cal. Health & Safety Code §§44300 to 44394 (West 2016).
46. Colo. Rev. Stat. §24-72-204.5 (2016).
47. The Connecticut Judicial Branch Law Libraries have created a webpage with links to useful information about the Connecticut FOIA: Connecticut Judicial Law Libraries, *Connecticut Law About Freedom of Information*, http://www.jud.ct.gov/lawlib/law/foi.htm (last visited Dec. 18, 2016) and the state's Office of Governmental Accountability has a Freedom of Information Commission that posts declaratory rulings and court decisions regarding the state FOIA online: Ct.gov, Freedom of Information Comm'n, *The FOI Act*, http://www.ct.gov/foi/cwp/view.asp?a=4161&Q=488540&foiNav= (last visited Dec. 18, 2016).
48. Conn. Gen. Stat. §1-225 (2016).

District of Columbia

District of Columbia Open Meetings Act[49]: D.C. Code §§2-571 to -280
District of Columbia Freedom of Information Act[50]: D.C. Code §§2-531 to -540

The District of Columbia maintains webpages for both its OMA and its FOIA.[51] The webpages have links to relevant reports, appeals, contact information, and other relevant materials relating to and interpreting the administrative transparency laws. The District of Columbia created an independent agency, D.C. Office of Open Government, to investigate complaints about public bodies failing to comply with the OMA and FOIA. That organization maintains a website with instructions on how to access records using the laws, and also advisory opinions, reports, and complaints.[52] The District of Columbia does not maintain a central FOIA office, but each agency accepts requests, which can be filed online through a central portal for many agencies.[53]

Delaware

Delaware Open Meetings Law: Del. Code tit. 29, §§10004, 10006 (this section is within the Delaware FOIA statutory scheme)
Delaware Freedom of Information Act: Del. Code tit. 29, §§10001 to 10007

Delaware's open meetings provisions are located in the state's FOIA. The open meetings provisions include open records provisions for video conferences.[54] If a resident believes the FOIA has been violated, they may contact the attorney general, who has 20 days to respond and decide whether a violation has occurred.[55] Delaware maintains a website containing links to online FOIA portals for various state agencies and agency guidance from state agencies that have created such guidance.[56]

---

49. OpenDC, *Open Meetings Act*, http://www.open-dc.gov/open-meetings-act (last visited Dec. 18, 2016).
50. DC.gov, *Freedom of Information Act (FOIA)*, http://dc.gov/page/freedom-information-act-foia (last visited Dec. 18, 2016).
51. *See supra* notes 49 and 50.
52. OpenDC, *Homepage*, http://www.open-dc.gov (last visited Dec. 18, 2016).
53. *See* DC.gov, *Welcome to the District of Columbia Government's Freedom of Information Act Public Access Website*, https://foia-dc.gov/palMain.aspx (last visited Dec. 18, 2016).
54. Del. Code Ann. tit. 29, §10004(e)(2) (2016).
55. *Id.* §10005(e).
56. Delaware.gov, *Freedom of Information Act: Request or Petition Any Agency for Information or Documents Subject to FOIA*, http://delaware.gov/topics/foia.shtml (last visited Dec. 18, 2016).

Florida[57]

Florida Sunshine Law (Open Meetings Law)[58]: Fla. Stat. §§286.011 to .01141
Florida Public Records Law: Fla. Stat. §§119.01 to .19

Florida's Constitution grants a right to public records access and open meetings of collegial public bodies.[59] In addition to the constitutional information access guarantees, there is also the state's Sunshine Law, which includes the open meetings law provisions and the public records law, located in Chapters 286 and 119, respectively.

Georgia

Georgia Open Meetings Act: Ga. Code §§50-14-1 to -14-6
Georgia Open Records Act: §§50-18-70 to -18-77

Hawaii[60]

Hawaii Sunshine Law: Haw. Rev. Stat. §§92-1 to -13
Hawaii Uniform Information Practices Act: Haw. Rev. Stat. ch. 92F, §§92f-1 to -43

Hawaii's Sunshine Law, which is the state's open meetings law, contains a declaration of policy that is especially pro-transparency.[61] The Uniform Information Practices Act governs access to public records, The state's administrative procedures act, in §91-2, also requires that every agency cre-

---

57. *See* Brechner Center for Freedom of Information, Univ. of Fla., *Florida, Government in the Sunshine: A Citizen's Guide*, http://www.brechner.org/citizen_guide_08.pdf (last visited Dec. 18, 2016).
58. Florida's state attorney general maintains a Sunshine Law webpage with links to a Sunshine Law manual, etc. Florida Attorney General, *Open Government: The "Sunshine" Law*, http://myfloridalegal.com/pages.nsf/Main/DC0B20B7DC22B7418525791B006A54E4 (last visited Dec. 18, 2016).
59. Fla. Const. art. I, §24.
60. Hawaii's Office of Information Practices administers both the Sunshine Law and the Uniform Information Practices Act, and maintains a website with training tools, guides, annual reports, forms, formal and informal opinions, and also the state's Records Report System which contains tens of thousands of state agency records.
61. *See* ch. 92, Haw. Rev. Stat. §92-1:

    In a democracy, the people are vested with the ultimate decision-making power. Governmental agencies exist to aid the people in the formation and conduct of public policy. Opening up the governmental processes to public scrutiny and participation is the only viable and reasonable method of protecting the public's interest. Therefore, the legislature declares that it is the policy of this State that the formation and conduct of public policy—the discussions, deliberations, decisions, and action of governmental agencies—shall be conducted as openly as possible. To implement this policy the legislature declares that:

    (1) It is the intent of this part to protect the people's right to know;
    (2) The provisions requiring open meetings shall be liberally construed; and
    (3) The provisions providing for exceptions to the open meeting requirements shall be strictly construed against closed meetings.

ate freedom of information methods that make records available for public inspection and allow people to make records requests.[62]

Idaho

Idaho Open Meeting Law[63]: Idaho Code §§74-201 to -208
Idaho Public Records Act[64]: Idaho Code §§74-101 to -126

The Idaho Open Meeting Law includes meetings conducted via telecommunications devices like phones and video conferencing devices.[65] The state's Public Records Act contains a number of exemptions found within §§74-104 to -114 of the law.

Illinois[66]

Illinois Open Meetings Act: 5 Ill. Comp. Stat. §§120/1.1 to /7.5
Illinois Freedom of Information Act: 5 Ill. Comp. Stat. §§140/1.1 to /11.5

Both in-person meetings and those held electronically by any means are covered by the Illinois OMA. The state's FOIA requires all of the state's public bodies to designate FOI officers to process FOIA requests in a timely fashion.[67] FOIA requests should not allow commercial enterprises to overburden public resources or unduly disrupt the work of a public body.[68] The Illinois FOIA also contains over 40 statutory exemptions, listed in §§140/7 to /7.5.

Indiana[69]

Indiana Open Door Law: Ind. Code §§5-14-1.5-1 to -14-1.5-8
Indiana Access to Public Records Act: Ind. Code §§5-14-3-1 to -14-3-10

---

62. *Id.*
63. Idaho Public Television maintains a website answering common questions about the state's open meeting law, and provides a link to the attorney general's *Open Meeting Law Manual*, http://idahoptv.org/dialogue/openidaho/meetingsfaq.cfm (last visited Dec. 18, 2016).
64. Idaho Public Television maintains a website answering common questions about the state's public records law, and provides a link to the attorney general's *Open Public Records Law Manual* http://idahoptv.org/dialogue/openidaho/recordsfaq.cfm (last visited Dec. 18, 2016).
65. Idaho Code §74-203(5) (Michie 2016).
66. Illinois' state attorney general maintains a website called "Ensuring Open and Honest Government," which contains educational materials for OMA and FOIA administrators and the public, advisory opinions and binding opinions, dispute mediation and requests for review. http://foia.ilattorneygeneral.net/.
67. 5 Ill. Comp. Stat. §§140/3.5.
68. *Id.* §140/1.
69. Indiana has a Public Access Counselor to assist the public in using the state's public access laws. They maintain a website containing sample request letters, advisory and informal statutes, presentations, and other useful tools. *See* IN.gov, Public Access Counselor, *Homepage*, http://www.in.gov/pac/ (last visited Dec. 19, 2016).

The Indiana Open Door Law, the state's OMA, explicitly excludes chance gatherings from the scope of the open access provisions.[70] Additionally, Indiana state law creates a special position, public access counselor, within the state government who is tasked with training officials and educating the public on the state's public access laws, and to help both groups in administering the laws and accessing public information, respectively.[71] This officer must be a practicing attorney.[72]

Iowa[73]

Iowa Open Meetings Law: Iowa Code §§21.1 to .11
Iowa Open Records Act: Iowa Code §§22.1 to .14

The state's Open Records Act contains the "Iowa Fair Information Practices Act," which requires all state agencies to create rules for administering open records processes, delineating confidential and public records, and dealing with personally identifiable information in agency records.[74] The Iowa Public Information Board Act (2013) created the Iowa Public Information Board and tasked it with ensuring compliance and enforcement of the state's records access laws (Chapters 21 and 22) by providing efficient, informal dispute resolution services.[75]

Kansas[76]

Kansas Open Meetings Act: Kan. Stat. §§75-4317 to -4320f
Kansas Open Records Act: Kan. Stat. §§45-215 to -253

Under the Kansas Open Records Act, agencies can refuse to satisfy requests when production of the records would place an unreasonable burden on the agency or in instances where agencies believe that repeated requests are designed to disrupt essential agency functions.[77]

---

70. Ind. Code §5-14-1.5-2 (2016).
71. *Id.* §5-14-4-10.
72. *Id.* §5-14-4-9.
73. The Iowa Public Information Board maintains a website containing tools, guides and information about using the state's open meetings and public records laws. *See* Iowa Public Information Bd., *Welcome to IPIB*, https://www.ipib.iowa.gov/ (last visited Dec. 19, 2016).
74. Iowa Code §22.11 (2016).
75. *Id.* §23.1 (the entire Act is Iowa Code §§23.1 to 23.12).
76. Kansas' state attorney general maintains an open government website covering the Kansas OMA and Kansas Open Records Act that contains annual reports for both Acts and a method for submitting Kansas OMA complaints and enforcement actions.
77. Kan. Stat. Ann. §45-218 (2016).

Kentucky

Kentucky Open Meetings Act: Ky. Rev. Stat. §§61.800 to .850
Kentucky Open Records Act: Ky. Rev. Stat. §§61.870 to .884

Louisiana

Louisiana Open Meetings Law[78]: La. Stat. §§42:11 to :28
Louisiana Public Records Law[79]: La. Stat. §§44:1.1 to :67.2 (criminal enforcement provisions are in La. Rev. Stat. §§14:132 to :133)

Louisiana has a constitutional provision declaring a right to "observe the deliberations of public bodies and examine public documents, except in cases established by law"[80] in addition to its statutory open meetings and open records laws.

Maine

Maine Open Meeting Law: Me. Rev. Stat. tit. 1, §§403 to 406 (this section is within the Maine Freedom of Access Act statutory scheme)
Maine Freedom of Access Act: Me. Rev. Stat. tit. 1, §§400 to 414

In 2011, the Maine Freedom of Access Act (FOAA) was amended to require the state to acquire information technology services and software that maximize public access to records and records exportability.[81] In 2012, the Maine legislature created the Office of Public Access Ombudsman within the attorney general's office to review and mediate FOAA complaints and assist the public and government entities with the FOAA's statutory requirements.[82] The law also establishes a Right-To-Know Advisory Committee to ensure the compliance and uphold the integrity of the state's FOAA.[83] Additionally, Maine has an electronic access to public information law called the InforME Public Information Access Act,[84] which

78. Louisiana's state attorney general maintains separate webpages for the state's open meetings law and open records law. The open meetings law webpage provides an overview of the statutory requirements: Louisiana Dep't of Justice, *Open Meetings Law*, http://www.ag.state.la.us/Article.aspx?articleID=21&catID=0 (last visited Dec. 19, 2016).
79. Louisiana's state attorney general's webpage for the open records law contains an overview of the law's statutory requirements, and also provides links to an FAQ about the law and a sample record request form: Louisiana Dep't of Justice, *Public Records Law*, http://www.ag.state.la.us/Article.aspx?articleID=20&catID=0&printer=1#Definitions (last visited Dec. 19, 2016).
80. La. Const. Ann. art. XII, §3.
81. Me. Rev. Stat. Ann. tit. 1, §414 (West 2016).
82. *Id.* §413.
83. *Id.* §411.
84. *Id.* §531.

establishes an online portal for individuals and business to obtain public information from the state.[85]

Maryland

Maryland Open Meetings Act[86]: Md. Code Ann. §§3-101 to -501
Maryland Public Information Act[87]: Md. Code Ann. §§4-101 to -601

Effective 2014, the Maryland state legislature created a State Open Meetings Law Compliance Board.[88] The board reviews OMA complaints and draft opinions, study compliance and make recommendations accordingly to the state legislature, and educate public bodies on how to abide by the law. Effective 2015, the Maryland legislature amended the Maryland Public Information Act (PIA), establishing a State Public Information Act Compliance Board to resolve fee disputes over $350[89] and creating the Office of Public Information Ombudsman in the attorney general's office to mediate PIA disputes.[90]

Massachusetts

Massachusetts Open Meetings Law[91]: Mass. Gen. Laws ch. 30A, §§18 to 25
Massachusetts Public Records Act: Mass. Gen. Laws ch. 66, §10 (exemptions are located in Mass. Gen. Laws Ann. ch. 4, §7(26))

Massachusetts' Open Meetings Law establishes a director of open government position in the attorney general's office and creates an open meeting law advisory commission.[92] Exemptions to the Massachusetts state Public Records Act are not located in the act itself, rather they are located in the

85. *Id.* §533.
86. Maryland's state attorney general's office created a Maryland OMA webpage with links to resources including sample forms, checklists, and complaint procedures overviews. *See* Maryland Attorney General, *Maryland Open Meetings Act*, https://www.oag.state.md.us/Opengov/Openmeetings/ (last visited Dec. 19, 2016).
87. Maryland's state attorney general's office created a Maryland Public Information Act webpage with links to a brief overview of the law and contact information for the state Public Information Ombudsman. *See* Maryland Attorney General, *Maryland Public Information Act*, https://www.oag.state.md.us/Opengov/pia.htm (last visited Dec. 19, 2016).
88. Maryland Attorney General, *Maryland Open Meetings Compliance Board*, https://www.oag.state.md.us/Opengov/Openmeetings/board.htm (last visited Dec. 19, 2016).
89. Md. Code Ann. §4-1A-01 (2016).
90. *Id.* §4-1B-03.
91. Massachusetts' state attorney general's office maintains an open meeting law webpage with links to resources and FAQs. *See* Massachusetts Attorney General, *The Open Meeting Law*, http://www.mass.gov/ago/government-resources/open-meeting-law/ (last visited Dec. 19, 2016).
92. Mass. Gen. Laws ch. 30A, §19 (2016).

definitions of statutory terms section of the state's laws, with the definition of "public records."[93]

Massachusetts' Toxics Use Reduction Act[94] requires facilities using toxic substances to provide reports and plans to reduce the use of toxic materials to the state Department of Environmental Protection.[95] Massachusetts residents are required access to review those reports and plan summaries, and residents within 10 miles of a regulated facility may petition the department to study the plans and determine their adequacy, and then report their findings back to the petitioners in writing.[96]

Michigan

Michigan Open Meetings Act[97]: Mich. Comp. Laws §§15.261 to .275
Michigan Freedom of Information Act[98]: Mich. Comp. Laws §§15.231 to .246

Incarcerated felons may not make public records requests in Michigan.[99]

Minnesota[100]

Minnesota Open Meeting Law: Minn. Stat. §§13D.01 to .08
Minnesota Data Practices Act: Minn. Stat. §§13.01 to .99

Minnesota's Open Meeting Law explicitly permits the use of social media without notice so long as the social media use is limited to exchanges with all members of the general public.[101] (E-mail is not considered social media under this provision.) Minnesota's Data Practices Act is a broad, sweeping law applying to all government agency data regardless of its physical form,

---

93. *Id.* ch. 4, §7(26).
94. *Id.* ch. 21I, §§1-23.
95. *Id.* ch. 21I, §11.
96. *Id.* ch. 21I, §18(A)-(B).
97. Michigan Attorney General, *Open Meetings Act Handbook*, https://www.michigan.gov/documents/ag/OMA_handbook_287134_7.pdf (last visited Dec. 19, 2016).
98. A 2007 pamphlet about Michigan's FOIA containing a summary of the statute and related case law is available online. *See* Michigan Attorney General, *Michigan's Freedom of Information Act*, https://www.michigan.gov/documents/ag/FOIA_Pamphlet_380084_7.pdf (last visited Dec. 19, 2016).
99. Mich. Comp. Laws §15.232(c) (2016).
100. Minnesota Department of Administration's Information Policy Analysis Division (IPAD) maintains a website providing "assistance and advice on Minnesota's public access and privacy laws to the public and government" including the open meetings and public records provisions of the state's laws. http://www.ipad.state.mn.us/ (last visited Dec. 19, 2016).
101. Minn. Stat. §13D.065 (2016).

and specifying policies and rules for particular types of data in its many statute sections.

Mississippi

Mississippi Open Meetings Act: Miss. Code §§25-41-1 to -41-17
Mississippi Public Records Act: Miss. Code §§25-61-1 to -61-17

The Mississippi Ethics Commission provides OMA orders online.[102] Like several other states, the statute explicitly states that the provisions of the open meetings law do not apply to chance meetings or social gatherings.[103] The Mississippi Public Records Act ensures access to electronic records,[104] including "sensitive" software that stores, processes, or retrieves information that is exempted from disclosure.[105]

Mississippi's Ethics Commission is tasked with enforcing the OMA[106] and Public Records Act[107] and hearing complaints. Thus, when either Act is violated, complaints must be filed with the Ethics Commission. The Ethics Commission created a set of Model Public Records Rules as "best practices" for complying with the state's Public Records Act and they encourage all of the state's public bodies to abide by them.[108] The model rules also establish the factors that the Ethics Commission considers when evaluating public records disputes.

Missouri[109]

Missouri Sunshine Law for Open Meetings: Mo. Stat. §§610.010 to .035
Missouri Sunshine Law: §§610.023 to .035

Missouri has created a single Sunshine Law scheme that encompasses both open meetings laws and government records access laws. Both the open meetings and public records access provisions are combined in a single

102. Mississippi Ethics Comm'n, *Open Meetings Act*, http://www.ethics.state.ms.us/ethics/ethics.nsf/webpage/A_meetings?OpenDocument (last visited Dec. 19, 2016).
103. Miss. Code Ann. §25-41-17 (2016).
104. *Id.* §25-61-2.
105. *Id.* §25-61-10.
106. *Id.* §25-41-15.
107. *Id.* §25-61-13.
108. Mississippi Ethics Comm'n, *Mississippi Model Public Ethics Rules, With Comments* (Mar. 5, 2010), http://www.ethics.state.ms.us/ethics/ethics.nsf/PageSection/A_records_model_PR_rules/$FILE/Mississippi%20Model%20Public%20Records%20Rules%20%283-5-10%29.pdf?OpenElement.
109. Missouri's state attorney general maintains a Sunshine Law webpage with links to both open meetings and public records access resources, including sample forms for both requests and complaints, open meetings notices, and an overview of the state's transparency laws. *See* Missouri Attorney General, *Sunshine Law*, https://www.ago.mo.gov/missouri-law/sunshine-law (last visited Dec. 19, 2016).

statutory scheme, sharing some provisions, including enforcement processes and procedures.

Montana

Montana Open Meetings Law: Mont. Code §§2-3-201 to -3-221
Montana Public Records Act: Mont. Code §§2-6-1001 to -6-1503

Montana's Constitution provides its citizens with a right to know, stating: "No person shall be deprived of the right to examine documents or to observe the deliberations of all public bodies or agencies of state government and its subdivisions, except in cases in which the demand of individual privacy clearly exceeds the merits of public disclosure."[110] Montana's public records laws were recently amended, with the former sections repealed[111] and new sections slated as 2-6-1001 et seq.[112]

Nebraska[113]

Nebraska Open Meetings Act: Neb. Rev. Stat. §§84-1407 to -1414
Nebraska Public Records Law: Neb. Rev. Stat. §§84-712 to -712.09

Nevada

Nevada Open Meetings Law[114]: Nev. Rev. Stat. §§241.010 to .040
Nevada Public Records Act: Nev. Rev. Stat. §§239.001 to .030

New Hampshire

New Hampshire Right-To-Know Law: N.H. Rev. Stat. §§91-A:1 to -A:9 (encompasses both open meetings and open records provisions, and N.H. Rev. Stat. §91-A:2 specifies open meeting terms and processes)

---

110. Mont. Const. art. II, §9.
111. They were formerly Mont. Code Ann. 2-6-101 et seq. (2016).
112. Mt.gov, *Montana Code Annotated 2015*, http://leg.mt.gov/bills/mca_toc/2_6_15.htm (last visited Dec. 19, 2016).
113. Nebraska's state attorney general's website contains a link called "Public Records & Open Meetings" with links to resources about both of these information access laws. *See* Nebraska Attorney General, *Homepage*, https://ago.nebraska.gov/ (last visited Dec. 19, 2016).
114. Nevada's state attorney general hosts an online open meetings law portal: http://ag.nv.gov/About/Governmental_Affairs/OML/ (last visited Dec. 19, 2016) with OML opinions, complaint forms, and other OML resources.

New Jersey

New Jersey Open Public Meetings Act (Open Public Meetings Act, the Sunshine Law, or the Senator Byron M. Baer Open Public Meetings Act[115]): N.J. Stat. §§10:4-6 to :4-21

New Jersey Open Public Records Act: N.J. Stat. §§47:1A-1 to :1A-13

The New Jersey Open Public Records Act establishes a Government Records Council in the Department of Community Affairs.[116] This body is tasked with providing an informal mediation program to resolve government records disputes and hear complaints filed by people who are denied access to records by a government records custodian. It also issues advisory opinions, and prepares guidelines and educational materials for the public and for records custodians so that everyone can utilize and maintain the provisions of the law.[117]

New Jersey's Worker and Community Right-To-Know Act[118] is an OSHA-approved (Occupational Safety and Health Administration) program for public-sector employers, including state and local governments and public schools that requires the proper labeling of hazardous substances. This law is similar to the OSHA labeling requirements.[119] Additionally, the New Jersey Pollution Prevention Act requires EPCRA-like plans to be submitted to the state environmental agency by facilities generating non-product output hazardous substances or any hazardous substances not intended for use as a product. This set of statutes mirrors already existing federal programs.

New Mexico[120]

New Mexico Open Meetings Act: N.M. Stat. §§10-15-1 to -15-4

New Mexico Inspection of Public Records Act: N.M. Stat. §§14-2-1 to -2-12

The New Mexico attorney general's office has an open government division that is tasked with representing the state's public body in legal proceedings related to the OMA and Inspection of Public Records Act (IPRA). This office also drafts OMA and IPRA opinions and advisory letters, and educates

115. N.J. Stat. Ann. §10:4-6 (West 2016).
116. *Id.* §47:1A-7.
117. *Id.*
118. *Id.* §§34:5A-1 to :5A-31.
119. 29 C.F.R. 1910.1200 (2016).
120. New Mexico's Open Government Division of the state attorney general's office maintains a website for the state's OMA and Inspection of Public Records Act (IPRA), which contains compliance guides for each law and complaints, as well as opinions and advisory letters on open government topics.

the state's public bodies and members of the public about the transparency laws via training seminars and compliance guides.

New York[121]

New York Open Meetings Law: N.Y. Pub. Off. Law §§100 to 111
New York Freedom of Information Law: N.Y. Pub. Off. Law §§84 to 90

New York's Department of State Committee on Open Government gives advice, provides training classes, and writes advisory opinions for the government, public, and news media on the state's open meetings law and FOI law.

The Sewage Pollution Right-To-Know Act requires public notification of raw or partially treated sewage discharges into New York waters. Public notification must occur within 24 hours of a discharge.[122] Under the statute, the state's Department of Environmental Conservation also produces an annual statewide sewage discharge report detailing discharges and remedial responses taken. Similar sewage discharge information access statutes exist in other states.

North Carolina

North Carolina Open Meetings Law: N.C. Gen. Stat. §§143-318.9 to -318.18
North Carolina Public Records Law: N.C. Gen. Stat. §132.11

North Dakota[123]

North Dakota Open Meetings Law: N.D. Cent. Code §§44-04-19 to -04-32 (some of these provisions also apply to open records)
North Dakota Open Records Law: N.D. Cent. Code §§44-04-18 to -04-18.26

North Dakota's Constitution establishes a public right to open meetings access[124] and public records access.[125] The public records provisions explicitly require state agencies to use electronic databases and information storage systems that allow for records access.[126] The specifically enumerated

121. The New York State Committee on Open Government of the New York Department of State maintains a website with a variety of open meetings law and FOI law resources, including case law, guidance, and advisory opinions. *See* http://www.dos.ny.gov/coog/index.html (last visited Dec. 19, 2016).
122. You can sign up for discharge notification alerts on the New York Department of Environmental Conservation's website at http://www.dec.ny.gov/chemical/90315.html (last visited Dec. 19, 2016).
123. North Dakota's state attorney general maintains a webpage called "Open Records & Meetings" containing manuals, legal opinions, guides, forms, and checklists for compliance with the state's.
124. N.D. Const. art. XI, §5.
125. *Id.* art. XI, §6.
126. *Id.* §44-04-18(3).

exemptions in the state's open access laws include the disclosure of electronic mail addresses.[127]

Ohio[128]

Ohio Open Meetings Law: Ohio Rev. Code §121.22
Ohio Public Records Law: Ohio Rev. Code §§149.43 to .435

Ohio refers to its open meetings and public records law, collectively, as the state's Sunshine Laws. Ohio's attorney general's office provides a free mediation program to help records requesters and the state's public bodies resolve disputes and reach a mutually agreed-upon solution with a member of the attorney general's public records unit.[129] The agency also publishes an annual Sunshine Laws manual, which is available online.[130]

Oklahoma

Oklahoma Open Meetings Act: Okla. Stat. tit. 25, §§301 to 314
Oklahoma Open Records Act: Okla. Stat. Ann. tit. 51, §§24A.1 to .30

Oklahoma's OMA includes video conferencing in its scope, allowing public bodies to whole meetings by video conference when each member of the public body is visible and audible and members of the public are also allowed to attend and receive electronic materials available in the video conference forum.[131] Oklahoma's Open Records Act contains many exemptions in its various statute sections.[132]

Oregon[133]

Oregon Public Meetings Law: Or. Rev. Stat. §§192.610 to .695
Oregon Public Records Law: Or. Rev. Stat. §§192.410 to .505

---

127. *Id.* §44-04-18.21.
128. Ohio's state attorney general maintains a Sunshine Laws webpage with videos, manuals, request forms, and training opportunities. *See* http://www.ohioattorneygeneral.gov/Legal/Sunshine-Laws (Dec. 19, 2016).
129. Public Records Mediation Program, http://www.ohioattorneygeneral.gov/Legal/Sunshine-Laws/Public-Records-Mediation-Program (last visited Dec. 19, 2016).
130. Ohio Sunshine Laws: An Open Government Resource Manual (2016), http://www.ohioattorneygeneral.gov/Files/Publications-Files/Publications-for-Legal/Sunshine-Law-Publications/2015-Sunshine-Laws-Manual.aspx (last visited Dec. 19, 2016).
131. Okla. Stat. tit. 25, §307.1 (2016).
132. *Id.* tit. 51, §§24A.1 to .30.
133. Oregon's state attorney general maintains a webpage dedicated to the state's public records and meetings law with manuals, online records request forms, and a link to the Public Records Law Reform Taskforce, formed in 2015. *See* http://www.doj.state.or.us/public_records/pages/index.aspx (last visited Dec. 19, 2016).

Oregon's Open Meetings Law only guarantees that a place to listen to any electronically or telephonically conducted meetings be established so the public can listen to the meeting.[134] (The statute does not grant video or participation access to these meetings.) Under Oregon's Public Records Law, public records do not include writing unrelated to the conduct of the public's business and contained on a privately owned computer.[135] The Oregon Public Records Law also allows agencies to provide a "condensation" from the records containing the non-exempt significant facts in lieu of providing the actual records.[136] A requester receiving a "condensation" can petition the agency's denial of the request.[137]

Pennsylvania[138]

Pennsylvania Sunshine Act: 65 Pa. Stat. and Cons. Stat. §§701 to 716
Pennsylvania Right-To-Know Law: 65 Pa. Stat. §§67.101 to .3104

Pennsylvania's Right-To-Know Law defines records broadly, as "[i]nformation, regardless of physical form or characteristics, that documents a transaction or activity of an agency and that is created, received or retained pursuant to law or in connection with a transaction, business or activity of the agency."[139]

Rhode Island[140]

Rhode Island Open Meetings Act: 42 R.I. Gen. Laws §§42-46-1 to -46-14
Rhode Island Access to Public Records Act: 38 R.I. Gen. Laws §§38-2-1 to -2-15

Rhode Island's public records law contains a provision guaranteeing that all records initially deemed public will continue to be accessible even if there is subsequent court action or investigations involving matters in the records.[141]

---

134. Or. Rev. Stat. §192.670 (2016).
135. *Id.* §192.410(4)(b).
136. *Id.* §192.423(1).
137. *Id.* §192.423(2).
138. Pennsylvania's Office of Open Records provides an online portal for the state's government transparency laws, with advisory opinions, forms, appeals, and educational materials about the statutes, as well as a Twitter feed. *See* http://www.openrecords.pa.gov/Pages/default.aspx#.V1B8LGalof8 (last visited Dec. 19, 2016).
139. 65 Pa. Stat. Ann. §67.102 (2016).
140. Rhode Island's Department of State has an online "Transparency" portal with a variety of open meeting information and charts, as well as some public records of interest (a state leases database, etc.) and links to public records access materials. *See* http://sos.ri.gov/divisions/Open-Government/Transparency (last visited Dec. 19, 2016).
141. 38 R.I. Gen. Laws §38-2-13 (2016).

South Carolina

South Carolina Open Meetings Law: S.C. Code §§30-4-60 to -4-90 (this section is within the FOIA statutory scheme)
South Carolina Freedom of Information Act: S.C. Code §§30-4-10 to 30-4-165 (some of these provisions also apply to open meetings)

South Dakota

South Dakota Open Meetings Law: S.D. Codified Laws §§1-25-1 to -25-10
South Dakota Sunshine Law: S.D. Codified Laws §§1-27-1 to -27-48

South Dakota has a statutorily mandated Open Meetings Commission.[142] The commission hears complaints and reviews investigatory files submitted to the state's attorney general. It issues written determinations regarding potential violations of the law.[143] The state maintains a website called OpenSD containing a variety of public records on topics ranging from payroll to boards and commissions.[144] When making a request for records in an electronic records system, note that the Sunshine Law requires public entities to provide a narrative description of every electronic records system, and an explanation of what information the entity needs from the requester in order to access records from the system.[145]

Tennessee

Tennessee Open Meetings Act: Tenn. Code §§8-44-101 to -44-111
Tennessee Open Records Act[146]: Tenn. Code §§10-7-503 to -7-516

Texas

Texas Open Meetings Act: Tex. Gov't Code §§551.001 to .146
Texas Public Information Act: Tex. Gov't Code §§552.001 to .353

---

142. S.D. Codified Laws §1-25-7 (Michie 2016).
143. *Id.* §1-25-7.
144. *See* http://open.sd.gov/ (last visited Dec. 19, 2016).
145. S.D. Codified Laws §1-27-48.
146. Tennessee's Office of Open Records Counsel is tasked with assisting requesters with open records requests. It maintains a website to those ends containing forms, reports, opinions, and other relevant information, as well as a link to Transparent Tennessee, an online source for many state records. *See* https://www.comptroller.tn.gov/openrecords/ (last visited Dec. 19, 2016).

Utah

Utah Open and Public Meetings Act[147]: Utah Code §§52-4-101 to -4-305
Utah Government Records Access and Management Act[148]: Utah Code Ann. §§63G-2-101 to -2-901

The Utah Open and Public Meetings Act includes e-mail, instant messaging, electronic chat, text messaging, or any other electronic communications.[149] The state's public records access law, the Government Records Access and Management Act (GRAMA) includes a provision requiring that the legislature create an online repository containing legislators' e-mails on the legislature's website.[150] GRAMA also creates a State Records Committee within the state's Department of Administrative Services to review and approve agencies' records retention and disposal policies and to hear and make determinations on appeals from agencies' GRAMA determinations and GRAMA-related disputes.[151]

Vermont

Vermont Open Meeting Law: Vt. Stat. tit. 1, §310 to 314
Vermont Public Records Act: Vt. Stat. tit. 1, §315 to 320

Virginia

Virginia Open Meetings Law: Va. Code §§2.2-3707 to .2.3714 (this section is within the Virginia FOIA statutory scheme)
Virginia Freedom of Information Act: Va. Code §§2.2-3700 to .2.3714

In addition to Virginia's general open meetings law and its FOI law, the state has a law requiring the state's DEQ to monitor sediment and fish tissue for contaminants and to make that information available to the public on the Internet for at least five years after the information is received by the agency.[152]

---

147. Utah provides a public notice website where people can look up notice of upcoming meetings in compliance with the state's Open and Public Meetings Act. *See* http://www.utah.gov/pmn/index.html (last visited Dec. 19, 2016).
148. Utah maintains several public records websites. The open records portal allows the public to deliver records request to Utah agencies online if you create an account with the state http://openrecords.utah.gov/ (last visited Dec. 19, 2016) and transparent.utah.gov is maintained by the state's Division of Finance, and it provides the public with financial records from that agency http://www.utah.gov/transparency/ (last visited Dec. 19, 2016) and Data.Utah.gov provides access to state data sets http://www.utah.gov/data/.
149. Utah Code Ann. §52-4-103(5)(a)-(e) (2016).
150. *Id.* §63G-2-208 (West) (for e-mails on or before Jan. 1, 2014).
151. *Id.* §§63G-2-501 to 63G-2-502.
152. Va. Code Ann. §62.1-44.19:6(3) (Michie 2016).

Washington[153]

Washington Open Public Meetings Act: Wash. Rev. Code §§42.30.010 to .30.920
Washington Public Records Act: Wash. Rev. Code §§42.56.001 to .56.904

West Virginia

West Virginia Open Governmental Proceedings Act: W. Va. Code §§6-9A-1 to -9A-12
West Virginia Freedom of Information Act: W. Va. Code §§29B-1-1 to -1-7

West Virginia's Ethics Commission has a Committee on Open Governmental Meetings charged with interpreting the state's Open Governmental Proceedings Act requirements, but the committee does not accept complaints about or enforce the open meetings law.[154] West Virginia's secretary of state's office recently created a FOIA database in compliance with the state's FOIA law[155] which is now available to the public.[156]

Wisconsin

Wisconsin Open Meetings Law: Wis. Stat. §§19.81 to .98
Wisconsin Open Records Law: Wis. Stat. §§19.31 to .39

Wisconsin's Constitution contains a provision requiring the legislature to keep a public journal of its proceedings and to have an open door policy, only allowing secrecy when the public welfare requires it.[157]

Wyoming

Wyoming Public Meeting Act: Wyo. Stat. §§16-4-401 to -4-408
Wyoming Public Records Act: Wyo. Stat. §§16-4-201 to -4-408

---

153. Washington's state attorney general's office oversees the Public Records Exemptions Accountability Committee (Sunshine Committee), a committee tasked with making recommendations about the Public Records Act. The committee's website contains open meetings and open records resources, including educational manuals and links to public records and open meetings information, as well as request forms for public records requests. *See* http://www.atg.wa.gov/sunshine-committee (last visited Dec. 19, 2016).
154. W. Va. Code §6-9A-1 (2016).
155. *Id.* §§29B-1-3(f), -1-3a(a).
156. West Virginia Secretary of State Natalie E. Tennant, *Freedom of Information Act Database*, https://apps.sos.wv.gov/FOIA/Requests/ (last visited Dec. 19, 2016).
157. Wis. Const. art. IV, §10.

Wyoming public records law was recently amended to explicitly exclude electronic communications solely between students attending school in Wyoming.[158]

IV. Example of Inconsistency: State Fracking Laws

The variety in state and local information access laws and provisions creates a national patchwork of environmental information access. This patchwork is full of inconsistencies that affect what kinds of notice and data people can obtain about environmental issues depending on what jurisdiction they are in. These disparities are especially clear in instances when states establish divergent transparency rules.

Fracking chemical transparency laws provide an example of the resulting muddle of information access that occurs without more standardized public records standards. Companies fracking on public lands are supposed to disclose chemical lists to FracFocus, but those fracking on private land have different disclosure obligations depending on which state they frack in.[159] Whether the fracking companies' chemical lists are available to the public depends on whether and to what extent the state grants them trade secret protection.

While around half of the states in the U.S. have either passed or proposed laws that require the disclosure of fracking chemicals to state environmental agencies,[160] half of the states do not require any data collection about the chemicals that fracking operations use. On the other hand, California's fracking chemical disclosure law not only requires fracking operations to disclose chemical information to the state's Division of Oil, Gas & Geothermal Resources, these disclosures must also be made public, even if the information is considered trade secret.[161] It also requires that the agency notify the public online about proposed fracking activities.[162] California's fracking chemical disclosure law varies greatly from Texas's law, which labels all fracking chemical information "confidential"[163] and North Carolina's fracking law, which actually criminalizes the disclosure of fracking trade secrets.[164]

---

158. Records and Recordation—Schools and School Districts—Communications, 2016 Wyo. Sess. Laws ch. 48 (H.B. 13).
159. Chemical Engineering News, *Tracking Fracking*, http://cen.acs.org/static/tracking-fracking.html (last visited Dec. 19, 2016).
160. *Id.*
161. Cal. Pub. Res. Code §3215 (West 2016).
162. *Id.* §3160.
163. Tex. Nat. Res. Code Ann. §134.031 (Vernon 2016).
164. Under the state's Trade Secrets Protection Act, N.C. Gen. Stat. §66-152 to-158 (2016). *See* Matthew Poppe, "North Carolina Enacts Criminal Penalties for Disclosure of Fracking Trade

As the variation in fracking law information access demonstrates, state laws do not create any sort of uniform environmental information access scheme. Thus, as an environmental information seeker, you must first determine what kind of information is available to you depending on what jurisdiction you are in before you prepare an open records request or otherwise try to access state or local environmental information.

Secrets," *Orrick Blog*, July 1, 2014, http://blogs.orrick.com/trade-secrets-watch/2014/07/01/north-carolina-enacts-criminal-penalties-for-disclosure-of-fracking-trade-secrets/.

Chapter 10: Finding Environmental Information in Other Countries: Foreign and International Law

Law in Action

Terry Thomas, a director at Wind Farm Solutions, a California wind energy products company, has been watching the recent United Nations Paris Climate Summit with the hopes of introducing his products in other countries as nations attending the Summit try to decrease their fossil fuel consumption. Terry is not sure which countries are attending the Summit or what kinds of agreements are being forged during the meetings, and he would like to know whether any of the agreements advocate wind energy as an alternative to fossil fuel use. Terry also wants to collect a survey of information about what kinds of energy use systems and laws are in place in several countries from each continent. Terry's not sure where to begin with all of this market research.

I. Introduction

Why would international and foreign environmental law matter to someone searching for environmental information relevant to where they live? Well, the simple answer is because the world is interconnected, and what happens in one country directly impacts another country's environment. Nations' borders do not isolate the environmental impacts of national undertaking. For example, Canada's mining practices impact Alaskan salmon fishing.[1] Nations are constantly meeting and signing international agreements and compacts to engage in cross-boundary environmental activities and efforts. Recent climate change meetings make it clear that global environmental progress requires international participation—the pollution decisions of nations on the other side of the globe, like China and India, directly impact us every day. We import harmful chemicals from other countries,

1. Brendan Jones, *A Canadian Threat to Alaskan Fishing*, N.Y. Times, Jan. 23, 2016, at SR6, http://www.nytimes.com/2016/01/24/opinion/sunday/a-canadian-threat-to-alaskan-fishing.html?emc=edit_th_20160124&nl=todaysheadlines&nlid=69056167.

too. When it comes to environmental information, national borders are somewhat arbitrary.

There are two major research areas for information abroad: international law and foreign law. These areas are not synonymous, but are instead, two separate research concepts: international law, the law governing relationships between nations; and foreign law, the domestic laws governing, and under the sovereignty of, each nation.[2]

II. Foreign Law

Finding the domestic laws of different countries can be difficult, as there is no uniformity in legal system organization, language, and method of dissemination in the world's national law and policy schemes. It helps to know whether a nation is a civil- or common-law country or a hybrid of the two. Is the country pluralist, with more than one source of authority? Is there a religious component to the nation's rules? Researching and understanding a nation's legal system provides context for the laws and rules that come out of that system. The Central Intelligence Agency's (CIA's) *World Factbook* provides a list of national legal systems and other background information about each nation's laws. It is available online, with all nations linked from an alphabetical list.[3]

Because only a minority of the world uses English as an official language, finding English versions of the foreign laws you seek may pose a challenge. If you find an unofficial English translation of foreign law (aka, a translation that was not created or approved by the nation's official government), keep in mind that the translation is approximate, and should only be relied upon to a certain extent. Luckily, the Internet has created easier access to translation services, and those services can help translate laws and provide understanding to those translations.[4]

Aside from the variations in legal system style and languages, for research purposes, it is important to note that there is also a great deal of variation in the amount of information governments provide online, and the technological capacities of their online systems. Some countries have excellent infor-

2. Frederic L. Kirgis, *Is Foreign Law International Law?*, 9 Insights (American Society of International Law) (Oct. 31, 2005), https://www.asil.org/insights/volume/9/issue/33/foreign-law-international-law.
3. CIA, *The World Factbook*, https://www.cia.gov/library/publications/the-world-factbook/fields/2100.html (last visited Dec. 19, 2016) [hereinafter *The World Factbook*].
4. A simple online search retrieves contact information for numerous translator services that will translate documents for a fee, as well as electronic online translators like Google Translate (https://translate.google.com/) that will translate documents free of charge. Be aware, however, that electronic translators do not guarantee correct translations.

mation systems, but others are quite limited, and in some instances, totally unavailable via the Internet.

III. Finding Foreign Laws, Generally

In the past, there were several major and well-regarded hardcopy resources that collected foreign laws in encyclopedic or loose-leaf format. These resources, for the most part, have migrated online and are no longer updated in their paper format. Even Flores & Reynolds' *Foreign Law Guide*, a staple for foreign legal research, is now available on the Internet. Thus, this book will present the online versions of these resources, noting the instances where paper versions preceded these Internet offerings. There are also organizations and institutions that gather links to all nations' laws in one place on the Internet. These online collections of national laws are great starting points for foreign law research. Here are some overviews of several of the best print and online resources for foreign law:

Publication	Online Status	Description
Foreign Law Guide (Flores & Reynolds, now the Brill Online Foreign Library)	Paper from 1990s-2000, acquired by Brill in 2012 and now online via Brill	*Foreign Law Guide* (or *FLG*, or Flores and Reynolds) is considered the "bible" of foreign law research. It contains legislation citations, English translations where available, and also links to relevant secondary sources for approximately 190 jurisdictions.
Germain's Transnational Law Research: A Guide for Attorneys	Paper since 1991 and it is available through a few universities, such as George Mason University, Catholic University, and George Washington University	This single-volume loose-leaf is a regularly updated volume designed for the legal community, providing foreign law information along with starting points for international litigation and other common legal uses for foreign law research.
The Modern Legal Systems Cyclopedia[a]	Paper from 1984-2005, now available online in HeinOnline's *World Constitutions Illustrated*	This was once a 22-volume encyclopedia, and it is available in paper in some law libraries, and online. Last updated in 2005, it likely contains outdated information for some countries.

World Constitutions Illustrated library in HeinOnline[b]	Online via HeinOnline	This constantly growing collection of foreign law information contains the constitution for every nation, numerous foreign law books and periodicals, country studies, and the CIA's *World Factbook*. HeinOnline is a subscription service, so this excellent resource may only be available to you via select law libraries.
GlobalLex	Online	This online database is maintained by the Hauser Global Law School Program at New York University School of Law and contains vast foreign law, international law, and comparative law resources.
Yale, Country by Country Guide to Foreign Law Research[c]		This website is not as in-depth at GlobalLex, but it does provide links to the best research guides and resources for each country, including the CIA's *World Factbook* link for each nation and each nation's *Library of Congress Country Study*.
The World Factbook[d]	Online via the CIA	Created for U.S. policymakers in 1981, and now available online, it provides overview information for every geographic entity in the world, such as government, economy, military, energy, and transportation structures, as well as identifying any transnational issues. It is an excellent resource for a quick overview of any nation, albeit a resource with a strictly U.S. perspective.

a. Lyonette Louis-Jacques, *The Modern Legal Systems Cyclopedia, Law News From the D'Angelo Law Library* (Sept. 11, 2011), http://news.lib.uchicago.edu/blog/2011/09/10/the-modern-legal-systems-cyclopedia/.

b. HeinOnline, *World Constitutions Illustrated: Contemporary & Historical Documents & Resources*, https://heinonline.org/HeinDocs/WorldConstitutionsIllustrated.pdf.

c. Yale Law School Lillian Goldman Law Library, *Country by Country Guide to Foreign Law Research*, http://library.law.yale.edu/research/guides/country-guide (last visited Dec. 21, 2016).

d. *See The World Factbook*, *supra* note 3.

Because environmental laws and policies tend to be scattered across various parts of the national law scheme for each sovereign nation, there is no comprehensive resource compiling only the environmental laws in every nation. Thus, these general foreign law resources are the best starting point for foreign environmental law research. All online international environmental law resources refer researchers to these major collections of foreign law research in the absence of more specialized environmental law databases.

Hypothetically, a major advantage of moving the paper version of foreign law resources online to an electronic version is that these new Internet repositories can be easily updated and are often more up- to date than paper books on shelves. Nevertheless, trusting that online foreign law resources are updated is not the best approach for researchers. Maintaining these foreign law websites is a gargantuan task, and legal information and links tend to become outdated very quickly without someone consistently keeping the files and links up to date. Keep the last updated date of the page you are viewing in mind when looking at legal resources on the Internet to make sure you are looking at the most current legal authority.

Another practical truth is that ultimately, the results of your foreign law research may not be as satisfying and certain as those in U.S. legal research. For one, it is likely that no libraries or U.S. research centers contain a full, exhaustive collection for the laws of another country, so it is hard to ensure you are getting everything, and that it is organized correctly. As one foreign law overview points out: "Very few jurisdictions publish legal information as thoroughly as the United States does."[5] Additionally, translating legal language is tough—you cannot be certain that the translation is clear and direct. Conducting foreign law research, however, can help you begin to formulate an understanding of the comparative environmental law in another nation, or their transparency scheme.

IV. Information Access Laws in Other Countries

Where most foreign law resources focus on the laws of foreign nations, from an environmental information access standpoint, it is important to go beyond the mere laws of a nation to discover what kind of environmental information can be obtained from a sovereign's government. Some nations lack a freedom of information (FOI) law, making it nearly impossible to get underlying information about environmental projects and issues within the country, however, other nations have special environmental information transparency laws and provisions that exceed the ordinary FOI access for environmental information. Be aware that such laws usually only provide

5. J.D.S. Armstrong & Christopher A. Knott, Where the Law Is: An Introduction to Advanced Legal Research 210 (4th ed. 2013).

access to the nation's citizens, however, so being a non-citizen may limit your ability to access the information.

As far as environmental law information goes, various nations tuck environmental information access laws and other environmental law provisions into all sorts of different places: statutes, regulations, constitutions, case law. While the United States does implement certain advanced environmental programs, like EPCRA's TRI and NEPA, other nations are engaging in innovative environmental programs and activities far different than those of the United States.

For instance, the Bolivian Constitution was amended in 2009 to include "Pachamama," or Mother Earth, as a deity, and in 2010 adopted a set of laws called *Ley de Derechos de la Madre Tierra*, which translates to "Law of the Rights of Mother Earth."[6] These laws bestow legal rights to nature. Costa Rica also has innovative environmental sustainability laws and policies. In 1997, Costa Rica was the first nation to create a "Pago Por Servicios Ambientales" program, where landowners receive compensations for using their land to engage in activities that promote the environment, like reforestation or sustainable forest management.[7] With so many unique perspectives on the role of the environment in the legal system, the amount of access and consideration given to environmental information varies greatly from nation to nation.

The unique considerations of nature and the environment in each country have led some countries to provide special legal priority to the access of environmental information. Recently China, whose environmental policies are often considered behind the times, created a legal venue to provide environmental information access to its citizens. China recently revised their environmental law, including a chapter giving citizens a right to obtain information about the environment and allowing more citizens to file environmental lawsuits against polluters.[8] The United Kingdom also has a special set of laws called Environmental Information Regulations that provide environmental information access.[9] These regulations are in addition to the U.K. Freedom of Information Act, and provide broader and more streamlined access to environmental information. Scotland also has special Environmental Information Regulations to parallel the U.K. policies.[10]

---

6. Ley de Derechos de la Madre Tierra, Law 071 of the Plurinational State of Bolivia.
7. Ley Forestal No. 7575.
8. *See* Tseming Yang, *The 2014 Revisions of China's Environmental Protection Law*, Risk Dialogue Mag., Oct. 16, 2014 (Swiss Re: Centre for Global Dialogue), http://cgd.swissre.com/features/The_2014_Revisions_of_Chinas_Environmental_Protection_Law.htm.l.
9. The Environmental Information Regulations 2004, SI 2004/3391 (Eng.).
10. The Environmental Information (Scotland) Regulations 2004 (2004 No. 520).

On the other hand, some nations do not even have basic FOI laws, restricting all access to government information. There are organizations that maintain websites listing the information access policies in different countries. For instance, journalism organizations like the Global Investigative Journalism Network keep lists of nations' information access policies in order to help journalists covering international stories get the information they need[11] and FOI nonprofits maintain similar lists to raise awareness of transparency issues across the globe.[12] David Banisar, legal counsel for Article 19, a nonprofit committed to increasing transparency worldwide as prescribed under Article 19 of the Universal Declaration of Human Rights, creates a regularly updated map reflecting international information access policies and posts it online, including in its Social Science Research Network.[13] Because nations' information access laws and policies vary drastically, the amount of obtainable environmental information also varies greatly from country to country (see Figure 10-1).

V. International Law

International law does not cover the unique legal schemes and structures within each sovereign nation. Rather, international law refers to the legal system that governs the relationships between countries, or international relations. International law involves not just sovereign nations, but also certain international organizations (like the United Nations, regional groups like the European Union, and intergovernmental special interest collaborations like the Organization of Petroleum Exporting Countries, better known as OPEC). International law is also called public international law, foreign-relations law, and transnational law.[14]

Although understanding the complexities of international law and the agreements of sovereign states is beyond the parameters of this environmental information access book, there are many educational resources for people doing international law research. These resources include benchmark treatises on international law, like Prof. Ian Brownlie's *Principles of Public International Law* (6th ed.) and *Oppenheim's International Law* (19th ed.).

11. Global Investigative Journalism Network, *Freedom of Information Laws*, http://gijn.org/resources/freedom-of-information-laws/ (last visited Dec. 21, 2016).
12. FreedomInfo.org, *93 Countries Have FOI Regimes, Most Tallies Agree* (Oct. 19, 2012), http://www.freedominfo.org/2012/10/93-countries-have-foi-regimes-most-tallies-agree/.
13. David Banisar, *National Right to Information Laws, Regulations and Initiatives 2015*, Article 19: Global Campaign for Free Expression (Dec. 1, 2015), http://papers.ssrn.com/sol3/papers.cfm?abstract_id=1857498.
14. Black's Law Dictionary, *International Law* (10th ed. 2014).

Figure 10.1 David Banisar's FOIA International Map

National Right to Information Laws, Regulations and Initiatives 2017

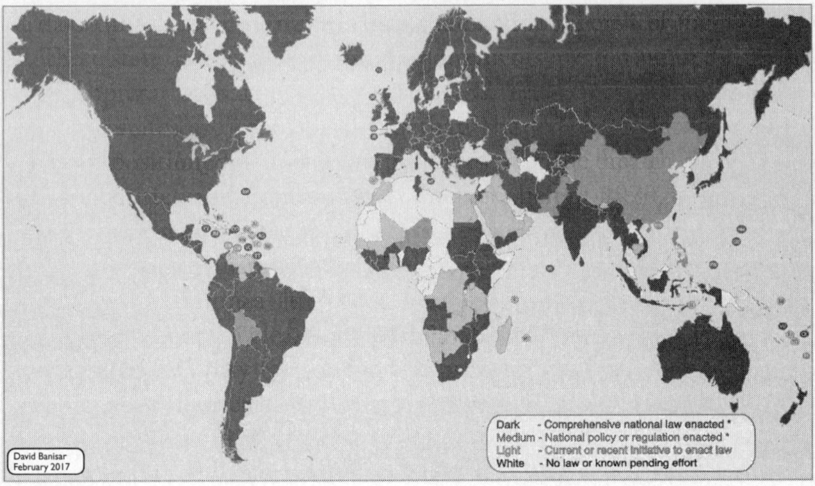

David Banisar, of Article 19: Global Campaign for Free Expression, created a map that shows the countries that have adopted national right to information laws, ordinances, or regulations. This map shows all the countries that have adopted such laws. David Banisar, National Right to Information Laws, Regulations and Initiatives 2017 (ARTICLE 19). Available at SSRN: https://ssrn.com/abstract=1857498.

There are also shorter overviews of international law concepts, like the *Public International Law in a Nutshell*, which explains basic concepts of public international law.

Generally, the agreements that nations come to under international law are memorialized in treaties. Countries agree to international treaties by expressing a willingness to undertake the legal rights and obligations within the treaty agreement. Countries demonstrate consent to be bound by treaties by signing, ratifying, accepting, approving, or acceding or some combinations of these acts. Simply signing a treaty, however, does not give a country any mandatory legal obligation under the agreement, rather, it just indicates a state's intention to move toward eventually being bound by the treaty at a later date. Until then, the nation's signature simply indicates that it will refrain from activities defeating the object and purpose of the treaty.

Additionally, there is no overarching enforcement mechanism for treaties. Each nation is sovereign and cannot, generally, be enforced against by an external entity. Although there is no compulsory judicial system or coercive penal system for nations failing to comply with treaties, there are tribunals to resolve international law issues. The International Court of Justice, established by the Charter of the United Nations, is a forum where U.N. Member states settle disputes in accordance with international law. Before the court can hear a case, each state involved must accept the court's jurisdiction, or else the states are not obligated to abide by the court's decisions. There are also specialized tribunals to handle certain types of disputes. For example, the International Tribunal for the Law of the Sea handles maritime disputes.[15]

Although the creation and enforcement of international agreements between sovereign nations is somewhat tenuous, treaties cover many environmental issues, conferring environmental responsibility on nations for things ranging from climate change, pollution and conservation to sustainable development. Many environmentally focused treaties are facilitated and overseen by the U.N. Environment Programme (UNEP).[16] While the binding nature of treaties is not certain, such agreements do dictate international best practices and lead to working papers, studies, and even the development of some foreign laws and programs dealing with and proposing innovations to deal with environmental issues. All of these materials can be very useful to those seeking information about international environmental topics.

VI. Finding Information About International Law

Like foreign law, many organizations and institutions have created online information repositories and research guides for international law. In contrast to foreign law resources, it is easier to find topical international law research guides. As with the foreign law resources, this is not an exhaustive list of resources, but some of the best tools for international environmental law searches.

15. These two paragraphs are abridged from the U.N. doc: United Nations, *Understanding International Law*, 2011 TREATY EVENT: TOWARDS UNIVERSAL PARTICIPATION AND IMPLEMENTATION, Fact Sheet #1, https://treaties.un.org/doc/source/events/2011/Press_kit/fact_sheet_1_english.pdf (last visited Dec. 21, 2016).

16. UNEP, *Homepage*, http://www.unep.org/ (last visited Dec. 21, 2016).

Publication	Online Status	Description
The American Society for International Law (ASIL) Electronic Resource Guide (ERG)[a] and Electronic Information System for International Law (EISIL)[b]	Online since 1997, ERG is a regularly updated chapter format and EISIL is a free online database providing links to web resources	ERG and EISIL work together to create a comprehensive international law research tool. ERG is formatted like a book, consisting of a table of contents and links to book chapters online, and EISIL augments ERG by providing links to online resources.
U.N. Treaty Collection (UNTC)[c]	Paper versions of treaty compilations and finding aids are discontinued and are now only available online	This database contains more than 500 major multilateral treaties and their accompanying information. It has a high-quality search mechanism providing full-text and field searches of the treaties.[d]
U.N. Research Guides[e]	Online	This online research guide helps anyone doing U.N.-related research navigate the special document types of the United Nations.
World Legal Information Institute[f]	Online	This resource is a nonprofit global information effort coordinated by the Australian Legal Information Institute (LII) featuring hundreds of international law resource databases compiled by a collection of LIIs.
Cornell Legal Information Institute[g]	Online	Cornell's LII offers one of the best free resources for U.S. legal information, and it also provides a great deal of international and foreign law materials online. This page contains links to continents' and countries' resources, and also a special list of links to environmental international law materials.

a. ASIL, Electronic Research Guide: An ASIL Information Resource (ERG), https://www.asil.org/resources/electronic-resource-guide-erg (last visited Dec. 21, 2016).
b. EISIL: An ASIL Research Tool, http://www.eisil.org/ (last visited Dec. 21, 2016).
c. U.N. Treaty Collection, *Depositary of Treaties*, https://treaties.un.org/Pages/Home.aspx (last visited Dec. 21, 2016).
d. *Id.*
e. Dag Hammerskjöld Library, *U.N. Documentation Research Guide*, http://research.un.org/en/docs (last visited Dec. 21, 2016).

f. World Legal Information Institute, http://www.worldlii.org/ (last visited Dec. 21, 2016).
g. Cornell University Law School Legal Information Institute, *Law by Source: Global,* https://www.law.cornell.edu/world (last visited Dec. 21, 2016).

Some of those specialized international environmental law resources are:

The ASIL ERG International Environmental Law Guide[a]	PDF available online, the ERG chapter containing links and information on environmental law, specifically	Links to treaties, search strategy discussions for environmental international law. Good discussion and overview of environmental international law and its broad scope.
International Environmental Agreements Database Project[b]	Online by the University of Oregon	A repository for environmental agreement information, including any agreement that attempts to "regulate human interactions with the environment." Contains thousands of agreements, and provides dates, links to agreements, UNEP information, members, agreement type, secretariat, and related data and articles.
GlobaLex Environmental Legal Research Page[c]	Last updated in 2013 (some of the links are outdated, but still an excellent resource over all)	Excellent international environmental law-focused guides providing a glossary of key treaties and abbreviations, and a environmental topic list with related treaties, and links to treaty webpages. Also contains links to environmental organizations' webpages.
U.N. Environment Programme (UNEP) Virtual Library[d]	Online resource of UNEP	All of UNEP's publications, including UNEP's Knowledge Repository, a searchable collection containing UNEP annual reports and other reports on international environmental topics.
ECOLEX[e]	Online collaboration information service by UNEP, U.N. Food and Agriculture Organization, and the International Union for Conservation of Nature	This database contains information on environmental treaties, judicial opinions, non-binding policy and guidance documents, national legislation, and judicial decisions.

a. Anne Burnett, e-*RG Electronic Research Guide: International Environmental Law,* https://www.asil.org/sites/default/files/ERG_ENVIROMENT.pdf (last visited Dec. 21, 2016).
b. Ronald B. Mitchell, International Envtl. Agreements Database Proj., *IEA Database Home,* http://iea.uoregon.edu/page.php?file=home.htm&query=static (last visited Dec. 21, 2016).

c. Heidi Frostestad Kuehl, UPDATE: *A Basic Guide to International Environmental Legal Research* (Aug. 2013), http://www.nyulawglobal.org/globalex/International_Environmental_Legal_Research1.html.
d. UNEP, *Virtual Library*, http://www.unep.org/library/ (last visited Dec. 21, 2016).
e. ECOLEX: THE GATEWAY TO ENVIRONMENTAL LAW, http://www.ecolex.org/start.php (last visited Dec. 21, 2016).

Wikipedia also contains a collected list of environmental law-focused treaties with links to the treaty materials.[17] While there is no assurance that these were compiled by a reliable source, the underlying treaties can easily be verified as environmental in nature by the researcher.

VII. Environmental Information Access in International Law

Several international treaties have bound signatories to provide access to governmental environmental information. These treaties engage their signatories to work toward environmental transparency in their nations, moving toward environmental information regulations like those in the United Kingdom and Scotland.[18]

The Rio Declaration of 1992 and the Aarhus Convention of 1998 are two major treaties that include environmental information transparency and access in their obligations. The Rio Declaration on Environment and Development is a list of 27 principles created to guide countries as they try to encourage sustainable development.[19] Principle 10 of the Rio Declaration states that "[a]t the national level, each individual shall have appropriate access to information concerning the environment that is held by public authorities, including information on hazardous materials and activities."

Over 170 countries have signed on to the convention.[20] In 2010, UNEP adopted the *Bali Guidelines*, special guidelines to further implement Principle 10 of the Rio Declaration.[21] Most notably, Principle 10 prompted the

17. Wikipedia, *Environmental Treaties*, https://en.wikipedia.org/wiki/Category:Environmental_treaties (last visited Dec. 21, 2016).
18. Sarah Lamdan, Nature Beyond the FOIA: Improving Environmental Information Access in the U.S., 29 GEO. ENVT'L L. REV. (forthcoming, Spring 2017).
19. Principle 10 of the Rio Declaration and also the Aarhus Convention's third "Environment for Europe" in 1995 agreed to new guidelines for public participation in environmental decisionmaking of which the first pillar was access to environmental information. *See* Rio Declaration on Environment and Development, princ. 10, U.N. Doc. A/CONF.151/5/Rev.1 (1992); Convention on Access to Information, Public Participation in Decision-Making, and Access to Justice in Environmental Matters, June 25, 1998, 2161 U.N.T.S. 447.
20. The American Presidency Project, *The President's News Conference in Rio de Janeiro* (June 13, 1992), http://www.presidency.ucsb.edu/ws/?pid=21079#axzz2gn1nPngp.
21. UNEP, *Bali Guideline Implementation Guide Published*, http://www.unep.org/civil-society/Implementation/Principle10/tabid/105013/Default.aspx (last visited Dec. 21, 2016).

Aarhus Convention, which is the first international, legally binding instrument specifying detailed obligations for effective implementation of Principle 10.[22]

The U.N. Economic Commission for Europe Convention on Access of Information, Public Participation in Decision-Making, and Access to Justice in Environmental Matters (popularly known as the Aarhus Convention) is a regional agreement involving 47 European nations that have ratified the agreement. The Aarhus Convention carries out Principle 10 of the Rio Declaration, creating mandatory obligations for implementation for all Members through guaranteeing access to environmental information (the right to obtain environmental information held by public authorities). The U.K. Environmental Information Regulations are the nation's implementation of this environmental treaty. Hopefully, as more nations implement the Rio Declaration and create more regional agreements like Aarhus, we will see more and more environmental information access to help international environmental seekers obtain environmental information all over the world.

22. European Commission, *What Is the Aarhus Convention?*, http://ec.europa.eu/environment/aarhus/ (last updated Oct. 3, 2016).

Chapter 11: Making Sense of Environmental Information

Law in Action

At a local meeting hall in Hattiesburg, Mississippi, town residents sat as EPA administrators presented the results of local groundwater tests to the group on a large projector screen. Residents knew that one of the industrial sites in their town was contaminated by chemicals from an old creosote factory, but beyond that they had very little information about the scope of the pollution, what it meant for their health and welfare, and how to improve the situation. As the U.S. Environmental Protection Agency (EPA) staff flashed slide after slide of technical, data-filled information on the screen, the residents looked on, simultaneously overwhelmed and perplexed. Citizens complained after the presentation: "You just wasted all these people's time, because they didn't get a thing from what you just said." Because the citizens did not have backgrounds in science, the complex format of the data presented did nothing to help them understand what impact this pollution might have on their lives. Although the citizens had successfully managed to obtain the information they needed, it was not useful to them in the form EPA presented it.[1]

I. Understanding Scientific Data

Finishing a research task with the correct records in hand is a rewarding experience. It is, however, often the case that when environmental information seekers receive the information they need, it is not immediately useful because the information contained therein is confusing and filled with complex formulae. Environmental information is usually full of scientific language, statistics, and jargon that the average person does not understand.[2]

1. This fictional scenario is loosely based on the article written by Ethan Contini-Field & Martha Merson, *Making Science Accessible to Environmental Advocates*, EARTHZINE (Apr. 12, 2013), http://earthzine.org/2013/04/12/making-science-accessible-to-environmental-advocates/.
2. The scientific complexity of EPA's Freedom of Information Act responses are well-documented. A 2009 EPA National Environmental Publications Information System project called the National Dialog on Access to Environmental Information led to EPA's Information Access Strategy, available at https://nepis.epa.gov/, which reflected the need for organizing EPA information and data to promote better understanding of the complex scientific information within EPA records (page 12 of the Access Strategy). Additionally, opening up many of EPA's publicly available records—clicking on the toxic release inventory database link or downloading one of the agency's thousand-paged environmental

Statistics and scientific data are unavoidable when dealing with environmental issues.[3] It is necessary to understand the underlying math and data to make any sort of environmental claim or argument. This chapter is designed to help laypeople (in this case, those of us who are not scientists, statisticians, and engineering experts) understand and use the information we have gathered from our research, open meetings, and freedom of information requests so that we can better engage with the scientific data at the core of many of those records.[4]

This chapter will help you begin to assess and understand the environmental information that is gathered using the information seeking strategies described in this book. While no single resource exists that can for all intents and purposes "translate" scientific information into easy to understand terms and concepts, there are certain tools and resources that can help us understand environmental information.

II. "Translating" or Breaking Down Environmental Information

A. *Understanding Environmental Terminology*

Several books have been published to help lawyers and laypeople understand the scientific jargon inherent in environmental law. The *Environmental Science Deskbook*, a publication in the Environmental Law Series published by Thomson Reuters, provides plain-English explanations of scientific and technical concepts used in environmental law.[5] This resource is edited by the American Chemistry Council and its contributors. There is also a resource published by the American Bar Association called *The Science and Technology Guidebook for Lawyers*, which explains science and technology through case law discussions to help attorneys understand "techno-science"

---

impact statement (EIS) file submissions—reveals pages of scientific data that are practically impossible to understand unless chemistry and engineering are your areas of expertise.

3. Statistics for Action Facilitator's Manual Draft (2010), http://www.toxicsaction.org/sites/default/files/tac/information/statistics-for-action-activities.pdf [hereinafter Statistics for Action].

4. Engaging with scientific materials is often especially important in the legal practice field. As Judge Posner recently surmised: "Innumerable are the lawyers who explain that they picked law over a technical field because they have a 'math block....'" "But it's increasingly concerning, because of the extraordinary rate of scientific and other technological advances that figure increasingly in litigation." Debra Cassens Weiss, *Posner: Lawyers Bad at Math Are an Increasing Concern; Inmate's Blood-Pressure Suit Shows Why*, ABA J. (Oct. 29, 2013), http://www.abajournal.com/news/article/posner_math_block_lawyers_an_increasing_concern_inmates_blood-pressure_suit.

5. James W. Conrad Jr., Environmental Science Deskbook (2016).

language.[6] Finally, the *Environmental Health and Science Desk Reference* defines and explains the terms used in environmental policy, science, and engineering fields.[7]

In addition to reference books that provide definitions, there are also online resources that can help decode the language within environmental information records. Environmental data contains terms like reporting limits, background levels, reference dose, and Occupational Safety and Health Administration Permissible Exposure Limits. Fortunately, both government agencies[8] and outside organizations provide explanations of what these measurements and terms mean, even though these explanations are not always easy to find. EPA even has a Terminology Services webpage that allows visitors to search for terms and acronyms and provides definitions and glossary entries for the words and phrases you enter into the search box.[9] The website also provides links to a vocabulary catalog that contains the 296 vocabularies used by EPA's various missions and operations and a taxonomies search tree and keyword search tool. Environmental science organizations, educational websites,[10] and advocacy groups also provide explanations for some of the environmental terminology used in EPA programs, as well. The United Nations Environment Programme (UNEP) also provides a glossary for environmental terminology used worldwide.[11]

Understanding environmental information also requires an understanding of some basic mathematical and scientific concepts. For instance, it helps to have an understanding of common units, limits, and levels. Metric measurements (liters, meters, grams), prefixes (kilo-, milli-, and micro-), and other measurements like acres, watts, and parts per million frequently appear in environmental data. There are many quick online references that

---

6. Cara Morris & Joseph R. Carvalko, The Science and Technology Guidebook for Lawyers (2014).
7. Frank R. Spellman & Revonna M. Bieber, Environmental Health and Science Desk Reference (2012).
8. *See, e.g.*, Disaster Info. Mgmt. Research Ctr., *Disaster Glossaries*, https://sis.nlm.nih.gov/dimrc/glossaries.html (last visited Dec. 24, 2016) (linking to the glossaries created by the various federal agencies relevant to environmental disaster terminology).
9. U.S. EPA, *Search: Terms & Acronyms*, https://iaspub.epa.gov/sor_internet/registry/termreg/searchandretrieve/termsandacronyms/search.do (last visited Dec. 24, 2016).
10. *See, e.g.*, McGraw Hill Online Learning Ctr., *Glossary A-D*, http://highered.mheducation.com/sites/0070294267/student_view0/glossary_a-d.html (last visited Dec. 24, 2016) (featuring the glossary from William P. Cunningham et al., Environmental Science: A Global Concern (7th ed. 2000)).
11. International Resource Panel, UNEP, *Glossary*, http://www.unep.org/resourcepanel/KnowledgeResources/GlossaryofTerms/tabid/133339/Default.aspx (last visited Dec. 24, 2016).

contain this information. Google even has an online unit converter that will convert measurements to different unit types.[12]

B. *The Contents of Environmental Records*

Beyond understanding basic terminology, it also helps to understand what the documents you are looking at actually contain. Familiarity with the formats, structures, and functions of the different types of data collections and documents created by the federal government is key for understanding what your search results will and will not contain, and even why the studies were conducted in the first place. Several common types of environmental records include[13]:

■ Environmental Testing Reports

Often, environmental information is generated by testing a sample of soil, air, water, or some other natural substance in a laboratory to see whether it contains toxic substances. To begin to understand where the environmental information from environmental testing comes from, and what it means, you can look to explanations and overviews about the studies' methodologies. EPA posts information about the methods their offices and laboratories approve of for various sampling types and toxins online.[14] For example, EPA's website has links to information about how they analyze drinking water, including information about the lowest concentration minimum reporting level calculator, which they use to assess the water.[15]

In form, testing reports usually begin with a narrative summary, contain several charts highlighting the "most important" data, maps of the testing locations, tables with the full lab results, descriptions of samples from the field (temperature, pH), and "chain of custody" reports.[16]

One limitation of environmental testing results reports is that they only cover the geographical areas where the testing occurred, the contaminants that engineers collected, and the substances studied by the chemists. The

---

12. Google, *Manage Calculator, Unit Converter, and Color Codes*, https://support.google.com/websearch/answer/3284611?hl=en#unitconverter (last visited Dec. 24, 2016).
13. These are merely brief overviews of the various types of environmental records. More information about each type of document is available on the various EPA.gov websites.
14. U.S. EPA, *Collection of Methods*, https://www.epa.gov/measurements/collection-methods (last visited Dec. 24, 2016).
15. U.S. EPA, *Lowest Concentration Minimum Reporting Level (LCMRL) Calculator*, https://www.epa.gov/dwanalyticalmethods/lowest-concentration-minimum-reporting-level-lcmrl-calculator (last visited Dec. 24, 2016).
16. *Id.*

absence of information about a particular location or substance does not mean that there are no environmental issues relating to them—it may be that they were not tested or accounted for in the study.[17] Also, because testing reports tend to be localized in nature, they are often found on state and municipal websites and in print in state and local agencies and libraries.

Figure 11.1 Environmental Testing Report

DEQ

State of Oregon Department of Environmental Quality

KLAMATH FALLS, OREGON
$PM_{2.5}$ PARTICULATE SITE VALIDATION STUDY

2000 - 2001

Conducted By

The Oregon Department Of Environmental Quality

Laboratories And Applied Research Division

Air Quality Monitoring Section

Sampling by: Pegeen Fitzpatrick

Report by: Mark Hansen, and Jeff Smith

August, 2001

Review by:__________ **Date:**__________

Review by:__________ **Date:**__________

Review by:__________ **Date:**__________

1

Table 4. Accuracy Data: FRM versus Survey at Benchmark Site (KFP)

All values in ug/m3.

Date	FRM	P&D Avg	FRM-Avg
7-Jun-00	3.5	4.7	-0.8
11-Jun-00	3.2	2.7	0.5
17-Jun-00	2.6	3.4	-0.8
23-Jun-00	3.9	5.4	-1.5
29-Jun-00	10.2	8.5	1.7
5-Jul-00	2.7	3.2	-0.5
11-Jul-00	5.6	6.4	-0.8
17-Jul-00	6.1	6.6	-0.5
23-Jul-00	3.9	4.2	-0.3
29-Jul-00	4.7	5.1	-0.4
10-Aug-00	4.6	5.3	-0.7
16-Aug-00	5.6	5.7	-0.1
22-Aug-00	8.2	9.3	-1.1
28-Aug-00	3.8	6.2	-2.4
3-Sep-00	2	5	-3
9-Sep-00	3.3	5	-1.7
15-Sep-00	5.3	4.4	-0.9
21-Sep-00	4.3	3.7	0.6
27-Sep-00	8.1	8.7	-0.6
3-Oct-00	6.6	7	-0.4
9-Oct-00	6.9	7	-0.1
17-Oct-00	12.9	12	0.9
21-Oct-00	2.9	3.1	-0.2
27-Oct-00	14.6	12.1	2.5
2-Nov-00	16.2	16.4	-0.2
8-Nov-00	5.9	5.6	0.3
14-Nov-00	19.8	19	0.8
20-Nov-00	32.4	30	2.4
26-Nov-00	21.8	21.1	0.8
2-Dec-00	24.4	23.6	0.8
8-Dec-00	53.6	48.1	5.5
14-Dec-00	6.4	7.5	-1.1
20-Dec-00	24.7	22.9	1.8
26-Dec-00	21.8	28.5	-6.7
1-Jan-01	36.7	40.5	-3.8
7-Jan-01	11.6	17.7	-6.1
13-Jan-01	7.8	10.1	-2.3
19-Jan-01	26.9	32.8	-5.9
25-Jan-01	4.5	6.6	-2.1
31-Jan-01	31.9	32.9	-1
6-Feb-01	2.7	6.3	-3.6
12-Feb-01	10.3	11.4	-1.1
18-Feb-01	1.9	3	-1.1
24-Feb-01	13	17.7	-4.7
8-Mar-01	7.5	9	-1.5
14-Mar-01	11.6	19.3	-7.7
20-Mar-01	11	14.1	-3.1
26-Mar-01	8.8	11	-2.2
1-Apr-01	6.3	8.5	-2.2
7-Apr-01	1.8	5.7	-3.9
13-Apr-01	4.6	7.8	-3.2
25-Apr-01	6.5	8.8	-2.3
1-May-01	3.1	9.3	-6.2
7-May-01	7	11.1	-4.1
13-May-01	5.9	11	-5.2
19-May-01	6	9.7	-3.7
25-May-01	7.1	10	-2.9
31-May-01	3	5.5	-2.5
Average =	10.34	11.85	-1.51
Count =			58
Correlation =			0.9699
Sigma =			2.46
Max diff =			7.7

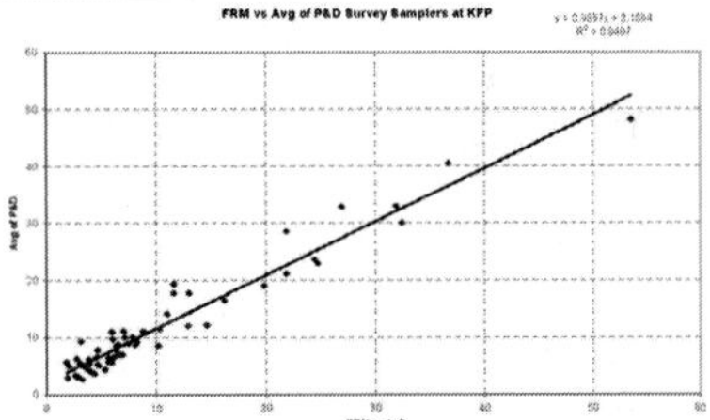

12

This Environmental Testing Report was published by the Oregon Department of Environmental Quality (DEQ) and contains results of particulate matter tests conducted in Klamath Falls, Oregon. The report explains the purpose and methodology of the study, how and why the testing sites were selected, and also provides graphical and tabular depictions of the test results, as well as maps and photographs of the testing locations. See http://www.deq.state.or.us/lab/techrpts/docs/KlamathFallsPM2.5%20Particulate%20Survey%20Report.pdf.

17. *Id.*

■ Environmental Impact Statements

Environmental impact statements (EIS) and reports (often drafted by government agencies pursuant to NEPA requirements[18] or similar state environmental laws) predict the ways that prospective projects may impact the environment and human health[19] without judging whether or not projects should be permitted or modified based on the findings. Rather, the impact statements are merely predictive documents, explaining what might happen if a plan is implemented.[20]

The types of information these statements contain include an explanation of the prospective project's purpose, explaining why the project is necessary and what geographical areas will be affected by the project. These explanations often contain data on the project's focuses. Projects can focus on anything from transportation demands and energy transport needs, to legislative efforts, efforts to link systems together, improve facilities that already exist, or to update or create infrastructure. Impact statements also contain lists of alternative ways that the project could be implemented, including the "no action" alternative, or doing nothing to change the status quo. For each alternative, the EIS compares the efficacy, feasibility, and potential effects on the environment and human health.[21]

There may also be a section discussing the environmental elements that will be affected by the project, which may include information about the condition of the environment in the area of the project—is it already a brownfield, or is it a nature reserve? Are there any major natural landmarks or preserves in the area, etc.?

The EIS may also include background information on the deliberations that went into the decisionmaking process surrounding the potential project. Final EIS filed pursuant to NEPA must contain summaries or full entries containing the comments made on the statement during the drafting process.[22] There may also be information about funding for the project, and

---

18. *See infra* Chapter 2 (discussing the requirements of NEPA).
19. Statistics for Action, A First Look at Technical Documents 5 (2013), http://www.toxicsaction.org/sites/default/files/a_first_look_at_technical_documents.pdf.
20. If the environmental impact statement is drafted because a statute like NEPA requires an agency to draft a predictive report before engaging in an activity or starting a project, the agency can decide to do whatever it wants with the information it gleans from EIS: it can proceed after considering the risks, modify the project to decrease risk, or put the project on hold. It is up to the agency, no matter what the impact statement predicts.
21. 39 C.F.R. §755.11(a)(2)(ii) (2016) (requiring analysis of alternatives, which must include the "no action" alternative).
22. *Id.* §755.11(b)(1)(ii).

about the people and organizations responsible for preparing the statement and conducting its underlying studies and technical reports.

■ Permit Records

Many federal and state environmental laws, including the Clean Water Act and the Clean Air Act, have in-depth permitting requirements for entities engaging in activities that affect the environment.[23] The permitting processes often include environmental studies, evaluations, and commentary from stakeholders that may benefit or be harmed by proposed activities like the building of a medical waste incinerator, dumping of a waste product, or restoration of a brownfield. Thus, the permitting processes, including permit renewal processes,[24] generate many records.

Most permit records are contained in or attached to the actual permit applications. The applications are, in essence, the proposals for new projects made to the municipality, state, or federal government. The applications are forms that the entity pursuing the permit fills out.[25] There are several general categories of activities that require permits:

(1) Release Permits

Certain permits allow individuals or entities to release a certain amount of pollution into waterways, the air, soil, or waste disposal facility. They may be called release permits, but may also be called permits for runoff, discharge, effluent, or some other waste release term. These permits usually only permit these releases for certain amounts of time.

(2) Site Permits

Certain permits allow individuals or entities to build things in certain locations or sites. Site permits require that construction plans contain provisions for dealing with environmental issues that come up during building projects, like sewage, trash, construction pollution, electricity, traffic, stormwater runoff, and other environmental effects inherent in construction.

(3) Other Permits

There are a variety of other types of permits that contain information that may be sought by people researching environmental issues. For instance, there

23. Many of these permitting schemes are discussed in Chapter 2.
24. Most permits are granted for a limited time and must be renewed periodically.
25. Many of the forms can be found on agency websites.

are permits that designate the amounts and types of fuel certain operations can burn, or how much water certain facilities can use. There are also many permitting processes for storing, treating, and selling hazardous substances.[26]

The components of various permits and the information found within the permit applications and renewals vary greatly by the type of permit and the type of agency overseeing the permitting process and granting the permit.[27]

III. The ADEPT Method: Analyzing Environmental Information

Once you obtain the testing reports, impact statements, permitting documents, or other studies or records containing the data you need, the next task is to break down the data into information you can understand so that you can formulate an assessment of the risks you may face.

There is an entire industry centered around risk communication—the art of explaining potential environmental and public health risks to the public.[28] One of the greatest challenges faced by these professionals is the task of explaining complicated scientific and engineering data to laypeople. EPA has created its own risk communication handbook to help public officials communicate environmental risks to potentially affected populations.[29]

As researchers have studied issues inherent in communicating risk to laypeople, they have identified some of the major issues related to understanding environmental and other hazard-related data. The barriers to understanding include the fact that non-experts often lack scientific comprehension skills and have trouble assessing the potential effects of environmental risks. Reports about environmental issues contain technical analyses that

26. See discussion of the Federal Insecticide, Fungicide, and Rodenticide Act, the Resource Conservation and Recovery Act, the Toxic Substances Control Act, and the Comprehensive Environmental Response, Compensation, and Liability Act in Chapter 2.

27. Generally, you can find information related to the proposed operation, building, or project in some level of detail, but until you see the permitting documents, it is hard to be sure of what they will contain. With permit documents, as well, it is also important to remember that the information given is provided by the entities and individuals hoping to get the permit, so the data contained in permits may be skewed favorably toward the granting of the permit, with the best predicted outcomes and the most optimistic data and study results rather than anticipating worst-case scenarios.

28. There are even degree programs dedicated to creating risk communication specialists, see Ashland Univ., *Health and Risk Communication*, https://www.ashland.edu/cas/communication-studies/programs/health-and-risk-communication (last visited Dec. 24, 2016).

29. Christine Reckelhoff-Dangel & Dan Petersen, EPA's Office of Research and Development, Risk Communication in Action: The Risk Communication Workbook (2007), http://nepis.epa.gov/Exe/ZyPDF.cgi/60000I2U.PDF?Dockey=60000I2U.PDF [hereinafter EPA Risk Communication Workbook].

are "often unintelligible to the average layperson."[30] To figure out what the environmental information records actually mean, on a practical level, you may have to:

- Translate jargon and technical terms into everyday language.
- Restructure concepts that are written in the passive voice, or in long, complex sentences.[31]
- Calculate and understand mathematical formulae and tables containing numeric and chemical information.

While there is no quick way to gain expertise into the worlds of environmental science and engineering, there are some strategies to help make the information more accessible and easier to understand. One set of strategies for gaining an understanding of environmental information is the ADEPT Method. Kalid Azad created this method for understanding complicated scientific and mathematical concepts while studying computer science at Princeton.[32] This method is a great way to break down and simplify difficult information and ideas so that you understand them. There are five parts to the ADEPT model[33]:

ADEPT Method Step	Description
Analogy	Tell me what it's like
Diagram	Help me visualize it
Example	Allow me to experience it
Plain English	Describe it to me with everyday words
Technical Definition	Discuss the formal details

This model for understanding is a tool for untangling and making sense of the scientific jargon and information often contained in the results of your environmental information seeking. Here is more detail about each step:

30. UNCERTAINTY IN RISK ASSESSMENT, RISK MANAGEMENT, AND DECISION MAKING 222 (V.T. Covello et al. eds., 1987).
31. EPA RISK COMMUNICATION WORKBOOK, *supra* note 29, at 8.
32. BETTEREXPLAINED, *About*, https://betterexplained.com/about/ (last visited Dec. 24, 2016). Of note, rather than writing a book or other formal print encapsulation of the ADEPT method concept, Kalid Azad, who developed the ADEPT method for understanding while studying computer science at Princeton University, created the BetterExplained website to explain and discuss the method.
33. BETTEREXPLAINED, *Learn Difficult Concepts With the ADEPT Method*, https://betterexplained.com/articles/adept-method/ (last visited Dec. 24, 2016) [hereinafter *Learn Difficult Concepts With the ADEPT Method*].

■ Analogy

Most new information you receive is a variation, extension, or combination what you already know. Compare the complex new information with things that you already understand. The analogy you make between the new information and what you know does not have to be perfect or 100% accurate, it just has to be sufficient to help you build an understanding of the information by relying on knowledge you already have. For instance, to understand the risk of climate change, some scientists analogize the environmental phenomenon with a medical issue that a patient is suffering from. In this analogy, climate scientists are like medical doctors diagnosing the patient, which is planet earth, determining that the earth has a "serious progressive disease" which can be treated with mitigation to stop the disease. Adaptation, on the other hand, is not an appropriate treatment, because adaptation is like putting band aids on visible wounds but allowing the disease to continue to progress.[34] This analogy makes it clear that it is better to attempt mitigation because, if it is successful, adaptation is then unnecessary.[35]

■ Diagram

Visual aids help people to better understand environmental information. Using visual aids—putting post-it notes on walls, drawing pictures and diagrams—helps by making visual depictions for complex concepts and providing a point of focus and understanding to build from. Visualizing abstract concepts is often the key to understanding the concepts, and the process of creating a visual representation of an abstract concept serves as an excellent early step in that understanding process. For instance, when the citizens in the Hattiesburg, Mississippi, town meeting were given a visual model of acceptable levels of toluene as a single drop and the levels of toluene actually in their water as three drops, the meeting attendees could more clearly imagine how far toluene levels were over the threshold in their local water supply. When all of the meeting attendees created their own visual representations of the toluene levels and shared them, the participants were finally able to actually see an illustration of the toluene pollution issue and picture it in their minds. One meeting participant described the benefit of creating visual diagrams, saying that even though the data has been around for years,

34. David Bergman, *Climate Change and the Medical Analogy*, TREEHUGGER (Apr. 30, 2014), http://www.treehugger.com/environmental-policy/climate-change-and-medical-analogy.html.
35. *Id.*

the meeting attendees did not truly "know," or understand the information before talking about the visual representations.[36]

■ Example

After analogizing and visualizing the information and concepts using ideas and pictures that you already understand, Azad recommends that you "experience the idea."[37] Rather than simply memorizing what others have already discovered, you should work through the data by yourself. Give yourself the gift of context by thinking critically about what the data means in real life. If you can envision the diagram or illustration that you created in a real-world situation, do that. If you can visit the site of the environmental issue, view images of it online, or see what the potential risks would lead to (view images associated with the same environmental risks in other places, for example) do it. Find a real-life example and try to imagine the experience of it. Azad recommends using Google images to find diagrams and depictions if you do not already have some. For example, if you are trying to assess what a mineral exploration project would look like on your property, an online search will provide images and notes about what that experience looks and feels like.

■ Plain English

Replace the complicated language in the environmental information records with your own words. Break down the complicated phrases, equations, and concepts into ideas that you understand. Make analogies to concepts that you are familiar with. Your words may vary from the understandings of others, and that is okay, as your understanding is a personal, unique experience. Be explicit, write it down, say it out loud—do not gloss over the complicated parts. Dive in and grasp an understanding of the material—your particular understanding. For instance, if you are looking at a report about toluene levels in a local well, describe in your own words what those risks mean, as if you are explaining them to your neighbor: "Toluene is a chemical found in crude oil and it's the stuff used in paint thinners and other solvents. If there is more than one part per million of the chemical toluene in the river, we should be worried, because toluene can damage our nervous systems, kidneys, and liver."[38]

---

36. Contini-Field & Merson, *supra* note 1.
37. *Learn Difficult Concepts With the ADEPT Method*, *supra* note 33.
38. Information gathered and synthesized from various EPA reports and websites.

■ Technical Definition

The final step in the ADEPT process is to convert your personal understanding, developed through the prior four steps, to the formal notation used in the environmental information world. Now that you have developed your own context, analogies, visual diagrams, and even language to describe the information, you should be able to transfer it back into its formal notation, going back and forth between your personalized understanding of the material and the material itself.

Sometimes, you will need to work backwards to the analogy step, or start with a rough analogy that you revise as you tackle the technical details. Here are some questions Azad recommends asking if you start with the technical details[39]:

- Can you explain the technical details in your own words?
- Can you solve an example problem, describing the steps in your own words?
- Can you create a diagram that represents how the concept fits together for you?
- Can you relate the concept to what you already know?

IV. Other Methods for Understanding Environmental Information

In addition to the ADEPT method, there are other methods for understanding mathematical concepts.[40] For numeric data, specifically, calling on "personal benchmarks," or measurements you are already familiar with, to visualize various quantities, to compare numerical information to things you are sure about, and to pay attention to how things are related. People visualize mathematical measurements in very personal ways. For instance, when asking a group of people what a kilogram weighs as much as—and the answers ranged from a pair of work boots, a pile of sweetener packets, or five apples. Relying on familiar concepts to grasp new concepts, making comparisons and forming analogies, helps to ground new ideas in things you already know.

39. *Learn Difficult Concepts With the ADEPT Method*, *supra* note 33.
40. Statistics for Action, *supra* note 3, at 2, 3.

Figure 11.2 Understanding Measurements

A kilogram? That weighs as much as ...

Imagining measurements in a familiar form can help you visualize the sizes, levels, and amounts of more abstract things like chemicals and elements.

A. *Using the Smart Moves Model*

The Smart Moves model, designed for helping people to understand mathematical concepts and understand sequential reasoning, can also help you to understand environmental data concepts.[41] Smart Moves urges people to approach numbers and mathematical information by:

- Slowing down and and taking the time to better understand the information, even if it takes extra time
- Playing with the numbers in different ways to "show it" and "say it"
- Using your senses to visualize and "touch" the concepts (for instance, counting out pennies and putting them in piles, moving them to represent mathematical change)
- Talking out loud, working in groups, and talking to peers
- Comparing the information, connecting it to what you already know
- Using "friendly numbers" (rounding numbers to the tens, etc. to make them easier to work with)

41. Michael Serra, Smart Moves: Developing Mathematical Reasoning With Games and Puzzles (2011).

- Seeking verification—consulting a peer or calculator, or contacting an expert.

Mathematical calculations need not be done only privately in your head, or alone in front of a calculator—math plays a public role in environmental issues, and it should be publicly used, discussed, and shared even if the numerical data is complicated or dry. You may have to reach out to scientists or the agency itself for help or create illustrations and visualizations to make mathematical concepts clearer and easier to talk about in public. Do not give up just because the information seems complicated and jargon-filled. Somebody out there documented the data you are looking at, so it is necessarily the case that somebody can help you understand that data, as well.

B. Getting Help From the Experts

In the era of quick and seamless online searching, it is sometimes easy to overlook the power of using the telephone or sending an e-mail. Reaching out to the authors of a study, the agency overseeing the work, or an expert in the area, can go a long way toward understanding a piece of environmental information. People generally enjoy discussing their areas of interest, study, and expertise, and are often happy to field questions or provide explanations of their projects, work, and interests. Do not hesitate to reach out to a scientist, nonprofit organization, or government office involved in creating or distributing the data you are trying to understand. If the person you contact cannot answer your questions, they can likely direct you to someone who can.

Additionally, if you are dealing with data from EPA or another federal agency, be aware that the agencies create fact sheets and hold webinars to explain some of the data, so you may be able to get explanations from the agency itself.[42] If you would like more information or a webinar on a specific data set or record, contact the agency to request those resources, as you may prompt the agency to create the necessary resource or webinar to explain the data. You may also be able to find more information about your data set by exploring other, non-agency resources. For instance, if you locate information about toluene pollution in a certain waterway on an EPA website, you may find more information about that pollution in a report posted on an academic or nonprofit organization's website, or in a paper in Google Scholar

42. Ethan McMahon, *I Know That Data Is Here Somewhere...*, Our Planet, Our Home (July 31, 2009), https://blog.epa.gov/blog/2009/07/datafinder/; *see also* City of Tuscaloosa Envtl. Servs. Dep't, A Reference Guide for Homeowners (2012), http://www.tuscaloosa.com/Assets/Government/Department-Documents/Environmental-Services/ESD-Documents/hhw%20brochure.pdf.

or a legislative Congressional Research Service report. These various information sources are discussed at length in Chapter 8.

V. Using Environmental Information to Assess Risks

After environmental information has been gathered, understood, and analyzed, the information gatherer, or related party, will often want to make environmental decisions based on the information. They may be deciding whether to file a citizen suit for an injunction pursuant to an environmental law, or whether they need to hire an attorney to defend themselves against such litigation. They may be deciding whether to buy a house in a particular area or organize public activism around an environmental event nearby. All of these decisions rely on assessing the risk associated with the environmental information. Risk can be defined by the "risk triplet," or three questions that capture the scope of the concept of risk:

- What can happen? (ex., What can go wrong?)
- How likely is it that that will happen?
- If it does happen, what are the consequences?

Risk is often measured by the likelihood of harm from a hazard, or the probability that something bad will happen. (For instance, exposure to a certain toxin creates a one-in-a-million chance of dying from cancer.) Risk is also thought of in terms of particular hazards, and certain activities are considered to be riskier than others. (The risks of ingesting mercury include birth defects and damaging the nervous system and kidneys.) Risk is also sometimes assessed by considering particular consequences. (What are the risks to air quality associated with smokestack pollution?) No matter how you define the term risk, there are several ways to conceptualize and evaluate potential risks. EPA has published factors to consider when assessing risks to human health[43]:

- Identify the hazards. Examine whether something could potentially harm humans or ecological systems, and if so, under what circumstances.
- Decide who might be harmed and how. Examine the relationship between exposure to the risk and the effects that it will have on health and safety using numerical and scientific data. EPA calls this determi-

43. U.S. EPA, *Human Health Risk Assessment*, https://www.epa.gov/risk/human-health-risk-assessment (last visited Dec. 24, 2015).

nation a Dose-Response Assessment.[44] Also examine the information available about the frequency, timing, and levels of contact with the potential risk. EPA calls this an Exposure Assessment.[45] These assessments will help you determine the nature of the risk.

- Evaluate the risks and decide on control measures. Determine the overall risk based on the potential hazards and harm threatened by the particular risky situation. Evaluate that risk and determine what plans could be implemented to control and minimize the risk and its harmful consequences. The various prevention and remediation measures may have different pros and cons; varying levels of cost effectiveness, impacts ranging from 100% eliminating risk versus merely decreasing risk, differing time frames for implementation and success, and other pragmatic considerations. Depending on your goals (Are you focused on balancing costs and benefits or would you prefer to maximize the risk reduction?), your preferred control measures may differ.
- Record your findings and implement them. After implementing a selected method of risk prevention following a risk assessment, the entity or individual implementing the plan should go back and review the risk assessment and methods periodically, to see whether the risk has changed (grown, decreased, etc.), whether the methods are effectively preventing the risk, and whether there are new, improved methods for preventing the risk. The risk assessment and methods should be amended and updated as needed.

EPA has a website dedicated to risk assessment containing links to many online environmental risk assessment guides and tools, These resources are created and maintained by the agency to deal with certain types of environmental risks in the United States.[46] Many of EPA's data resources for particular environmental laws covered in Chapter 2 were actually created for the purpose of risk assessment communication. One of EPA's website links connects to risk hotlines for various environmental issues in the United States and also to offices where the public can reach "risk assessment specialists" who will help you determine the actual risks associated with particular

44. U.S. EPA, *Dose-Response Assessment for Assessing Health Risks Associated With Exposure to Hazardous Air Pollutants*, https://www.epa.gov/fera/dose-response-assessment-assessing-health-risks-associated-exposure-hazardous-air-pollutants (last visited Dec. 24, 2016).
45. U.S. EPA, *Guidelines for Human Exposure Assessment*, https://www.epa.gov/osa/guidelines-human-exposure-assessment (last visited Dec. 24, 2016).
46. U.S. EPA, *Risk Assessment*, https://www.epa.gov/risk (last visited Dec. 24, 2016).

environmental events and issues.[47] EPA also has many assessment guidelines for assessing different types of environmental risks, although these are as filled with jargon and technical terminology as many of the sources you are trying to make sense of by reading this very chapter.[48]

Risk assessment can be a very personal determination, and it is not always a uniform judgment—what one person perceives as very risky, another may perceive as relatively free from risk. Indeed, as several environmental risk assessment scholars write: "Perceptions of risk play a prominent role in the decisions people make, in the sense that differences in risk perception lie at the heart of disagreements about the best course of action between technical experts and members of the general public, men vs. women, and people from different cultures."[49] These different perceptions are very visible in current environmental debates about issues like fracking and climate change—people's opinions on the risks associated with fracking and climate change vary widely despite the data available to the public.

Factors like past experiences and religious beliefs also affect the public's assessment of risks. Perception of risk is also swayed by varying levels of familiarity with mechanisms underlying the relevant technology, the visibility of the threat, whether the risk is within public control, and the weight of the risk against other societal concerns like property values and decline in community image.[50]

Media's depictions of potential risks and current events also affect how the public assesses risk.[51] What's more, the public often associates mere correlation or association with actual causality, drawing conclusions about what causes things that may not be based in evidence.[52] All of these variables should be considered when assessing and discussing risk, because they will, either consciously or subconsciously, shape your understanding of the risk at hand. Considering the personal perspectives and beliefs associated with

---

47. U.S. EPA, *About Risk Assessment: Getting Help With Risk Assessment Issues*, https://www.epa.gov/risk/about-risk-assessment#tab-3 (last visited Dec. 24, 2016).
48. U.S. EPA, *Risk Assessment Guidelines*, https://www.epa.gov/risk/risk-assessment-guidelines (last visited Dec. 24, 2016).
49. Paul Slovic & Elke U. Weber, Perception of Risk Posed by Extreme Events, Paper presented at Risk Management Strategies in an Uncertain World, Apr. 12-13, 2002 (Columbia University and Wissenschaftskolleg zu Berlin), http://www.rff.org/files/sharepoint/Documents/Events/Workshops%20and%20Conferences/Climate%20Change%20and%20Extreme%20Events/slovic%20extreme%20events%20final%20geneva.pdf.
50. National Research Council, Analysis of Cancer Risks in Populations Near Nuclear Facilities: Phase I, §5.2 (2012), http://www.ncbi.nlm.nih.gov/books/NBK201996/pdf/Bookshelf_NBK201996.pdf.
51. *Id.*
52. *Id.* §5.2.1.

perceptions of risks adds another layer of complexity to the environmental decisionmaking process.

VI. Communicating About Environmental Information and Risk

After you have assessed the risks associated with the environmental information you have gathered, you can communicate the risks to your target audience. Environmental information is rarely collected in a vacuum, and you have likely gathered these records in anticipation of litigation, or a public awareness campaign, or even scholarly writing. If you would like to use the information you have gathered for advocacy or policy purposes or present it to a wider audience, there are several things to consider when trying to communicate about your environmental issue after you have made sense of the information[53]:

- Determine who your target audience is. Is it a group of officials (a board or agency) with expertise in the area or a reporter with no technical knowledge or background of the environmental issue? If your target audience is members of the public, determine how much knowledge they have of the situation. You should also try to assess the general attitude of your target audience—do they agree with your proposition and just need to be supported or spurred into action, or do they disagree and need to be convinced? Are they aware of the situation, or do they need to be alerted to the environmental issue?
- Determine the format of your communication. Will you communicate verbally or in writing? You could have an interview or meeting and give a presentation, or conversely, you could draft written materials like a flyer, fact sheet, or newspaper ad. Both of these methods require preparation. Verbal communication may involve creating visual tools like a PowerPoint or charts, or even an online video. You may have to schedule a press conference or find the proper public forum, like a town meeting, and determine how much time you will need or be allowed to speak. Written communication requires you to select either an online or paper format (or both) and determine how much space you would like your communication to take, what visual formatting to use and how many pieces of information to bullet-point. Remember that often

53. Statistics for Action, *Communicating With Numbers*, http://sfa.terc.edu/using/options/communicating.html (last visited Dec. 24, 2016).

times, charts, graphs, and photos help people better understand the information you are presenting.

- Determine the best setting for your message. If you are engaging in environmental litigation, then you may be using the information in a complaint or other court motion or filing, or possibly a legal brief or memorandum. You may be using the information to testify at an administrative or legislative hearing or a public meeting. If no written or in-person forum exists for you to communicate your findings, you may want to contact a news reporter, write an opinion piece, or set up a community meeting to raise awareness of your issue.

If you can, create graphs and memorable messages to present key facts. Also meet with others; getting together, doing group activities and providing handouts and fact sheets can help you communicate environmental information to others. If you effectively share your environmental information with other stakeholders, you can help people make environmental decisions in their communities.

VII. Conclusion

Although the data you receive through your environmental information seeking efforts may initially seem impenetrable, with the right assistance and resources, you can break the information down into a helpful, usable form. For instance, in Hattiesburg, after community members were able to visualize and create their own illustrations of the water pollution issues facing their town, they were able to understand the information and share it with others.[54]

Beyond Hattiesburg, understanding and obtaining environmental information is a vital part of participation in environmental decisionmaking. Being able to collect and understand information about our local environs is vital for participation in environmental work and advocacy. An informed public is imperative for environmental protection and key to peoples' ability to safeguard public health.[55] In fact, some regard environmental information access as a primary tool for cultivating "environmental democracy."[56] In an ideal world, environmental information would be

54. Contini-Field & Merson, *supra* note 1.
55. William A. Wilcox, *Access to Environmental Information in the United States and the United Kingdom*, 23 Loy. L.A. Int'l & Comp. L. Rev. 121, 122 (2001).
56. Kofi Annan, *Foreword*, *in* The Aarhus Convention: An Implementation Guide (2000); Ban Ki-Moon, *Foreword*, *in* The Aarhus Convention: An Implementation Guide (2d ed. 2014).

streamlined, centrally located, and available to all in an easy to understand format. The real world, however, is very complex, containing an array of languages and perspectives, multiple creators of environmental facts, figures, and deliberations, within layers of bureaucracy and government systems. Knowing how to access information across those layers, using the various access resources created through legal authority, information science technology, and people and organizations dedicated to environmental work is an imperative step in environmental decisionmaking.

Appendix 1: Chronological List of Select Major Federal Environmental Statutes and Amendments

This list is illustrative as opposed to exhaustive. It contains some, but not all, amendments to major environmental statutes. Additionally, the CFR sections selected are not exhaustive, as many of these statutes generated numerous regulations by multiple agencies and large-scale environmental programs with many regulatory aspects.

Statute	Related U.S. Code	Public Law No.	Statutes at Large	CFR	Popular Name	Enacted Bill	Year
Yellowstone National Park Act	16 U.S.C. §§21-40c	Act of Mar. 1, 1872, ch. 24	17 Stat. 32	36 C.F.R. Parts 1.1-1.10 (general National Park Service provisions)	Yellowstone Act	S. 392	1872
Timber Culture Act (repealed in 1891)		Pub. L. No. 42-277	17 Stat. 605		TCA		1873
Forest Reserve Act (repeals Timber Culture Act)	16 U.S.C. §§471-539r	Pub. L. No. 51-561	26 Stat. 1095		FRA Creative Act	H.R. 7254, S. 5129 (includes a rider authorizing president to establish reserves)	1891
Yellowstone Game Protection Act (amended by National Park Protection Act)	16 U.S.C. §§24-31	Act of May 7, 1894	28 Stat. 73		National Park Protective Act/Lacey Act	H.R. 6442	1894
Forest Management Act of 1897 (implements Forest Reserve Act)	16 U.S.C. §§473	Act of June 4, 1897	30 Stat. 34, 36		The Organic Act of 1897	Part of the annual sundry civil appropriations act	1897
Rivers and Harbors Act	33 U.S.C. §§401, 403, 407-407a	Act of Mar. 3, 1899, ch. 425	30 Stat. 1151	19 C.F.R. Part 4.66b	Refuse Act		1899

Statute	Related U.S. Code	Public Law No.	Statutes at Large	CFR	Popular Name	Enacted Bill	Year
National Park Protection Act or Lacey Act of 1900	16 U.S.C. §§3371–3378	Act of May 25, 1900	31 Stat. 187	50 C.F.R. Part 16	Lacey Act	H.R. 6634	1900
Antiquities Act of 1906 (repealed in 2014, replaced by 54 U.S.C. §§320301-320303)	16 U.S.C. §§431-433	Pub. L. No. 59-209	34 Stat. 225	36 C.F.R. Parts 7.1-7.100, 12.1-12.12		H.R. 11,016, S. 4698	1906
Pure Food and Drug Act (replaced by the Federal Food, Drug, and Cosmetic Act of 1938)	21 U.S.C. §§1-15	Pub. L. No. 59-384	34 Stat. 768		Wiley Act		1906
Federal Insecticide Act	7 U.S.C. §§136 et seq.	Pub. L. No. 61-152	36 Stat. 331	40 C.F.R. Parts 150-180.2020	FIA FIFRA Insecticide Act of 1910	S. 6131	1910
Migratory Bird Act (replaced by the Migratory Bird Treaty Act)		Pub. L. No. 62-430	37 Stat. 828		Weeks-McLean Act	Rider to an appropriation bill for the Dep't of Agriculture	1913
National Park Service Act (repealed in 2014 and replaced by 54 U.S.C. §§100101-100102)	16 U.S.C. §§1-4	Pub. L. No. 64-235	39 Stat. 535	36 C.F.R. Parts 1-79	The Organic Act of 1916	H.R. 15522	1916
Migratory Bird Treaty Act	16 U.S.C. §§703-712	Pub. L. No. 65-186	40 Stat. 755	50 C.F.R. Parts 20-21	MBTA	S. 1553	1918

Statute	Related U.S. Code	Public Law No.	Statutes at Large	CFR	Popular Name	Enacted Bill	Year
Federal Water Power Act (amended in 1935)	16 U.S.C. §§791a-828c	Pub. L. No. 66-280	41 Stat. 1063	18 C.F.R. Subchapter B	Federal Power Act	H.R. 3184	1920
Mineral Leasing Act (replaced the General Mining Act of 1872)	30 U.S.C. §§181-287	Pub. L. No. 66-146	41 Stat. 437	43 C.F.R. Parts 3100-3930 generally deal with minerals management		S. 2775	1920
The Taylor Grazing Act of 1934	43 U.S.C. §§315-315r	Pub. L. No. 73-482	48 Stat. 1269	43 C.F.R. 4100.0-1-4100.0-9		H.R. 6462	1934
Fish and Wildlife Coordination Act	16 U.S.C. §§661-667e	Pub. L. No. 73-121	48 Stat. 401	50 C.F.R. covers wildlife and fisheries, generally	FWCA	S. 2529	1934
Historic Sites, Buildings, and Antiquities Act (repealed in 2014, replaced by 54 U.S.C. §§320301-320303)	16 U.S.C. §§461-467	Pub. L. No. 74-292	49 Stat. 666	36 C.F.R. Parts 7.1-7.100, 12.1-12.12	Historic Sites Act	S. 2073	1935

Statute	Related U.S. Code	Public Law No.	Statutes at Large	CFR	Popular Name	Enacted Bill	Year
Soil Conservation and Domestic Allotment Act (repealed)	16 U.S.C. §§590(e), 590(e)(1), 590(g), 590(h), 590(p)(1), 590(q)(3)	Pub. L. No. 74–461	49 Stat. 1148		Soil Conservation Act of 1935 (amended and renamed Soil Conservation and Domestic Allotment Act)	S. 3780	1936
Federal Food, Drug, and Cosmetic Act (amends Pure Food and Drug Act)	21 U.S.C. §§301-399f	Pub. L. No. 75-717	52 Stat. 1040	21 C.F.R. 1-1499 covers food and drugs, generally	FFDCA FDCA FD&C	S. 5	1938
Bald and Golden Eagle Protection Act	16 U.S.C.A. §§668-668c	Pub. L. No. 76-567	54 Stat. 250	50 C.F.R. Part 22	BGEPA	H.R. 4832	1940
Walrus Protection Act (replaced with Marine Mammal Protection Act)	48 U.S.C. §§248 et seq. (omitted as obsolete)	Pub. L. No. 77-218	55 Stat. 632			H.R. 1606	1941
Atomic Energy Act of 1946	42 U.S.C. §§2011-2023	Pub. L. No. 79-585	60 Stat. 755	10 C.F.R. Parts 960.1-960.5-2-11, 962.1-962.3	McMahon Act	S. 1717	1946
Federal Insecticide, Fungicide, and Rodenticide Act	7 U.S.C. §§136-136y	Pub. L. No. 80-104	61 Stat. 163	40 C.F.R. Parts 150-189	FIFRA	H.R. 1237	1947

Statute	Related U.S. Code	Public Law No.	Statutes at Large	CFR	Popular Name	Enacted Bill	Year
Whaling Convention Act of 1949	16 U.S.C. §916-916l	Pub. L. No. 81-676	64 Stat. 421	50 C.F.R. Parts 230.1-230.8		S. 2080	1950
Submerged Lands Act	43 U.S.C. §§1301-1315	Pub. L. No. 83-31	67 Stat. 29			H.R. 4198	1953
Atomic Energy Act of 1954	42 U.S.C. §§2011-2021, 2022-2286i, 2296a-2297h-13	Pub. L. No. 83-703	68 Stat. 919	10 C.F.R. Parts 960.1-960.5-2-11, 962.1-962.3	AEA	H.R. 9757	1954
Air Pollution Control Act	42 U.S.C. §7401 et seq.	Pub. L. No. 84-159	69 Stat. 322	40 C.F.R. 50.1-98.478 cover air programs, generally	APCA	S. 928	1955
Fish & Wildlife Act	16 U.S.C. §§742a-742j	Pub. L. No. 84-1024	70 Stat. 1119	50 C.F.R. covers wildlife and fisheries, generally		S. 3275	1956
Multiple-Use Sustained Yield Act	16 U.S.C. §§528-531	Pub. L. No. 86-517	74 Stat. 215	30 C.F.R. Parts 293.1-293.17	MUSY / MUSYA	H.R. 10572	1960

Statute	Related U.S. Code	Public Law No.	Statutes at Large	CFR	Popular Name	Enacted Bill	Year
Refuge Recreation Act	16 U.S.C. §§460k-460k-4	Pub. L. No. 87-714	76 Stat. 653	50 C.F.R. Parts 71.1, 71.2, 71.11, 71.12, 70.1-70.9, 32.1-32.8, 29.1-29.5, 31.1-31.2		H.R. 1171	1962
Clean Air Act of 1963	42 U.S.C. §§7401-7671q	Pub. L. No. 88-206	77 Stat. 392	40 C.F.R. 50.1-98.478 (cover air programs, generally)		H.R. 6518	1963
Federal Insecticide, Fungicide, and Rodenticide Act Amendments (amends FIFRA)	7 U.S.C. §§136-136y	Pub. L. No. 88-305	78 Stat. 190	40 C.F.R. Parts 150-189	FIFRA Amendments	S. 1605	1964
Wilderness Act	16 U.S.C. §§1131-1136	Pub. L. No. 88-577	78 Stat. 890	50 C.F.R. Parts 35.1-35.14		S. 4	1964
Land and Water Conservation Act	16 U.S.C. §§460d, 4601-4 to 4601-11	Pub. L. No. 88-578	78 Stat. 897	36 C.F.R. Parts 327.0-327.31	LWCF Act	H.R. 3846	1965

Statute	Related U.S. Code	Public Law No.	Statutes at Large	CFR	Popular Name	Enacted Bill	Year
Solid Waste Disposal Act (combined into a single bill with Motor Vehicle Air Pollution Control Act)	42 U.S.C. §6901-6908a	Pub. L. No. 89-272	79 Stat. 992	40 C.F.R. Parts 264.1-264.3	SWDA	S. 306	1965
Motor Vehicle Air Pollution Control Act (combined into a single bill with Solid Waste Disposal Act)	42 U.S.C. §§7521-7590	Pub. L. No. 89-272	79 Stat. 992	40 C.F.R. Parts 86.000-2-86.099-11		S. 306	1965
National Historic Preservation Act (repealed in 2014, replaced by 54 U.S.C. §§320301-320303)	16 U.S.C. §§470a-470w-6	Pub. L. No. 89-665	80 Stat. 915	36 C.F.R. Parts 7.1-7.100, 12.1-12.12	NHPA	S. 3035	1966
Fish and Wildlife Conservation Act	16 U.S.C. §§668dd-668ee	Pub. L. No. 89-669	80 Stat. 926	50 C.F.R. (covers wildlife and fisheries, generally)	Nongame Act	H.R. 9424	1966
National Wildlife Refuge System Administration Act (same sections as Fish and Wildlife Act, but focuses on administrative and management issues)	16 U.S.C. §§668dd-668ee	Pub. L. No. 89–669 §§4, 5	80 Stat. 927	50 C.F.R. Parts 25.11-25.13 (administrative provisions)		H.R. 9424	1966

Statute	Related U.S. Code	Public Law No.	Statutes at Large	CFR	Popular Name	Enacted Bill	Year
Air Quality Act of 1967 (amends Clean Air Act of 1963 and is amended by the Clean Air Act of 1970)	42 U.S.C. §1857	Pub. L. No. 90-148	81 Stat. 485	40 C.F.R. 50.1-98.478 (cover air programs, generally)		S. 780	1967
National Wild and Scenic Rivers Act	16 U.S.C. §§1271-1287	Pub. L. No. 90-542	82 Stat. 906	36 C.F.R. Parts 297.1-297.6	WSR ACT WSRA	H.R. 8416	1968
National Environmental Policy Act	42 U.S.C. §§4321-4370h	Pub. L. No. 91-190	83 Stat. 852	40 C.F.R. Parts 1500-1508 (CEQ regulations for implementing NEPA)	NEPA	S. 1075	1969
Occupational Safety and Health Act	29 U.S.C. §§651-678	Pub. L. No. 91-596	84 Stat. 1590	29 C.F.R. Parts 1910-1910.1450 App B (OSHA standards), 20 29 C.F.R. Parts 1900-1990, (OSHA generally)	OSH Act	S. 2193	1970

Statute	Related U.S. Code	Public Law No.	Statutes at Large	CFR	Popular Name	Enacted Bill	Year
Clean Air Act Amendments	42 U.S.C. §§7401-7671q	Pub. L. No. 91-604	84 Stat. 1676	40 C.F.R. 50.1-98.478 (cover air programs, generally)	CAA	H.R. 17255	1970
Federal Water Pollution Control Act	33 U.S.C. §§1251-1387	Pub. L. No. 92-500	86 Stat. 816	40 C.F.R. Parts 100-140, 401-471, 501-503	Clean Water Act CWA	S. 3122	1972
Federal Environmental Pesticide Control Act (amends FIFRA)	7 U.S.C. §§136-136y	Pub. L. No. 92-516	86 Stat. 973	40 C.F.R. Parts 150-189	FEPCA	H.R. 10729	1972
Marine Protection, Research, and Sanctuar-ies Act	33 U.S.C. §§1401-1445	Pub. L. No. 92-532	86 Stat. 1052	15 C.F.R. Parts 992.1-992.198, 33 C.F.R. Parts 324.1-325.10	Ocean Dumping Act, MPRSA	H.R. 9727	1972
Marine Mammal Protec-tion Act (amended in 1994)	16 U.S.C. §§1361-1423h	Pub. L. No. 92-522	86 Stat. 1027	50 C.F.R. Parts 18.1-18.129	MMPA	H.R. 10420	1972
Noise Control Act	42 U.S.C. §§4901-4918	Pub. L. No. 92-574	86 Stat. 1234	40 C.F.R. Parts 201.1-211.214	NCA	H.R. 11021	1972

Statute	Related U.S. Code	Public Law No.	Statutes at Large	CFR	Popular Name	Enacted Bill	Year
Coastal Zone Management Act	16 U.S.C. §§1451-1466	Pub. L. No. 92-583	86 Stat. 1280	15 C.F.R. 923.1-932.3, 49 C.F.R. 1105.9	CZMA	S. 3507	1972
Endangered Species Act (implements 1975 Convention on International Trade in Endangered Species (CITES) international treaty)	16 U.S.C. §§1531-1599	Pub. L. No. 93-205	87 Stat. 884	50 C.F.R. Part 17	ESA	S. 1983	1973
Forest & Rangeland Renewable Resources Planning Act	16 U.S.C. §§1601-1610	Pub. L. No. 93-378	88 Stat. 476	36 C.F.R. Parts 219.1-219.62	Resource Planning Act RPA	S. 2296	1974
Energy Reorganization Act of 1974	42 U.S.C. §§ 5801-5891	Pub. L. No. 93-438	88 Stat. 1233	10 C.F.R. Parts 1.1-171.25 (Nuclear Regulatory Commission regulations, generally)	ERA	H.R. 11510	1974
Public Health Service Act	42 U.S.C. §§300f-300j	Pub. L. No. 93-523	88 Stat. 1660	40 C.F.R. Part 22	Safe Drinking Water Act SDWA	H.R. 13002	1974

Statute	Related U.S. Code	Public Law No.	Statutes at Large	CFR	Popular Name	Enacted Bill	Year
Federal Coal Leasing Amendments Act (amends Mineral Leasing Act of 1920)	30 U.S.C. §§202a, 208-1	Pub. L. No. 94-377	90 Stat. 1083	30 C.F.R. Parts 2-1243, 43 C.F.R. 16-9230, 36 C.F.R. 228	FCLAA	S. 391	1976
Toxic Substances Control Act	15 U.S.C. §§2601-2695d	Pub. L. No. 94-469	90 Stat. 2003	40 C.F.R. Parts 700-789, 790-799	TSCA	S. 3149	1976
Resource Conservation and Recovery Act	42 U.S.C. §§6901–6992k	Pub. L. No. 94-580	90 Stat. 2395	40 C.F.R. Parts 239-282	RCRA	S. 2150	1976
Federal Land Policy and Management Act	43 U.S.C. §§1701-1787	Pub. L. No. 94-579	90 Stat. 2744	43 C.F.R. Part 4100	FLPMA	S. 507	1976
National Forest Management Act of 1976	16 U.S.C. §§1600-1614	Pub. L. No. 94-588	90 Stat. 2948	36 C.F.R. Parts 219.1-219.19	NFMA	S. 3091	1976
Surface Mining Control and Reclamation Act	30 U.S.C. §§1231-1244	Pub. L. No. 95-87	91 Stat. 445	30 C.F.R. Parts 700.1-955	SMCRA	H.R. 2	1977
Clean Air Act Amendments	42 U.S.C. §§7401-7671q	Pub. L. No. 95-95	91 Stat. 712	40 C.F.R. 50.1-98.478 (cover air programs, generally)	1977 CAAA	H.R. 6161	1977

Statute	Related U.S. Code	Public Law No.	Statutes at Large	CFR	Popular Name	Enacted Bill	Year
Federal Water Pollution Control Act (amends CWA)	33 U.S.C. §§1251-1376	Pub. L. No. 95-217	91 Stat. 1566	40 C.F.R. Parts 100-140, 401-471, 501-503	Clean Water Act	H.R. 3199	1977
Outer Continental Shelf Lands Act (amends same of 1953, 1969, 1976)	43 U.S.C. §§1331-1356	Pub. L. No. 95-372	92 Stat. 629	43 C.F.R. Part 2880, 30 C.F.R. Chapter 2 (minerals management, generally)	OSCLA	S. 9	1978
Federal Environmental Pesticide Control Act (amends FIFRA)	7 U.S.C. §136	Pub. L. No. 95-396	92 Stat. 819	40 C.F.R. Parts 150-189	FEPCA; Federal Pesticide Act of 1978	S. 1678	1978
Fish and Wildlife Conservation Act	16 U.S.C. §§2901-2912	Pub. L. No. 96-366	94 Stat. 1322	50 C.F.R. Parts 10, 13-18, 21-24	Nongame Act	S. 2181	1980
Comprehensive Environmental Response, Compensation, and Liability Act	42 U.S.C. §§9601-9675	Pub. L. No. 96-510	94 Stat. 2767	40 C.F.R. Part 307	Superfund; CERCLA	H.R. 7020	1980

Statute	Related U.S. Code	Public Law No.	Statutes at Large	CFR	Popular Name	Enacted Bill	Year
Federal Oil and Gas Royalty Management Act (mandates protection of the environment and conservation of federal lands in the course of building oil and gas facilities)	30 U.S.C. §§1701	Pub. L. No. 97-451	96 Stat. 2447	30 C.F.R. Part 1200, 25 C.F.R. Parts 211-212, 43 C.F.R. Part 3100 (federal oil and gas leasing, generally)	FOGRMA	H.R. 5121	1983
Emergency Planning and Community Right-To-Know Act	42 U.S.C. §§11001-11050	Pub. L. No. 99-499	100 Stat. 1613	40 C.F.R. Parts 300-311, 355, 373	EPCRA EPCRTKA SARA Title III	H.R.2005 (originally introduced as separate bill but enacted as part of CERCLA)	1986
Marine Protection, Research, and Sanctuaries Act (amends MPRSA of 1972)	16 U.S.C. §§1431-1445c, and 33 U.S.C. §1401-1402	Pub. L. No. 100–711	102 Stat. 4755	40 C.F.R. Parts 220.1-220.4	MPRSA; Ocean Dumping Ban Act	H.R. 4189	1988
Dolphin Protection Consumer Information Act	16 U.S.C. §1385	Pub. L. 101-627	104 Stat. 4465	50 C.F.R. Parts 216.90-216.95	DPCIA	S. 2044	1990

Statute	Related U.S. Code	Public Law No.	Statutes at Large	CFR	Popular Name	Enacted Bill	Year
Oil Pollution Act	33 U.S.C. §§2701-2762	Pub. L. No. 101-380	104 Stat. 484	40 C.F.R. Parts 110, 112, and 300 subparts C, D, E, 49 C.F.R. Part 194, 33 C.F.R. Part 154	OPA OPA 90	H.R. 1465	1990
Pollution Prevention Act	42 U.S.C. §§13101-13109	Pub. L. No. 101-508	104 Stat. 1388	48 C.F.R. 52.223-5	PPA	S. 585	1990
Wild Bird Conservation Act	16 U.S.C. §§4901-4916	Pub. L. No. 102-440	106 Stat. 2224	50 C.F.R. Parts 15.1-15.53	WBCA	H.R. 1627	1992
Marine Mammal Protection Act Amendments of 1994 (amends 1972 MMPA)	16 U.S.C. §§1361-1423h	Pub. L. No. 103-238	108 Stat. 532	50 C.F.R. Parts 18.1-18.129	MMPA	S. 1636	1994
Food Quality Protection Act	7 U.S.C §§136-136y	Pub. L. No. 104-170	110 Stat. 1489	40 C.F.R. 180.1-180.6 (tolerance levels for pesticide chemical residues in food, generally)	FQPA	H.R. 1627	1996

Statute	Related U.S. Code	Public Law No.	Statutes at Large	CFR	Popular Name	Enacted Bill	Year
Oil and Gas Royalty Simplification and Fairness Act of 1996 (amends FOGRMA)	30 U.S.C. §1701	Pub. L. No. 104-185	110 Stat. 1700	30 C.F.R. Part 1200, 25 C.F.R. Parts 211-212, 43 C.F.R. Part 3100 (federal oil and gas leasing, generally)		H.R. 1975	1996
Chemical Safety Information Site Security and Fuels Regulatory Relief Act (amends CAA)	42 U.S.C. §§7401-7431	Pub. L. No. 106-40	113 Stat. 207	40 C.F.R. Parts 51.15-51.50, 68.1-68.15	CSISSFRRA	S. 880	1999
Pesticide Registration Improvement Act (amends FIFRA, and has been reauthorized twice: PRIA II, 2007 and PRIA III, 2012)	7 U.S.C. §136w-8	Pub. L. No. 108-199	118 Stat. 3	40 C.F.R. Parts 150-189	PRIA I	H.R. 2673 S. 1664	2003
Maritime Pollution Prevention Act	33 U.S.C. §§1901-1915	Pub. L. No. 110-280	122 Stat. 2611	40 C.F.R. Parts 1043, 33 C.F.R. Parts 2-158, 46 C.F.R. Parts 2-162	N/A	H.R. 802	2008

Statute	Related U.S. Code	Public Law No.	Statutes at Large	CFR	Popular Name	Enacted Bill	Year
Executive Order 13514, Federal Leadership in Environmental, Energy, and Economic Performance (PDF)					EO 13514		2009
Executive Order 13693, Planning for Federal Sustainability in the Next Decade (revoked 2007 Executive Order 13423, Strengthening Federal Environmental, Energy, and Transportation Management)					EO 13693		2015
Frank R. Lautenberg Chemical Safety for the 21st Century Act (amends TSCA)	15 U.S.C. §§2601-2629, 42 U.S.C. §§280g-17, 47 U.S.C. §§254, 609	Pub. L. No. 114-182	130 Stat. 448		The Lautenberg Act	H.R.2576	2016

Appendix 2: U.S. Environmental Protection Agency Regional Offices: Contact Information

When in doubt, information seekers looking for U.S. Environmental Protection Agency (EPA) records and information can often get what they need by contacting the EPA office in their region. Note that different regions have varying numbers and types of offices, and that each region provides different types of contact information. These variations follow the diffused informational structure of the regional EPA office system, described in Chapter 2. An EPA Staff Directory and People Locator is available online to find specific EPA employees and Regions.[1] Additionally, each Region's home page (URLs are listed in the footnotes) contains a "contact" page link, where visitors can get contact information for various offices and subdivisions in the Region. Each Region has its own Freedom of Information Act (FOIA) office, but all of EPA's FOIA offices recommend making initial FOIA requests through FOIAonline.[2]

Region 1: New England[3]

Serving Connecticut, Maine, Massachusetts, New Hampshire, Rhode Island, and Vermont

EPA New England Customer Call Center: (888) 372-7341 or (617) 918-0101[4]

EPA Region 1 Offices:

EPA New England Region 1
5 Post Office Square, Suite 100
Boston MA 02109-3912

1. U.S. EPA, *Staff Directory*, https://cfpub.epa.gov/locator/index.cfm (last visited Dec. 25, 2016).
2. *FOIAonline*, https://foiaonline.regulations.gov/foia/action/public/home (last visited Dec. 25, 2016).
3. U.S. EPA, *EPA Region 1 (New England)*, https://www.epa.gov/aboutepa/epa-region-1-new-england (last visited Dec. 25, 2016).
4. If calling from outside New England.

Long Island Sound Office
Stamford Government Center
888 Washington Boulevard
Stamford CT 06904-2152
(203) 977-1541

EPA New England Regional Laboratory
11 Technology Drive
North Chelmsford MA 01863-2431

EPA Region 1 FOIA Office:

Regional Freedom of Information Officer
EPA Region 1 (OARM01-6)
5 Post Office Square, Suite 100
Boston MA 02109-3912
(617) 918-1102

Region 2[5]

Serving New Jersey, New York, Puerto Rico, U.S. Virgin Islands, and Eight Tribal Nations

Region 2 Citizen Hotline/Help Desk: (877) 251-4575[6]

EPA Region 2 Offices:

EPA Region 2
290 Broadway
New York NY 10007-1866

Main Regional Office
290 Broadway
New York NY 10007-1866

Edison, New Jersey Environmental Center
2890 Woodbridge Avenue
Edison NJ 08837-3679

---

5. U.S. EPA, *EPA Region 2*, https://www.epa.gov/aboutepa/epa-region-2 (last visited Dec. 25, 2016).
6. This number is toll free from locations within Region 2.

Caribbean Environmental Protection Division
City View Plaza II, Suite 7000
#48 Road, 165 km 1.2
Guaynabo PR 00968-8069

Hudson River Field Office
187 Wolf Road, Suite 303
Albany NY 12205

Western New York Public Information Office
EPA
186 Exchange Street
Buffalo NY 14204

Virgin Islands Field Office
Tunick Building, Suite 102
1336 Beltjen Road
St. Thomas VI 00801

Long Island Sound Office
888 Washington Boulevard
Stamford CT 06904-2152

EPA Region 2 FOIA Office:

Regional Freedom of Information Officer
EPA Region 2
290 Broadway, 26th Floor
New York NY 10007-1866
(212) 637-3668

Region 3: Mid-Atlantic[7]

Serving Delaware, District of Columbia, Maryland, Pennsylvania, Virginia, and West Virginia

EPA Mid-Atlantic Customer Call Center: (800) 438-2474 or (215) 814-5000[8]

7. U.S. EPA, *EPA Region 3 (Mid-Atlantic)*, https://www.epa.gov/aboutepa/epa-region-3-mid-atlantic (last visited Dec. 25, 2016).
8. If calling from outside Region 3.

EPA Region 3 Offices:

EPA Region 3 Office
1650 Arch Street
Philadelphia PA 19103-2029

EPA
Environmental Science Center
701 Mapes Road
Fort Meade MD 20755-5350
(410) 305-2600

Chesapeake Bay Program
410 Severn Avenue, Suite 112
Annapolis MD 21403-2567
(410) 267-5700

EPA
Wheeling Office
1060 Chapline Street, Suite 303
Wheeling WV 26003-2995

EPA Region 3 FOIA Office:

Regional Freedom of Information Officer
EPA Region 3
1650 Arch Street (3CG00)
Philadelphia PA 19103
(215) 814-2050

Region 4: Southeast[9]

Serving Alabama, Florida, Georgia, Kentucky, Mississippi, North Carolina, South Carolina, and Tennessee

EPA Southeast Customer Call Center: (800) 241-1754 or (404) 562-9900[10]

---

9. U.S. EPA, *EPA Region 4 (Southeast)*, https://www.epa.gov/aboutepa/about-epa-region-4-southeast (last visited Dec. 25, 2016).
10. If calling from outside Region 4.

EPA Region 4 Offices:

EPA Region 4
Sam Nunn Atlanta Federal Center
61 Forsyth Street, SW
Atlanta GA 30303-8960

(404) 562-9900
Toll Free: 1-800-241-1754

EPA Region 4 FOIA Office:

Regional Freedom of Information Officer
EPA Region 4
AFC Building, 61 Forsyth Street SW, 9th Floor (4PM/IF)
Atlanta GA 30303-8960
(404) 562-9891

Region 5[11]

Serving Illinois, Indiana, Michigan, Minnesota, Ohio, and Wisconsin

EPA Region 5 Customer Call Center: (800) 621-8431 or (312) 353-2000[12]

EPA Region 5 Hotline E-Mail Address: r5hotline@epa.gov

EPA Region 5 Offices:

EPA Region 5
77 West Jackson Boulevard
Chicago IL 60604

Cleveland Office:
EPA Region 5
25063 Center Ridge Road
Westlake OH 44145-4114

11. U.S. EPA, *EPA Region 5*, https://www.epa.gov/aboutepa/epa-region-5 (last visited Dec. 25, 2016).
12. If calling from outside Region 5.

EPA Region 5 FOIA Office:

Regional Freedom of Information Officer
EPA Region 5
77 West Jackson Boulevard (MI-9J)
Chicago IL 60604-3590
(312) 886-6686

Region 6: South Central[13]

Serving Arkansas, Louisiana, New Mexico, Oklahoma, Texas, and 66 Tribes

EPA Region 6 Offices:

EPA Region 6 Main Office
1445 Ross Avenue, Suite 1200
Dallas TX 75202
(800) 887-6063

Houston Laboratory
10625 Fallstone Road
Houston TX 77099
(281) 983-2100

EPA Border Office
4050 Rio Bravo, Suite 100
El Paso TX 79902
(915) 533-7273

Oklahoma Field Office
EPA
1645 South 101 East Avenue, Room 134
Tulsa OK 74128-4629

EPA Region 6 FOIA Office:

Regional Freedom of Information Officer
EPA Region 6
1445 Ross Avenue (6MD-OE)
Dallas TX 75202-2733
(214) 665-7202

13. U.S. EPA, *EPA Region 6 (South Central)*, https://www.epa.gov/aboutepa/epa-region-6-south-central (last visited Dec. 25, 2016).

Region 7: Midwest[14]

Serving Iowa, Kansas, Missouri, Nebraska, and Nine Tribal Nations

EPA Midwest Customer Call Center: (800) 223-0425 or (913) 551-7003[15]

EPA Region 7 Offices:

EPA Region 7
11201 Renner Boulevard
Lenexa KS 66219

EPA Region 7 FOIA Office:

Regional Freedom of Information Officer
EPA Region 7
11201 Renner Boulevard
Lenexa KS 66219
(913) 551-7003

Region 8: Mountains and Plains[16]

Serving Colorado, Montana, North Dakota, South Dakota, Utah, Wyoming and 27 Tribal Nations

EPA Region 8 Call Center: (800) 227-8917 or (303) 312-6312[17]

EPA Region 8 E-Mail Address: r8eisc@epa.gov

EPA Region 8 Offices:

EPA Region 8
8OC-EISC
1595 Wynkoop Street
Denver CO 80202-1129

14. U.S. EPA, *EPA Region 7 (Midwest)*, https://www.epa.gov/aboutepa/epa-region-7-midwest (last visited Dec. 25, 2016).
15. If calling from outside Region 7.
16. U.S. EPA, *EPA Region 8 (Mountains and Plains)*, https://www.epa.gov/aboutepa/epa-region-8-mountains-and-plains (last visited Dec. 25, 2016).
17. If calling from outside Region 8.

EPA Region 8 FOIA Office:

Regional Freedom of Information Officer
EPA Region 8
1595 Wynkoop Street
Denver CO 80202-1129
(303) 312-6306

Region 9: Pacific Southwest[18]

Serving Arizona, California, Hawaii, Nevada, American Samoa, Commonwealth of the Northern Mariana Islands, Federated States of Micronesia, Guam, Marshall Islands, Republic of Palau, and 148 Tribal Nations

EPA Pacific Southwest Customer Call Center: (866) 372-9378 or (415) 947-8000[19]

EPA Region 9 Offices:

EPA Region 9
75 Hawthorne Street
San Francisco CA 94105

Southern California Field Office
600 Wilshire Boulevard, Suite 1460
Los Angeles CA 90017

San Diego Border Office
610 West Ash Street, Suite 905
San Diego CA 92101

Pacific Islands Office
300 Ala Moana Boulevard, Room 5124
Honolulu HI 96850

18. U.S. EPA, *EPA Region 9 (Pacific Southwest)*, https://www.epa.gov/aboutepa/epa-region-9-pacific-southwest (last visited Dec. 25, 2016).
19. If calling from outside Region 9.

EPA Region 9 FOIA Office:

Regional Freedom of Information Officer
EPA Region 9
75 Hawthorne Street (OPA-2)
San Francisco CA 94105
(415) 947-4251

Region 9 Library Hours and Contact Information:

Monday through Thursday (except federal holidays)
9:00 AM - noon (PST)
1:00 PM - 4:00 PM (PST)
Closed on Fridays
Phone: (415) 947-4406
E-mail: library-reg9@epa.gov

Region 10: Pacific Northwest[20]

Serving Alaska, Idaho, Oregon, Washington, and 271 Native Tribes

EPA New England Customer Call Center: (800) 424-4372 or (206) 553-1200[21]

EPA Region 10 E-Mail Address: epa-seattle@epa.gov

EPA Region 10 Offices:

EPA Region 10
1200 6th Avenue
Seattle WA 98101

Alaska Operations Office
222 West 7th Avenue #19
Anchorage AK 99513-7588
(907) 271-5083

Idaho Operations Office
950 West Bannock, Suite 900
Boise ID 83702
(208) 378-5746

20. U.S. EPA, *EPA Region 10 (Pacific Northwest)*, https://www.epa.gov/aboutepa/epa-region-10-pacific-northwest (last visited Dec. 25, 2016).
21. If calling from outside Region 10.

Oregon Operations Office
805 SW Broadway, Suite 500
Portland OR 97205
(503) 326-3250

Washington Operations Office
300 Desmond Drive SE, Suite 102
Lacey WA 98503
(360) 753-9437

Hanford Project Office
309 Bradley Boulevard, Suite 115
Richland WA 99352
(509) 376-6865

EPA Region 10 FOIA Office:

Regional Freedom of Information Officer
EPA Region 10
Office of Ecosystems, Tribal and Public Affairs
1200 6th Avenue ETPA-124
Seattle WA 98101
(206) 553-8665

Index

Notes

Notes

Notes

Notes

Notes

Notes